Another party. "Haven't seen you in a while. What'dya do again?" "Used to be an editor" I said, but it felt like a lie. I wanted to say "I manipulate audio and video material and combine them with graphics and information" but it sounded both pretentious and vague. "I work in media integration." She sucked back a gob of guacamole and continued to munch on chips while managing to say: "You mean, like on the Internet?" I began to rant. "Well, yes and no. I still work in broadcasting and movies, but they aren't the same as they used to be. It used to be simple. TV as you know it is only a few years away from totally disappearing; hell—the death of celluloid film projection is on the horizon!" She was losing interest, perhaps looking for a VC, her big doe eyes focused on the far part of the room. "You know, I'm a pioneer in the union of Silicon Valley and Hollywood." She stopped munching chips and looked up at me, touched my arm, and said: "My website gets 3500 hits a day, but I've still got bugs in the streaming code. Do you think you could take a look at the Javascript?"

"I'm an editor . . ." I say.

"I know. Do you know Java?"

"I tell stories."

"Is that a 'no'?"

"I take the data captured from the real world in whatever media it arrives, and I put it together in some kind of meaningful way, usually involving the creative synthesis of new images and sound, or modified images and sound, or even old images and sound that I have sleuthed out of the media mines and utilized for my own benefit. And sometimes I even merge this stream of audio and video with text, for passive entertainment or interactive education, or both. I'm a writer. I'm an artist. I'm a technologist . . . I'm, I'm . . ."

"You don't know Java, do you?" I shook my head, "I use a computer, but I'm not a programmer. It's just a media tool."

I think she called me a tool as she walked away. I began to see my life as having purpose. My role perhaps to surf the tsunami of recorded data wherever it was, and put bits of everything together to say something to the world. I'm a media machine, a media sculptor who doesn't see the lines of production and post-production, of acquisition and distribution. I reject these terms as archaic!

A golden light began to shine from a flat panel plasma display behind my head, and I smiled. I had clarity with what it was I did. I'm an editor in the new millennium.

This is what this book is about.

N O N L I N E A R

a field guide to digital video and film editing

Fourth Edition

by Michael Rubin

4th Edition revised with
Ron Diamond

a book from playground productions • santa cruz, california
TRIAD PUBLISHING COMPANY • GAINESVILLE, FLORIDA

Printed in Canada

First Edition November 1991
Second Edition June 1992
Third Edition October 1995
Fourth Edition August 2000
10 9 8 7 6 5 4 3 2

Library of Congress Cataloging-in-Publication Data

Rubin, Michael, 1963–
    Nonlinear : a field guide to digital video and film editing / by Michael Rubin. -- 4th ed. / revised with Ron Diamond.
        p. cm.
    Includes index.
    ISBN 0-937404-85-3
    1. Motion Pictures--Editing.   2. Video tapes--Editing.   I Diamond, Ron.   II. Title.

TR899 R83 2000
778.5' 235--dc21   00-034363

Every effort has been made to provide accurate and up-to-date information about the edit systems and equipment described in this book, utilizing information supplied by the manufacturers. The opinions given, however, are solely those of the author.

Screens and interfaces of editing systems © by their manufacturers; the names of the editing systems and equipment are trademarks of their manufacturers, as follows:

Avid, Media Composer, Symphony, AVR, OMF, MediaLog, MediaSuite Pro and ePublisher are trademarks of Avid Technology, Inc. Apple, QuickTime, Final Cut Pro, ADB, FireWire, G3, G4 and Macintosh are trademarks of Apple Computer, Inc. Premiere, PostScript are trademarks of Adobe Systems, Inc. Avio and Casablanca are trademarks of Draco Systems. Panasonic is a trademark of Matsushita Electric Industrial Co., Ltd. CL-550 is a trademark of C-Cube Microsystems. NuVista+ and Vidi/o are trademarks of Truevision, Inc. DVI, Pentium, and ActionMedia are trademarks of Intel Corporation. Indigo, Extreme, Silicon Graphics, Silicon Studio are trademarks of Silicon Graphics, Inc. Virtual Master, CMX 600, 6000, Chyron, Cinema, 3400, and 3600 are trademarks of Chyron/CMX. LaserVision and Laserdisc are trademarks of Pioneer Electronics Corp. PostBox is a trademark of Pioneer Electronics, Inc. Play, Trinity, Preditor are trademarks of Play, Inc. Montage, and Montage Picture Processor, are trademarks of Pinnacle Systems. PostScript, Photoshop, PageMaker, Illustrator are trademarks of Adobe, Inc. RAZOR is a trademark of in:sync, inc. KeyKode is a trademark of Eastman Kodak Company. DCode is a trademark of Denecke, Inc. 68010-68040, PowerPC are trademarks of Motorola Corp. Moviola is a trademark of the J & R Film Company. EditBox, Harry, Paintbox, Henry are trademarks of Quantel Corp. E•Pix, NightSuite are trademarks of Adcom Electronics. Ediflex, Ediflex Digital are trademarks of Ediflex Systems, Inc. Star Wars, Death Star, Darth Vader, Droid are trademarks of Lucasfilm, Ltd. editDROID is a trademark of LucasArts Entertainment, Inc. TouchVision, D-Vision, D-Vision Pro, D-Vision OnLINE, are trademarks of Discreet Logic. Spectra Laser Edit System and Electronic Laboratory are trademarks of Laser-Pacific Media Corp. Betacam, Betacam SP, TRV900, Digital Betacam, iLink and Sony are trademarks of Sony Corporation of America. Oscar and Academy Award are trademarks of A.M.P.A.S. Any and all other trademarks used within are acknowledged as property of their respective owners.

Published and distributed by
Triad Publishing Company
PO Drawer 13355
Gainesville, FL 32604

This book was written and published entirely on a Macintosh G3 (OS 9) with 256MB RAM—using Adobe InDesign versions 1.0 and 1.5; Illustrations in Macromedia Freehand and Adobe Photoshop; Collaborations were facilitated by Adobe Acrobat version 4.0 and Microsoft Word.

ACKNOWLEDGMENTS

Can you believe it has been about a decade since this guide first evolved? Through all the years, NONLINEAR has been a collaboration with the best people I have known in a variety of fields.

For their technical and historical expertise, and many wonderful memories, thank you all: Tom Beams, Stan Becker, William Butler, Jerry Cancellieri, Josiah Carberry, Shawn Carnahan, Gabriella Cristiani, Dick Darling, Andy Delle, Robin Dietrich, David Donaldson, Robert Doris, Herb Dow, Adrian Ettlinger, Bennett Goldberg, Leigh Greenberg, Richard Greenberg, Patrick Gregston, Ralph Guggenheim, Seth Haberman, Marvin Hall, Augie Hess, Paul Hoffman, Barbara Koalkin, Don Kravits, Robert Lay, Keith Lissak, Harry Marks, Phil Mendelson, Gary Migdal, Harry Mott, Maureen O'Connell, Tom Ohanian, Brian Pinkerton, Mark Pounds, Robert Purser, Thomas Dolby Robertson, Diane Rogers, Tony Schmitz, Arthur Schneider, Steve Schwartz, Tom Scott, Larry Sherwood, Walter Shires, William Stein, Leon Silverman, Martha Swetzoff, Bob Turner, Randy Ubillos, Nancy Umberger, Michael Walker, Susan Walker, Steve Weisser, Kenneth Yas, and—especially—to Dean Godshall (technical support master), Scott Johnston, Derek McCants, and Ron.

Ron is here too, and wanted his own paragraph to specially thank a number of folks (it goes without saying that I, too, am grateful to the following people). Take it away, Ron:

For their expertise, inspiration and support (which is in many cases ongoing), I'd like to thank these kind and talented people: Jay Ankeney, Alberto Arce, John Broaddus, Steve Cohen, Tad Davis, Chris Emhardt, Eve Gage, Patrick Gregston, Will Hall, Ben Hershleder, Ann Holbrook, Joanne Izbicki, Marilyn Madderom, Patty Montesion, Rick Moore, Andy Parke, Lugh Powers, Miriam Preissel, Scott Redlich, Tim Vandawalker—and Mike, keeper of the Gestalt-O-Meter™ *. . as well as my family, whom I love.*

Thanks, Ron. Also, no edition of this tome would be complete without my continued gratitude to Mary Sauer and Steve Arnold, who brought me into this business in the first place.

Finally, I want to thank my family—Lorna and Mel; Sara; Gabrielle, Jacques, Joshua; Danny, Louise, Maida, Asa; Jerry, Judie, Jeff; Gordon and the Petroglyph gang; all my friends who tolerate this behavior; my precious and inspiring son Jonah; and *most importantly* to my love and wife Jennifer♥ , (who I owe *big* time) and who constantly reminds me that at the end of the day, all this technology isn't really worth squat.

C O N T E N T S

PREFACE TO THE FOURTH EDITION XI
PREFACE TO THE FIRST EDITION XIII
INTRODUCTION XV

1. FIRST THINGS

WHAT IS LINEAR EDITING? 3
    So What's Wrong with Linear Editing? 4
    How Can Nonlinear Editing Help? 4
    Is There Anything Wrong with Nonlinear Editing? 5
BUZZWORDS 11
    Nonlinear 11
    Real Time 12
    Frame Accurate 13
    Broadcast Quality 14
    Random Access 16
    Digital 17
    Desktop 17

2. BACKGROUND, 19

NOVICE'S INTRODUCTION TO PRODUCTION 21
THE FILM STORY 24
THE VIDEO STORY 32
    The Ghost of RT-11 38
HISTORY AND MOVING PICTURES 39
A BRIEF HISTORY OF ELECTRONIC EDITING.. 41

3. FUNDAMENTALS, 77

ONLINE AND OFFLINE.. 79
TIMECODE 83
TYPES OF TIMECODE (VITC, LTC AND BEYOND) 86
THE EDL.. 92
    The Evolution of the EDL.. 95
FILM EDGE NUMBERS... 98
    Keykode® 99
TELECINE AND 3:2 PULLDOWN 102
CUTTING FILM USING VIDEOTAPE 108
    Preparation for Telecine 113
TELECINE LOGS 118
COMPONENT AND COMPOSITE VIDEO 120
    Component/Composite Video Format Table............................... 128
    Videotape Quality 129

Color Space, Gamuts and the Kitchen Sink.............................. 130
Cabling 136
    Cable Connectors 138
    Balanced and Unbalanced Audio 141
    FireWire®... 143
The Digital World 145
Digital Video 151
Compression 156
    MPEG Profile/Level Table 162
    Compression Made Simple.. 163
    Color Sampling and 4:2:2.. 165
RAID Technology 167
Digital Media Characteristics 169
Bandwidth... 171
    Video Compression, Bandwidth Table....................... 174
Pixels 175
Aspect Ratio Basics... 178
Light and Color 180
Typography 187

4. Editing Primitives, 191
What is an Editing Primitive? 193
Editing Time 194
Editing Flowcharts 196
Computers 202
File System Organization 207
Nomenclature (reels, scenes, takes) 211
Primitives of Editing... 216
3- and 4-Point Edits 231
Synchronization... 232
Effects 233
Editing Ergonomics 237
Monitors and Windows 241
Timelines 244
Fundamental Edits 251
Digitizing and Capturing... 253
Videotape Assemblies.. 257
Printouts.. 260

5. Distribution, 265
A Brief Review of Broadcasting 267
The Economics of Revolution..................................... 277
Why Shows Cut Film (Standards Conversion and HDTV) 280

Brief History of HDTV 283
Video on the Web 287
DVD.. 301
Video-to-Film Transfers.. 309
Electronic Cinema... 312

6. Systems, 315
Types of Editing Systems 317
Turnkey Systems 321
System Interface Historical Overview 324
The Best System 349

7. The Real World, 351
Practical Editing Methods.. 353
Deconstructing Production.. 359
Pricing.. 360
Telecine Trouble.. 362
Color Bars and Monitor Adjustment 364
Demo Reels 366
Tape Formatting 370
Editor's Bag of Tricks 371

8. Theory, 375
Horizontal and Vertical Nonlinearity....................................... 377
Interview with the Author 388

Appendices
Film Footages 392
Binary Counting.. 393
First Decade of Nonlinear Cinema... 395
Acronyms and Abbreviations 398
Index.. 404

Navigation Note: Occasionally, at the bottom of some pages, you will find *directional footnotes* ▶▶ Check this out: p. 416
to lead you to other key sections that may define, support or augment the information on that page. These are by no means the full extent of additional resources, and do not replace the index, but they may prove useful to the busy reader.

PREFACE TO THE FOURTH EDITION (2000)

The profession of editing has many similarities to, but does not completely coincide with, the *joy* of editing. There is something about watching recorded pieces of reality (or even a make-believe version of reality) and taking them apart and putting them together again. Telling a story in pictures and in sound. Any story. Any pictures and sounds. Interest in the Internet is no wonder. Building websites (particularly the "websites" of the future) is deeply intertwined with the joys of editing and post-production.

The future lies in the convergence of media: the useful control over images (still and moving) and sounds. The technology at this epicenter of media convergence must handle video and audio editing and synthesis. Technological wizardry allows more and more of these tasks to become integrated into a single "swiss-army-knife" of media control; economics makes this tool more and more available to the individual. And who is the *de facto* benefactor of all this control? Who holds the profession most like the MEDIA CONVERGENCE MASTER of the future? While he or she may come from any one or series of disciplines, I suggest that the profession *most like* this future job description is **nonlinear editor**. The editor sits squarely at the focus of media, art and technology. Today, I suggest, this is a particularly good place to be.

Post-production is not often written about with detached reverence. Sometimes it appears an embarrassment because it is less romantic than production. Production is where the *stars!* are. "Post" is a back room somewhere. Editing is more like watching TV than shooting a movie, even if it is perhaps the place where the movie is "made."

In 1932 Ansel Adams was a pioneer of photography as something other than pictorialism and "pretty pictures." He and the so-called "Group f64" were quiet, creative and technological radicals. The art community didn't look at photography as art. Painting: *that* was art. Photography was mechanical. It was simply documenting a moment in the world, with no interpretation. It was a result of technology, of optics, of chemistry. The early creative uses of photography during the prior decades tried to imitate the impressionistic painting style of the day that was so popular. Group f64's vision was in demanding that photography didn't need to imitate painting. In the same way that the first television shows were often filmed versions of radio shows, and the first websites were little more than Internet newspapers and billboards, each new technology is often forced to make a baby step from the technology it ostensibly replaces. The technology invariably exists before the vision of how to utilize it emerges. We don't yet know what kind of creative expression is possible with the technologies we are developing.

Ansel Adams called his style of working "pre-visualization." He worked out his vistas and ranges of light in detail in advance of opening the shutter on his 8x10" negative hand-built camera. While he was famous for this, it is less widely known that he did extensive work in the developing and printing of images to make them look so wonderful. His hallmarks were heavy dodging and complex exposure settings on different parts of the enormous negative. He and his printing assistants were masters of these exposures. He was famous for pre-visualizing, but his secret was "fix it in post."

If you are an editor today, you have access to the greatest quantity of affordable technology in history, with which you can edit your personal slice of the media pie. There is a lot of media out there, and it is BEGGING to be edited.

The Internet requires a great deal of graphic design sensibilities, which not all editors have. But those who can manage both the sense of story, pace, focus and time (from audio and video), to layout, readability and excitement (from publishing and design) will find this new medium interesting. Clearly, the more it evolves, the more moving images and sound can become a part, and the more editors will find it friendly and familiar territory. The Web will no doubt become something greater than a cool frame for streaming media: it will become its *own* medium.

And that's why this book is here. We love editing. We are unafraid of technology. We grow more and more able to afford the tools of our trade and hobby. This is a golden age. We contributed to its ascendance. We have a few useful observations and fundamental information for you that doesn't really go out of style, but continues to grow in relevance.

I hope this book and our website provide for you a Rosetta stone, a launching point, as well as an enjoyable romp in politics and business (a necessary side hobby of many an editor). Please let us know what we might improve, and we'll see what we can do. Until then, hang on, enjoy the ride, thank you and good luck.

MHR
July 2000

PREFACE TO THE FIRST EDITION (1 9 9 1)

At this year's National Association of Broadcasters show I met many videotape editors who thought the new digital editing systems they saw were the first nonlinear products. I met film editors who didn't know the difference between hard disks and laserdiscs. As I walked the exhibit floor, it became apparent that editing and editing systems were about to get ... confusing ... if they weren't already.

For over six years I have been actively involved with nonlinear editing, and only today it seems the editing community is sitting up and saying, "Hey, look at this nonlinear thing ..."

If you are finally discovering nonlinear editing (or only now find it ready for you), you are probably in need of a perspective on these systems: where they came from, how they do what they do, how to work them, and which is the right one for you. So where can you turn for help?

No person in the business can give you an objective overview. Although I promise to try, it probably won't be absolutely evenhanded. No one is unbiased and few know all the systems well.

Not the *trade press*—they have magazines to sell and advertisers to please.

Not the *manufacturers*—they really don't know much about any editing systems other than the ones they make (and sometimes don't know that much about their own).

Facilities? They tend to get informed once they've owned a piece of equipment for awhile (often knowing it better than the manufacturer) but will have only a very selective knowledge about any other systems.

Producers? Their principle concern, and rightfully so, is cost. Their judgments are important, but their bias is toward economic and logistic features more than editing functionality.

Editors? Most who edit nonlinearly have fallen in love with the first system they learned and rarely want to venture out much farther. There is a growing body of editors who started on film, experimented with videotape (but generally rejected it), and then discovered nonlinear video. Of this group, most have edited on two or three different systems at different times, giving them a somewhat diverse knowledge of system functionality. These are good people to talk with. You will find their recommendations divided between the system they like the most and the system that most productions want them to use.

No system is perfect. Everybody wants different things and every system offers different things. The "right" features are in at least half a dozen different editing systems but no one system has them all. Today there are at least twelve significant manufacturers of nonlinear electronic editing systems. By the time you read this, there may be twelve others. Though the specifics of each system may be different, there is much background that is fundamental to all of them.

This book does not presume to teach anything about the aesthetics of editing. It arose simply from a need to demystify the myriad editing equipment choices an editor has today, to bring to editors a unified terminology and a basis for learning more about systems not yet developed. Technology happens whether we like it or not. Consequently, this book was designed for (1) film editors who need to understand more about the world of "electronic film" editing—to develop some understanding of the techniques, vocabulary, and technologies of the video world; (2) videotape editors who want to investigate the stylistic power and history of nonlinear editing; and (3) producers and directors who are baffled by their electronic options.

As the decade progresses, videotape and film editors will be more and more in the same ship—learning new ways from each other about editing in the future. I hope this book will begin the foundation for that new community.

MHR
November 1991

INTRODUCTION

No book can replace actually working in post-production and with nonlinear editing. The purpose of this book is to familiarize the reader with fundamental concepts associated with nonlinear post-production. It should be read in conjunction with class training or real-world experience.

For learning how to operate any of the available nonlinear systems, manufacturers sponsor classes, the Editors Guild supports training, and most universities and community colleges (and even high schools) offer programs. However, technical background is often passed over in the classroom setting for a more direct "how-to" application of knowledge.

NONLINEAR is not meant to be read cover-to-cover like a novel. It is completely random-access and, well, nonlinear. The reader should feel free to peruse sections and find topics that are of particular interest.

Many topics are repeated in different chapters and in slightly different ways. Film, video and computer people often have vastly different backgrounds, and some will need more explanation on certain topics than others. NONLINEAR assumes the reader has only a rudimentary knowledge of either film, video or computer concepts and terminology.

New in this edition comes a relationship to the web. Check out the nonlinear4 website—at **www.nonlinear4.com**—where updated information, resources, and greater detail on many topics is provided.

This "primer" is neither intended to teach you how to edit on any particular system nor to convince you which system is "best." It will discuss the history of the nonlinear systems and give an overview on how they work. When appropriate, I will offer my opinions on the good and the bad, hopefully without proselytizing and without malice.

ADDITIONAL NOTES

ON TERMINOLOGY

There is (still) no uniform terminology in the hybrid world of digital editing. Since many individuals approach this world from neither a film nor video background, user vocabularies often reflect a pidgin from these disciplines, plus street *computerese*. The problem would not exist if we had a central editing school teaching one set of terms; but since there is none, this handbook of nonlinear editing will use what we feel is a consistent and appropriate set of terms. These definitions and spellings are not necessarily better than others used elsewhere, but are generally accepted. After all, it is not so important which term is used, but that the term is understood by everyone.

For our purposes, this book will use the following conventions: *online* and *offline*, meaning editing types, are one word, as is *nonlinear*. *Pre-* and *post-production* are hyphenated. *On-line*, also with a hyphen, describes the state of being connected to a system or network. *3:2* pulldown will be called such, even though it is technically more accurate often to refer to the pulldown cadence with the increasingly common "*2:3*" designation. And even though the phrase normally refers to technical specifications, *broadcast quality* will also be used to describe the image quality associated with SDTV and HDTV television broadcasts.

ON PRODUCTS

To make this book as useful as possible, actual systems are named and sometimes used in illustrations. It should be recognized, however, that these systems are always changing, upgrading, modifying their pricing, and releasing surprise features to anxious trade-show audiences. This book is not reviewing nor endorsing any specific products, and uses products in an illustrative way only. Although some editing features are patented, most are general enough to appear in many systems.

Only parts of a couple of chapters actually name individual nonlinear products, the prevailing ones at the time of this writing. All are certain to change over the months and years following this publication. The background history is culled from dozens of interviews and countless articles with key players in the nonlinear game—inventors, company presidents, facility staff, and editors. Although it often reflects many of their biases, it it is intended to portray a fair view of the evolution of these products.

Chapter 1

First Things

"In the '60s I thought that editing was
the moment of law and order—like the
police—on the body of the movie . . . but
in the early '70s I met [editor] Kim Arcalli,
who made me discover that editing could be
a fantastic moment, a creative moment . . ."

— Bernardo Bertolucci (1991)

director: Last Tango in Paris, The Last Emperor

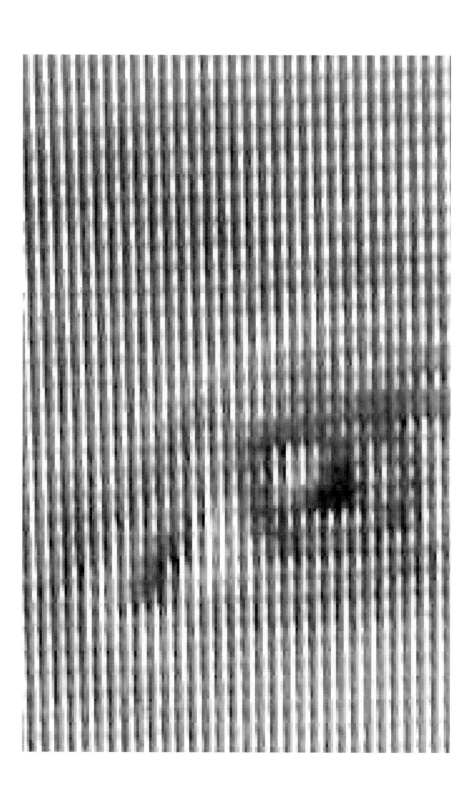

WHAT IS LINEAR EDITING?

You're having a party. Your job, starting early that morning, is to prepare the "party tape"—a cassette of music that you can pop in the audio cassette deck and that will run throughout the evening. You collect up your favorite albums (the *source* for your party tape), choose your favorite songs, and start recording. What do you do first?

Insert a cassette tape into your tape deck, cue it up to the beginning, and then load up the first CD into the disc player. Let's say it is Madonna. *Ray of Light*. Press *record* (on the tape). Hit play (on the CD). You're rolling.

Five minutes and 21 seconds later, the song ends. You press stop at just the right time, and then unload the CD. One down, 28 to go. Load the second album, Ritchie Valens. Re-cue the tape after the Madonna song, and prepare to start the next CD at the beginning of *La Bamba*.

Whoops. You missed. Stop recording and re-cue everything. On the second try you get it right. While it is recording you run to the kitchen to put beer in the fridge. By the time you get back into the living room, *La Bamba* has ended and the next song on the album has already started.

No problem. Stop the recording. Stop the album. Rewind the tape to cue it at the end of the second song.

By afternoon you have a terrific party tape. The *list* of songs, and their durations, looks like this:

Ray of Light	5:21
La Bamba	2:04
Johnny B. Goode	2:18
All I Wanna Do	4:33
Rock Lobster	6:51
The Flat Earth	6:35
Livin' La Vida Loca	4:03
Stairway to Heaven	8:01
(plus 21 others)	

What you have done, in a very real way, is perform a traditional, *linear* (albeit audio-only) edit session. That's what a party tape is. Linear editing is simple and very understandable. Compact discs were your *sources*, the *edit list* is printed above, and the party tape is your *edited master*.

NB: If you've already got an MP3 player, and use it for sonifying your parties, then you're already nonlinear. Feel free to skip this whole party tape analogy and jump ahead.

So What's Wrong with Linear Editing?

Nothing, really. Except this: let's pretend that you are checking over your party tape an hour or so before everybody arrives, and you decide that you'd rather not use the second song, *La Bamba*. It used to be your date's favorite song with Somebody Else. Now what? RE-EDIT.

All you have to do is record a new song over the old one. No problem. You pick Springsteen's *Born to Run*, but find that it is almost 5 minutes long. If you start at the end of Madonna, it will roll over *La Bamba* (2:04) and the beginning of the next song, *Johnny B. Goode*. So that won't work. You really like Chopin's Minute Waltz, but, of course, it is only a minute long. You'd still have 1:04 of *La Bamba* left over, hanging out the end. What do you do?

Short of finding an alternate song that is EXACTLY THE SAME LENGTH AS THE OLD SONG (which is extremely unlikely), you have only a few options: record the new song, and then record all the following songs *again*, from their original discs; or copy (dub) the first tape onto a *second* tape, record the new song in the right place on the first tape, then re-record the dub back onto the first tape in the right place.

The first option will take all day because it means re-locating and re-cueing every song. The second option is much faster, but unless the tape and cables are digital, the music will lose a couple generations of quality from all the re-recording. For your party, you probably will opt to skip all of this, turn the volume down low, and hope your date doesn't notice—because LINEAR re-editing is sometimes more trouble than it is worth.

How Can Nonlinear Editing Help?

If the party tape had been created on a nonlinear editing system (of which an MP3 player is one), or programmed into a multi-disc CD player, you would only re-program the order of the songs you wanted, skip *La Bamba*, add other songs, do whatever you wanted. Ten CDs, or whatever the maximum number of discs you can load at once, determines how much music you can listen to without using any tape, or before needing to re-load.

Nonlinear editing is a method for editing that allows editors to work in any order or style they want. Unlike traditional (linear) videotape editing, an editor can insert frames into an edited sequence, shoving everything that follows farther down in time, or remove frames and have the following material slide up. Unlike an edited master tape in video editing, each shot is not locked to time—it exists only as a "source in" and "source out."

▶▶ Nonlinear editing primitives: p. 218

As with editing film, a nonlinear work can be roughed together, and then slowly honed and sculpted into a final cut. In linear videotape editing, however, this isn't a practical way to work. And since making certain changes can be so much trouble, editors (and clients) are sometimes forced to skip those changes, and ultimately the final product suffers.

At the utopian ideal of nonlinear editing, one has instant access to *all* source footage, and there is no drawback to modifying *any* particular edit—whether it is the second shot in a list or the last. Film editing is nonlinear. But accessing source material is not necessarily quick, and the handling of trims and coding of dailies is relatively laborious. For many kinds of projects, particularly those with relatively small quantities of source material, this ideal of nonlinearity exists today. For other projects, larger projects with a lot of source material, true nonlinearity is still only approximated, but to acceptable limits.

Digital nonlinear systems possess degrees of "nonlinearity" because (1) they many not (for a variety of reasons) be able to load all source material at one time and (2) they may need to render effects in order to see them. Both technical and economic limitations impact the creative experience.

Is There Anything Wrong with Nonlinear Editing?

While digital nonlinear does provide many advantages over the older, traditional methods of editing film and video, there are nevertheless various mitigating factors to keep in mind. Are these valid reasons not to work in nonlinear post-production? Not necessarily; they are simply things to pay attention to when moving into a different way of working.

Turnaround time

The sheer simplicity of linear video editing can mean much greater expediency for situations requiring very rapid turnaround — day-of-air news stories, for example. Quick-and-dirty "machine-to-machine" editing lets you rush into the edit bay at 5:30PM, quickly lay down a series of shots, sound bites and narration in the right order directly onto the master tape, and still have some hope of finishing before the start of the 6 o'clock news.

However, the most common nonlinear editing workflow consists of three discrete "modes"—first, **input** (either *digitizing* or *capturing*), then **editing**, and finally **output**. The often-lengthy procedure of having to digitize everything first before beginning any actual "work" can be quite frustrating when you're anxious to get on with the actual editing. And if you first do an offline edit at low resolution followed by re-digitizing at high resolution, then you've added even more steps to the process.

These trade-offs are in flux, however: tape formats that can be digitized at faster-than-real-time speed; cameras and recorders that eliminate tape completely, in favor of random-access source media; "hybrid" nonlinear systems that integrate the task of digitizing into a traditional linear editing workflow; and shared servers that eliminate the need to do a finished output entirely. Nevertheless, there continue to be quick-and-dirty "I need it done *now*" situations that might still be better suited to more venerable methods, particularly when the editing itself is something that demands little experimentation.

Reliability

Traditional film and video editing equipment benefits from a long-established and well-understood design that is *far* simpler than that of contemporary nonlinear gear. Problems are relatively trivial to diagnose, and replacement parts are readily available and generally cheap. A "catastrophic failure" in a Moviola or KEM, for example, might require all of a new light bulb or rubber belt in order to be quickly back up and running again.

By contrast, a digital nonlinear editing system is an *extraordinarily* complicated and fragile combination of intricate computer hardware, software and digital data. Furthermore, a constant tinkering with the system's design prevents it from ever reaching a well-understood equilibrium. Crashes and other problems can be notoriously difficult to isolate—particularly any that are intermittent and can't be readily reproduced on command.

Cost factors

Some nonlinear systems, particularly higher-end professional systems, can be much more expensive to rent or buy than the gear traditionally used for, say, editing a feature film: Moviolas, flatbeds, synchronizers, *etc.*

Yet one of the ways in which nonlinear editing was originally "sold" to the major film studios during the 1980s–1990s was by convincing them that any higher equipment costs would be recouped through shortened post-production schedules. This turns out to be only partly true. An ultimately *creative* endeavor such as the production of a feature film may entail a fair amount of trial and error, in order to see what works best and

▶▶ Back-ups: p. 206

what doesn't, and to provide time for reflection and coming up with additional new ideas. Sometimes referred to as the "gestation period" of a film, it has little to do with the time it takes the editor to physically make a given splice. The modern nonlinear cutting room for a theatrical feature requires additional assistants to keep the operation running, as well as to do the eventual conforming of film workprints for screenings and previews.

The additional cost for the digital nonlinear approach can thus be a mixed blessing. Projects that might have benefitted from extra time for finessing could, in the final analysis, end up shortchanged.

Mentorship

When it is cost-prohibitive for an editor and assistant to each have separate nonlinear editing systems, they may be forced to share one by working opposite shifts, and perhaps rarely even seeing each other in person. This sharing has proven devastating to the traditional mentoring process and can result in greater frustration and even burnout for those newer to the business and most eager to advance.

There is something truly powerful in closely observing a mentor who has years of experience. The editing unions believe it is important to foster mentorship in the teams that make films. Admittedly, there comes a point when you need to move on and edit on your own; but until then, it can be a fabulous experience.

Visualization and preparation

Once the footage has been digitized, nonlinear "edits" certainly happen faster than the equivalent on video or film, but that's assuming that you already have a pretty good concept of what edits you want to make. The relative inertia that characterized the older media had a beneficial side effect: since it did take longer to do the physical editing, people became more adept at forming a detailed mental picture and/or written structure first, in order to avoid reliance on time-consuming, brute-force experimental execution to evaluate an idea. In addition, an excessive number of physical splices (for varied attempts at creating a particular film edit), or "traced" EDLs (for offline video editing), did carry a stigma of sorts, and acted as a incentive to make lasting decisions sooner rather than later.

By contrast, digital nonlinear editing imposes fewer obvious penalties for lack of preparation. Sequences in timelines leave no trace of intermediate experimentation. Moreover, any individual action carries the illusion of being "only a mouse click away" (though this may add up to hundreds or even thousands of extra steps, and numerous creative dead-ends over the course of a project).

As a result, those who have only edited nonlinearly may be less likely to develop certain imaginative or organizational skills, and be more dependent on avoidable trial and error to see what works and what doesn't. There is power in both raw discovery by experimentation as well as learning though premediation (and education). Neither is better than the other, and both are important.

Creative indecision

While it might be too much trouble to make a certain set of changes late in a video or film project (which is then declared "good enough" by default), nonlinear editing has unlimited malleability. It's seemingly never too late in the game to make changes.

But, at some point, every project needs to be finalized. It's important to recognize the point at which continued tweaking no longer makes the results genuinely better, but simply "different." Directors, producers and editors without this innate discipline can find themselves drowning in a sea of indecision. Clients are also more apt to ask for continued changes, knowing the relative ease with which they can be accomplished.

Random access

There's a hidden but invaluable "serendipity" factor in having to shuttle through reels of seemingly unrelated video or film source material to locate the footage that (you think) you want: you're exposed to other possibilities you might not have otherwise considered, and which could turn out to be even better than what you'd originally had in mind.

Your constant exposure to large quantities of the raw footage promotes a greater familiarity with *all* the material available to choose from. It also facilitates the ability of your subconscious mind to foment new creative ideas and solutions to editorial problems.

Picture quality

Ideally, during post-production, the editor and director can view a picture that's as good or better than what will ultimately be shown to the audience, as was usually the case with the old-fashioned linear sys-

▶▶ Productivity tip: p. 358

tems. Editing systems now tend to reverse the equation: digitized images (even at high-resolution) fall short of the picture quality of celluloid film. They're compromised by the lesser quality of a video transfer, and during offline editing, suffer artifacts from the data compression used to make up for finite storage capacities.

This reduced picture quality can be a real limitation because you're making editorial decisions regarding details that can't quite be seen. In dramatic material, for example, subtle facial expressions (such as eye movements) can help you decide which performances to use, and when to cut from one shot to the next. Editors and directors aware of this and related trade-offs might insist on first watching their "dailies" footage projected on film. Others gamble that the limited resolution of the digital system doesn't obscure anything too important, and hope that there are no unpleasant surprises once the "cut" is seen at truly high resolution on a very large screen.

Changing perception of editorial skills

Because of the simplicity of traditional film equipment, learning to use it was relatively easy and afforded the luxury of a much longer (arguably lifelong) process of learning the aesthetics and craft of editing—the ability to harness a series of raw performances in order to tell the best story, and an understanding of the myriad subtleties involved in character development, motivation, story arcs, and the like.

By contrast, editing now diverts more effort and attention toward the purely technical: having to keep up with a complex, never-ending parade of software and hardware configurations and the demands of newer tangential skills such as graphics, animation and compositing. Thus, there's now an occasional tendency—even by those making hiring decisions—to equate technical know-how with being a good *editor*. For decades, a comprehensive reservoir of storytelling experience was widely considered the best qualification for the job. It will be a number of years still before the technical skills are so commonplace that this is once again true.

Specialization of video/audio equipment and skills

The democratization of so-called "broadcast quality" on the nonlinear desktop has fundamentally changed the post-production equation, putting some online finishing capabilities into what were traditionally (and are still often doggedly referred to as) "offline" edit bays, and sometimes creating confusion as to which equipment and personnel are genuinely best suited to each task. Some very common examples:

      ⊘ Many rooms used for online nonlinear editing routinely lack very simple amenities, such as real-time waveform/vectorscope monitoring (for measuring video levels); timebase corrector (TBC) remote controls (for convenient and real-time adjustment of those levels); and video processing amplifiers, or "proc" amps (to prevent video levels that are still excessively low or high from making it into the final master output).

      ⊘ Nonlinear edit bays are often marginally suited to quality control of audio, by virtue of their often-excessive ambient noise levels, inadequate less-than-full-range monitor speakers, marginal acoustics, and a lack of experience on the part of some who operate those rooms. Assistant editors with limited familiarity in the proper set-up of mixing boards or knowledge of the sonic idiosyncrasies of VTR audio tracks unwittingly find themselves "in charge" of the ultimate final audio quality nevertheless—simply because editing systems are capable, in a technical sense, of capturing audio at a sufficient sampling quality from the very outset of a project.

      ⊘ Though the multi-track audio layering capabilities of nonlinear systems tend to reduce the need for a separate "sweetening" bay, they still don't replace some of the finessing of the audio (such as judicious use of audio compressor/limiters and other types of processing, and the knowledge of when and how best to apply them) that help make an output sound like a properly finished and cohesive mix.

In addition, since the "traditional" finishing workflow tends to be more parallel in nature, it is sometimes more expedient as well. For example, while one person is executing a series of edits or setting up a DVE move, another person might be typing lower-third titles into a stand-alone character generator, even as someone else is EQing an audio track. In comparison, the "all-in-one" editing workstation can be a bottleneck—there's only one operator performing the tasks in series (and using one set of ergonomically-generalized physical controls) so it may take a greater number of hours to complete.

▶▶ Editing ergonomics: p. 237

B   U   Z   Z   W   O   R   D   S

NONLINEAR   REAL-TIME
RANDOM ACCESS   FRAME ACCURATE
BROADCAST-QUALITY   DIGITAL   DESKTOP

If you've been reading the trade magazines on video or editing any time in the past dozen years, you've seen the "buzzwords" listed above over and over. They appear in informative articles about nonlinear editing systems and in the advertisements for them. Sales people, especially, string the words together to make their system sound particularly powerful and impressive. Unfortunately, the words have been used and misused so often that they are on the verge of becoming almost meaningless adspeak.

Do you know what they mean? Once you understand these terms, you will be better able to understand what a magazine or manufacturer has to say. Identifying the best system for your needs still won't be easy, but you will be off to a better start.

Oddly enough, these terms not only can describe an editing system, they are generally the defining characteristics of this type of product. So, while the words as used may well be right, they don't *necessarily* differentiate the various products from one another.

"NONLINEAR"

As you now know, traditional videotape editing is, by definition, *linear*. It means that first you make the first edit, commit to it, then make the second edit, and so on, starting at the beginning of your project and ending at the end. Changing any edit prior to your current edit is somewhat difficult, and the farther back toward the beginning you go, the more trouble you may have.

"Nonlinear" is the key word describing digital editing systems that are designed for cutting in a film-like style, and not needing to record to videotape (except for distribution, perhaps). "Nonlinear" is a good term for the *general* category of editing systems discussed here (more so than "random access" or "frame accurate," for example). It serves to separate these technologies from the rather wide field of *linear* electronic systems that preceded them (and still exist, mostly for online) since the 1960s.

▶▶ The Video Story: p. 32

By doing away with recording the edited master to videotape, any editing system, regardless of its source capacity, output resolution, human interface, or total speed rendering, can be nonlinear.

"REAL-TIME (PREVIEW)"

For the first many years of analog nonlinear editing, the way the editing sequence was viewed was through a "preview" or "real-time preview." This was a variation of a common video editing term: a preview is a rehearsal of an edit prior to committing that edit to a record tape. Film editors have no equivalent concept for the preview—you cut something and you watch it. You don't like it? Change it. Now. Later. Whenever. But because videotape is linear, you want to watch an edit before you record it . . . because once it is recorded it is more difficult to modify.

Digital nonlinear post-production means not needing *any* record tape, and talking about "previews" today is somewhat obsolete. On the other hand, the term "real-time preview" has been redefined from a key word of "horizontal" nonlinearity (separate shots are presented to the editor as a single cohesive unit, as if they were recorded onto tape) to a key word for "vertical" nonlinearity (composited effects).

Digital systems generally have functionality with *layers of video tracks*. Ironically, when multiple layers of video (and effects) were built and composited together in analog systems, the editor saw the results of her work as she performed them—in "real time." But **digital** effects are enormously complex for computers to handle, and often the editor must wait to see the results of compositing until the system chews on the data in each layer and puts them together. *When modern systems preview these vertical layers in **real time**, this book refers to it as "vertical" nonlinearity.* When a digital system maintains two or more "streams" of video, it means that many effects can be played in real time, without **rendering**, much in the way analog systems had managed effects. Cost-savings in terms of computer horsepower often translate into diminished "real-time" previews of effects. It's the classic price-performance trade-off.

Many of today's digital nonlinear systems are both horizontally *and* vertically nonlinear. In other words, they offer some degree of compositing effects functions and previews (from little to mind-blowing). Since *all* digital nonlinear systems possess real-time *horizontal* previews, the term "real-time preview" now only refers to a system's capacity to preview composited effects in layers of more than one video track without the need to render.

▶▶ Horizontal and vertical nonlinearity: p. 377

"FRAME ACCURATE"

This is a tough term. It has many meanings to many people and is more misleading than any other marketing phrase.

The term never existed in the world of film. Editing film *is* frame accurate. You might have asked, on occasion, if a negative cutter matched the workprint *frame accurately* . . . but that is another story.

The term showed up when editors began offline for videotape, and timecode was the key. If I am using a computer to control a piece of video-tape, and I am looking at a picture on a monitor, I generally want to know "What frame am I looking at?" If I have striped timecode along my video-tape, my computer will display that timecode on my screen.

BUT, there are different kinds of timecode and each has different fea-tures (and costs). *Vertical Interval Timecode* (VITC) is extremely accu-rate, but it used to be relatively expensive. No matter what frame I am on, whether I slowly shuttle forward, then backward, then fast, then slow, it always knows the exact timecode of the frame I see on the monitor. *Longitudinal Timecode* (LTC) is less expensive and potentially less accu-rate. With LTC it is possible to confuse the computer by marking an edit point after shuttling around erratically, in non-play speeds. If for any reason the machine misreads the LTC, you may try to make an edit on frame 3, but the computer might actually record the edit on frame 4 or 5. Your machine and editing system are thus *NOT FRAME ACCURATE.*

This is the original meaning of frame accurate: *the computer knows the video timecode of the frame that you edit on.* Always. Accurately.

Some nonlinear systems use LTC as their cost-effective timecode, which can lead to small errors; consumer DV camcorders generate their own timecode—which makes them extraordinarily more accurate than con-sumers are used to, but can still lead to mistakes if used improperly. Systems using anything but VITC are potentially not frame accurate. Unfortunately . . . **this is often NOT what people mean when they say "frame accurate" with regard to nonlinear systems.**

Nonlinear editing systems were originally created to give film editors the power of modern videotape editing, but with a style that they were used to. The process was thus highly creative, like film editing, but involved no head/tail trims and allowed faster access to material, like videotape editing.

When nonlinear projects start with a film shoot, they often want to finish on film as well, even though they might air on television. Since film speed in the USA is 24 frames per second (fps), and videotape speed

▶▶ Timecode: p. 83   ▶▶ 3:2 pulldown accuracy: p. 111

here is 30 fps, there is a problem of mathematics when editing something on videotape that began and must end on film. Without going into detail here, whether or not a system accurately does this math—allowing a cut on videotape (or in a digital system) to "frame accurately" conform back to film—is the crux of this term.

There are two principal kinds of accuracy: *visual accuracy* and *temporal accuracy*. If a cut is visually accurate, it is frame per frame identical to the frames chosen by the editor on the editing system, but it is still possible that the ultimate timings of cuts are off. If a cut is temporally accurate, it will have the same duration as the digital cut, but it will only be visually accurate to a point; isolated frames can be added or lost to keep the timings in sync.

It is impossible, in the 24/30 fps world, for a digital edit to be identical to both the film cut *and* the video cut. Only one can be truly identical, and the other will be an extremely close approximation, based on a mathematical formula. Type of accuracy and method of conversion varies from system to system. All systems are *frame accurate*, depending on your own particular set of defining parameters. Just make sure you're asking for what you want.

"BROADCAST QUALITY"

Since video information was first broadcast — that is, transmitted through the air to receivers located in your house — it has been well understood that the *strength*, and thus the *quality* of that broadcast signal degraded substantially by the time it got to you. Consequently it was very important for those doing the broadcasting to have a very strict set of engineering guidelines to maintain the technical quality of the shows you see on TV. These strict rules were based on the parameters of the standard video waveform: signal-to-noise ratios, amplitudes of colors, sync and blanking (don't ask), *etc*. While the exact guidelines are perhaps unimportant here, what should be noted is that the term "broadcast quality" is not an arbitrary criterion for images and sound—it is a very well-documented objective series of limits to insure that the shows you get are designed correctly for broadcast (and by the way, looking good was kind of secondary). For almost forty years, the SMPTE-set rules, adopted by the FCC, defined what you could and could not broadcast.

The relationship of image quality to broadcast quality is incidental. So while broadcasters were concerned with getting tapes with broadcast-quality video signals on them, it was the networks, stations, and advertis-

▶▶ Online and offline: p. 79

ers who were most concerned about the actual quality of the images. Each network each came up with its own image quality criterion; but eventually these limitations were dropped. It is marketing-speak to use the term "broadcast quality" to relate to image quality. Still, common usage tends to toss around the *image quality of broadcast-quality video* to actually define broadcast-quality video. This cannot be emphasized enough: *broadcast quality is the quality of the signal and not the image*.

WITH THAT HAVING BEEN STATED FOR THE RECORD . . . the question arises: can you take a less-than-high-quality image and place it into a broadcast-quality signal? (*Yes.*) And what **is** the expected image quality of broadcast-quality video? (*High.*)

Well, it used to be that if the picture quality was deemed *low*, it would not be considered for broadcast. Thus, video images had to be recorded on high-quality equipment in order to be appropriate for TV broadcasting. But the highest-end technology of the 1970s was often less able than today's camcorders to deliver this high-quality product. With the advent first of Y-C camcorders (*e.g.,* S-VHS or Hi-8) and then the substantially higher-quality MiniDV and DVCPRO, the quality of video that the public would accept on TV seemed to be less a factor of the "image quality" *per se*, so much as the content. If some consumer had a home video of a train wreck, the actual image quality certainly wasn't going to stop it from going on the air. Then, with network show after show of consumers' home videos, the resistance of the TV-viewing public to lower-quality images began to drop. Which brings us to *digital images*.

The standard definition of a frame of video is considered to have 525 lines of resolution, refreshing on the screen at 60 fields per second (30 frames per second). But only about 480 of those lines of resolution show up on the screen. In computer terms, an image of "full" resolution (the

GENERATION LOSS. Has it come to this? Has digital become so pervasive that people have begun to forget the idea of the old analog problem of **generation loss**. One reason image resolution and video signals had to be so strictly controlled and high to begin with was that there was so much generation loss over the production, post- and distribution path—film to 1" composite video (1st generation), to an edited master (2nd generation), to a color-corrected and formatted-for-air master, and so on, to an analog broadcast signal, to a consumer monitor. When post-production and distribution are in digital formats, there is simply less generation loss in the path; consequently you have a less-stringent requirement on the starting point resolution of this path. (With digital formats, minimizing the compressions/decompressions and digital-to-analog transfers back and forth keeps images in higher quality.)

▶▶ Image quality-compression table: p. 174

rough equivalent of the "broadcast resolution" criterion), is about 640 pixels across by 480 pixels down, with each pixel having one color out of a spectrum of 16 million (or 24 bits). When you work out the math, you get a frame of about 900K and a second of video (without audio) at about 27MB. Since 27MB/second (216Mbps) is tough for many computers to handle, compression techniques are used to make the video "smaller." Of course, for the most part, making the video smaller also means degrading the image—making it look worse. But many viewers cannot distinguish the uncompressed 27MB/second video from the same image compressed to 2MB/second. And one system's image quality with 2MB/second video (68K/frame) will likely look a little different from another's, depending on what the images are, the method of compression, and so forth; so even describing an image as a given "bytes per frame" is not necessarily a fair comparison.

The video system manufacturers are left with a quandary: *how do you objectively describe a resolution?* Even if it's good enough to broadcast, it is inappropriate to describe it as "broadcast quality." At first, they did it anyway until the officials got mad. Now, there is a whole array of subjective terms that describe image qualities. "Online quality" is a sloppy substitute, roughly corresponding to the images you might see in a professional online bay. Comparisons to tape formats are slightly more concrete: "Betacam SP quality" and "S-VHS quality" are two examples of subjective descriptions. These are not optimal ways to describe image qualities, but they are considered acceptable today as long as users understand the obvious limits and "softness" of the description.

As an end-user of a nonlinear system, realize that there are two ways to describe an image's quality: via *subjective* descriptions and via *objective* ones. "Broadcast-quality video" is an objective description, but of a video signal; as far as image quality goes, **there is simply no such thing as a required broadcast quality**. And now with HDTV broadcasting standards, this term will subside slightly, perhaps. It will all just be "standard definition." Regardless, subjective comparisons of image quality are always possible, and are certainly the norm in the marketing of digital video systems.

"RANDOM ACCESS"

This is an odd term that is unusually common. All editing is random access—that's what editing is. You access THIS scene, then you go to THAT scene . . . then look around for some other scene. . . .

◀◀ Limitations of random access: p. 8

What is generally meant by "random access" is that there is *quick* access to randomly selected points in the dailies and/or the edited master. If you want to go from the beginning to the end of a particular scene, it's really not random access if you have to shuttle through the middle. It is random access if you just "pop" there. Compact discs, laserdiscs, and computer disks all offer serious random access to source material. As long as dailies or created edits are NOT recorded to videotape, that system is going to be indisputably random access.

For the ideal in random access, all source material should be available upon demand. Unfortunately, cost and practicality tend to limit source capacity as much as technical limitations of the various systems. Nonlinear systems, which all provide for a degree of random access considerably greater than other kinds of editing (traditional film or tape), are often referred to as "random access" or "RA" editing systems.

It is important to note that "nonlinearity," "random access," and "film-style" have unique meanings, and can only be used interchangeably in rare cases. In general, nonlinearity is a kind of editing, random access is a kind of cueing, and film-style is a kind of interface.

"DIGITAL"

Since the beginning of the 1990s, and the release of the first EMC2 and Avid nonlinear systems, the descriptor *digital* joined the buzzword ranks. The term refers to the fact that with the newer generation systems, source material is stored digitally on the random-access media, thus differentiating these systems from the analog tape- and laserdisc-based progenitors.

"Digital" is *not* synonymous with "nonlinear." There have been ample nonlinear systems that were not digital; and there are many digital systems that are not nonlinear. The original nonlinear systems achieved nonlinearity via multiple copies of analog (tape or disc) source material under the control of a central computer. Today's digital nonlinear systems simply remove the need for multiple copies of source material. On the other hand, there are currently many digital systems and equipment that are not nonlinear editors; "digital" only refers to the format of the media being manipulated. Linear video editing systems, for example, may use digital tape formats in the post-production process or compositing systems may digitize high-quality video; all are accurately described as "digital" systems.

▶▶ Digital video: p. 151

"DESKTOP"

When the Macintosh was released in 1984 it included a new kind of bitmapped display (new for personal computers, anyway) with icons instead of the familiar computer text CRT; Apple termed this interface "the Desktop." In 1985 Aldus released a software product called *PageMaker* that brought professional publishing powers to a person with a Macintosh computer. In doing so, Aldus coined the term "desktop publishing" to describe their new revolution.

Aldus was making a kind of double reference with their new term. First, the Aldus software literally ran on the Macintosh "desktop," and since the product was designed for and released on the Mac, it was quite realistic to call PageMaker a "*desktop* publishing system." Second, but of greater importance today, they were truly taking a set of tasks and procedures—from typesetting to page layout—and integrating them into a single tool that could be performed on a personal computer, in essence, bringing publishing to a single "desktop." In this sense, the word has come to imply that a personal computer sits on your desk and is run by a single individual.

Since the revolution in publishing brought about by PageMaker, Quark Xpress, and other desktop publishing tools, people have sought to expand on the term's meaning by using it as a suffix for other industries: in particular, **desktop video.**

Today "desktop" is used to describe a system that runs with low-cost software on a personal computer. Ironically, as personal computers have evolved, a "desktop" system for either publishing or video can very often involve many pieces of equipment (that have trouble fitting on all but the biggest desks); they *can* be reasonably expensive and generally difficult to use.

Today the true meaning of the *desktop*, now that it no longer implies the exclusive Macintosh domain, involves a software plus hardware set-up that runs on off-the-shelf personal computers rather than dedicated platforms; it is a tool that while perhaps professional, is still within the grasp of the average user; and most importantly, it is a break from the production methodologies of the previous era, where tasks that had once been diverse are now integrated.

Unfortunately, the *desktop* is sometimes used as a type of diminutive term for professional equipment, but ultimately the word tends to relate more to individual empowerment and liberation.

Chapter 2

Background

"Tele-vision? The word is half Greek and
half Latin. No good will come of it."

— C.P. Scott (1928)

newspaper editor: Manchester [England] Guardian

N O V I C E ' S   I N T R O D U C T I O N

If You Have Ever Worked On a Production . . .
NO NEED TO READ THIS

In the early years, folks going into nonlinear editorial were coming from either a film or a video background. But today, people are getting into nonlinear work with absolutely *no* background in editing. While this book would not presume to teach the novice how to edit, it has been suggested that a small explanation of the production and editing domains would be helpful. As promised, it will start at the very, very beginning . . .

Let's say you're shooting and editing a movie. A big movie. Lots of movie stars. You know how when you watch it on the big screen, it all seems to happen in *real time*. When you're shooting the movie, it doesn't.

So. You and I are in a scene. We're sitting in a restaurant having dinner. (I'll call you "Derek" for the moment.) Here's a piece of our script for scene 24:

                    MIKE
        Nice food.

                    DEREK
        You gonna eat those egg rolls?

                    MIKE
        No. You can have 'em. I'm stuffed.
                    (beat)
        Say, what time is it?

                    DEREK
        (looking at watch) 7:21.

                    MIKE
        Shouldn't Jennifer have called by now?

                    DEREK
        Uh oh, maybe something's happened.
        Let's get out of here.

        (They nod, and exit)

This is the scene. Simple enough? What do you think it would take to shoot it? If you're thinking 20 seconds, think again. Even though the scene will cut to about 20 seconds (and appear to take 20 seconds of real time), it might take all day to shoot. *Why?* Unlike those sitcoms on TV where you see four big cameras rolling around shooting the scene, most scenes in a film are shot with only one camera. Realize that these cameras cost a fortune,* and the company shooting the movie is renting them. So you usually need to get by with one camera.

Next, unless you are going to play the entire scene in a wide shot, which you could certainly do if you wanted it to look like you were in the audience of a stage play, you are probably going to want to cover the action from a few different viewpoints. You may want a close up on Mike when he says "Shouldn't Jennifer have called by now" because you want to see the concern in his eyes. You might want a close up on Derek when he says "Uh oh" and a close shot of a watch when he looks at it. To build suspense, you may want a viewpoint that is far away, with our heroes tiny in the frame, silently playing with their food before you punch in on Mike's close up once he realizes that Jennifer is missing. The point is that directors don't shoot a scene for you to hear the characters simply reciting dialog. They often want to use the photography to build emotion. This is done by shooting from certain angles, by using close ups and long shots, by using different lighting conditions, by moving the camera or not moving the camera, and so on.

So, to *really* shoot the scene, you may begin by placing the camera at the table, where Derek should be sitting, and shooting the entire thing with the camera up close on Mike's face. Derek doesn't even need to be there. He could be off getting lunch. Each time you move the camera to a new spot, it is called a **set-up**. And when you slate each set-up, it is given a unique scene number. The overall scene number for this dinner sequence is "Scene 24." So the first set-up might be called 24A. Now, once you set up the camera, and the lights, and have Mike sitting there, you could shoot him doing the dialog once and it would use about 20 seconds of film.

Done? Are you kidding? That was just Take 1. Mike performed the scene once. But the director may want more emotion, and so they do it again. Take 2. They may shoot scene 24A three, five, a dozen times until the director is satisfied she "caught it." The director may not use every take she shot, but will circle a few good performances on a sheet of paper and "print" those circled takes. Now, the director moves the camera and the whole thing goes again, this time with scene 24B: the close up on Derek. This could go on for hours. Then the camera may be moved again, this

---

*It's not atypical for a good 35mm motion picture camera to rent for $450/day, and lenses, tripods, filters and other accessories can easily raise the amount to $1,000. Film and processing, of course, make it even more expensive . . .

time for a long shot. Both actors will be required in this shot. 24C. And so on. The over the shoulder shot: 24D. The reverse angle over Mike's shoulder back to Derek: 24E. The birds eye view looking straight down at the table: 24F. The Steadicam shot that starts on their last line and follows them out the door: 24G. The aerial photography: 24H. And that's just the shots with the director and the actors. It is possible that these people are all too important to hang out all day to do this. So at some point, another group of people might shoot the outside of the building at night, the close up of the watch, and, of course, the eggrolls. This alternate group is the 2nd Unit (they do *2nd Unit Photography*). And all of this: a total of 45 minutes of film, must be edited together into the simple 20 second sequence you see on the screen.

The film goes to the lab to get developed (as a negative) and then printed (into positive) and then synced with the sound. It's not altogether unlike dropping your vacation pictures off at the Fotomat: you take in the film, and the next day you get prints and negatives back. The editor will get all this raw material, and from it must construct the final product. But how?

When delivered, it is likely that the eight set-ups (24A through H), each with 3 or 4 chosen takes will be assembled together, end on end, for a long, repetitive, boring kind of show (called "dailies," because of the frequency of this ritual). The editor must watch this show, think about how the scene *should* look when done, perhaps talk with the director about her vision for the finished scene, and then go to work—pulling the best parts out of all the takes, and putting them together in a compelling and interesting way, all the while making the work invisible to the average viewer. The finished cut needs to flow *naturally*. There is no one way (and certainly no "right way") to edit a sequence together. And remember, regardless of what was intended in the writer's mind, regardless of what the director was hoping to shoot, the only thing that really matters is what was captured on the film—for that is **all** the editor has to work with.

Another point worth noting: most films are not shot in order, that is, the order of the scenes as they appear in the script. On the first day of shooting, it is common to begin somewhere in the middle of the script (based on logistics and scheduling of actors, equipment, locations, and so on). This means that editors traditionally cut scenes in *shooting* order and not *scene* order. It's just another of the many joys of being an editor.

▶▶ Scene/take numbering: p. 211

Although this handbook is not specifically about the editing of film or video, it is likely that many individuals working in videotape are unfamiliar with film, and vice versa. The following two sections present a brief overview of these two worlds of editing. Not all film or video projects will be exactly as described here, but it should provide you with a general understanding of the concepts and language involved.

T H E   F I L M   S T O R Y

Francis Ford Coppola once described film editing as cutting up a dictionary and writing a novel with the pieces. At a mechanical level, editing is quite simple—cutting and taping.

Let's begin by examining the flow chart on page 31. Film is shot on *location*, along with production sound which is usually recorded on 1/4" audio tape or DAT tape. The negative film you put in your Nikon 35mm camera is not all that different from the film loaded into a motion picture (like a Panavision) camera: it is 35mm wide, the film stock comes in small cans, usually from Kodak, Agfa, or Fuji.

When you get your snapshots back from the Photomat, you know how the film has those little numbers on the edges? 1...1A...2...2A...and so on through your roll. These are called *latent edge numbers*.

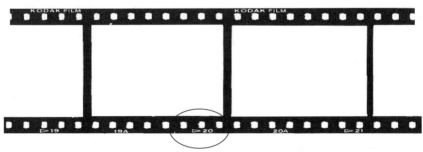

On motion picture film, these *edge numbers* (also known as *key numbers*), are divided into two parts: the *prefix*, which indicates what roll of film was used; and a *footage*, which increments once for each foot of film—usually 16 frames. These key numbers are extremely important because every frame of film must be uniquely labeled. Reference to a key number is the only way a specific frame of negative can be located.

There is also a manufacturer's code on the edge of the film. This tells people what kind of film stock was used, what company manufactured the film, and so on.

▶▶ Edge numbers: p. 98

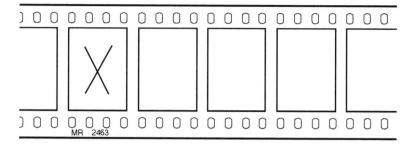

Another difference between your snapshot film and movie film is that your roll is about 3 or 4 feet long, giving you room for 24 to 36 exposures (frames). Movie film rolls are longer, usually 1000 feet each.

The shape of the frame on a roll of film is NOT determined by the film stock, but rather by the CAMERA—its mechanics and optics. The familiar shape and size of snapshot frames is different than the shape of 35mm frames shot in motion picture cameras.

Different kinds of motion picture cameras can even place the image on the frame in different positions and in different proportions. Panavision and Vistavision offer just two of the different ways frames can be placed on 35mm film.

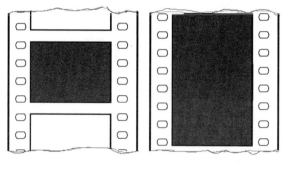

Aspect ratio is the length-to-width proportion of a film frame, regardless of the actual size of the frame. Clearly, two frames can have the same aspect ratio but the area of one can be vastly different from the other. More area means higher resolution (or higher quality) of image.

These two rectangles (at left) have the same aspect ratio but the larger one has more than four times the area.

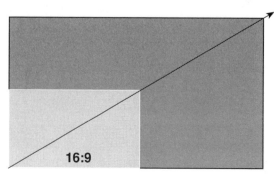

16:9

▶▶ Aspect ratios: p. 177

The *sprocket holes* (also called *perforations*) along the side of the film allow for toothed spools to move the film along in cameras (as well as in all film equipment).

For every frame on traditional 35mm film, there are four perforations along the side (hence the expression "4-perf film"). If you want to make the images a little smaller, you can modify your camera to shoot frames that only have three perforations along the side, called "3-perf film." Doing this makes the same 1000-foot roll of film hold 25% more frames, and although the aspect ratio is changed and film area is smaller, it often provides acceptable image quality at less cost.

When a film roll is finished it is sent to the lab for processing, just like your own 35mm rolls. You are aware how expensive 36 exposures can be to develop and print; imagine millions and millions of exposures! Film productions would rather not waste money on prints of useless shots, so even though they develop every roll they only print selected scenes.

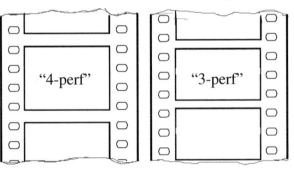

While each roll is being shot on a location, someone (often a Camera Assistant) takes notes as to what scenes are on each roll, how many takes were shot of each scene, and how much footage was used on each take. When a shot is completed, the director will often ask that it be "selected" or "circled," meaning that he liked it or might like it, and wants it to be printed. The Camera Assistant actually circles the take number on the "camera report" and this report goes with the roll of film to the lab.

Also during the shoot, the Script Supervisor takes notes on a copy of the script. These *continuity* notes describe each camera set-up, highlight circled takes, and "line" the script with vertical lines depicting each take. At the end of each day, these continuity notes along with the camera reports are sent to the editorial department.

At the same time, the film and camera reports are sent to the lab for processing. The lab develops all the negative and makes a positive print of the *circled takes*. The negative that was not printed is considered no good (or N.G.) and is placed in a vault as "B-neg"; it will probably never be used but could be printed later if needed. The negative that was printed is vaulted also, but kept separately from the B-neg. No one wants to handle the negative very much. You don't want to risk ruining it with scratches or dirt or anything. Remember, after you have shot a 50-million-dollar movie, after everybody goes home, all you really have to show for it is the negative. It is delicate. It is irreplaceable.

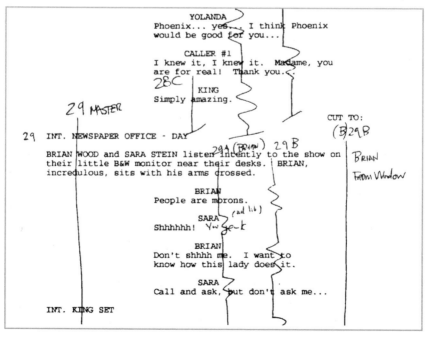

Meanwhile, the 1/4" audio tape that was recorded on location (probably on a Nagra brand tape recorder) is transferred to another magnetic stock, this one with sprockets on the sides so that it is in the same format as the picture. This stock is called *sprocketed mag.*

In the editing department, the print from the shoot (usually called *workprint*) and the sprocketed mag (now referred to as *track*) are delivered usually the next day. First, one of the editorial assistants will *sync* the picture and sound. This involves finding each scene's slate, locating and marking the "clap frame" and then doing the same on the mag. Once the clap point has been found in both picture and sound, the two strips of sprocketed film are lined up, marked, labeled, evened out, and built into larger rolls of dailies.

Then the film dailies are *coded.* Coding workprint and track provides a numeric reference to locate which track goes with which picture. Workprint and track are fed into a table-top coding machine (Acmade is a popular brand) that applies *code numbers* (also sometimes called "rubber numbers") to the celluloid, one number per foot.

Now the workprint has two sets of numbers on it: the original key numbers that came on the negative and can be read on the positive print, and the new code numbers that allow the editor to sync picture and sound, and identify every frame on the roll.

The code numbers and key numbers have no relationship to each other in value or position; they are completely different sets of numbers. So assistants build extensive *log books* that refer key numbers to code numbers, and code numbers to scene numbers and they make labels for every scene to go with the rolls of film.

For each day that film is shot, it is sent to the lab, developed and printed overnight, and returned to the production the next day.

When there is time, the editor, director, cinematographer, and anyone else of sufficient importance gather to watch dailies (or *rushes*) in a screening room. They screen all the previous day's film, perhaps discuss shots the director likes or how he'd suggest approaching the edit, and then other acting or technical matters are evaluated.

Finally, the editing can begin.

The dailies are "broken down" into small rolls, each containing one take. Racks of these little rolls are brought to the editor, who begins watching the film on a machine called an *upright.* The most widely used upright editing machine is manufactured by Moviola. The upright "Moviola" was

invented in 1924, and has remained virtually unchanged since then.

Film is fed into the Moviola; it
can be stopped and played, forwarded
and reversed, at play and slow speeds.
The editor uses foot pedals to control
the machine. With a grease pencil, also
known as a *china marker* (marks made are
sometimes called "chinagraph marks")
the editor marks frames, then pulls the
film from the machine, and with a splicing block to hold the film, cuts it between
frames.

Film dailies are cut apart and taped
into sequences. The leftover portions of
the dailies, the *head and tail trims*, are hung up on small hooks and draped
into a large cloth basket—known as a "trim bin." Each piece of film must
be labeled so that it can be later found if needed. Editors might have many
bins surrounding them while working.

Many editors work with two Moviolas: one for watching and cutting
dailies, and one for playing and re-cutting the developing sequences.
Sequences are strung together until their total length is about 1,000 feet—
these are called "reels." A thousand-foot reel is about the largest manageable load of film, although reels can be built larger, up to 2,000 feet, if
necessary.

An average feature film starts with about 150,000 feet of selected dailies (equivalent to about 27 hours of film), and is edited down over the
course of about 12 weeks to just under 2 hours (about
12 reels of film). Some movies begin with 500,000
feet or even a million feet on rare occasions.

A typical film room might also have
a flatbed editing table for editing or
screening. The most popular manufacturers of flatbeds are KEM and
Steenbeck. Some editors do not work
on Moviolas, but prefer the flatbed
approach. Other editors find the flatbeds cumbersome.

Either way, after months of film editing, a final cut is chosen, and the much-re-spliced workprint is sent to the *negative cutter*, along with a copy of the log book. Using the key numbers to locate negative, the negative cutter carefully cuts and splices the original film negative, using a hot-splicer to melt the ends of the shots together. The workprint is used as a guide to cutting the negative, to insure total accuracy. Negative cutting for a film takes a week or two.

The final negative is later matched back with the final mixed tracks of audio (which, when completed, is printed as a *single optical track negative*) and dubbed to an "answer print" positive to check and modify color balance between shots. Answer prints may sometimes go through a few trials until the color timer gets all the colors right, and everyone likes how the film looks.

When these choices are finalized, an *interpositive print* or IP is created. This is a positive print of the film on negative stock, and is basically a protection copy for the original negative. From the IP, a handful of *dupe negatives* are produced, from which all *release prints* of a film will be created. The very first release print is also used as a check print, to verify the color timings after all the generations of copies have been completed. If the colors are still okay, all the remaining release prints are copied and sent to theaters around the country (and the world).

If this movie were for television, the final process would be slightly different. Rather than make all the expensive prints and dubs, the original negative of the cut film would be sent to a videotape post-production facility and *telecined* (transferred) to broadcast quality videotape, usually Digital Betacam, but today probably a high definition (HDTV) format. Until electronic post-production for television began changing the broadcast industry in the mid-'80s—by allowing for source negative to be telecined, edited on video, and delivered on both film and video—this cut-negative telecine was the primary way film-originated material was shown on television.

▶▶ HDTV: p. 280

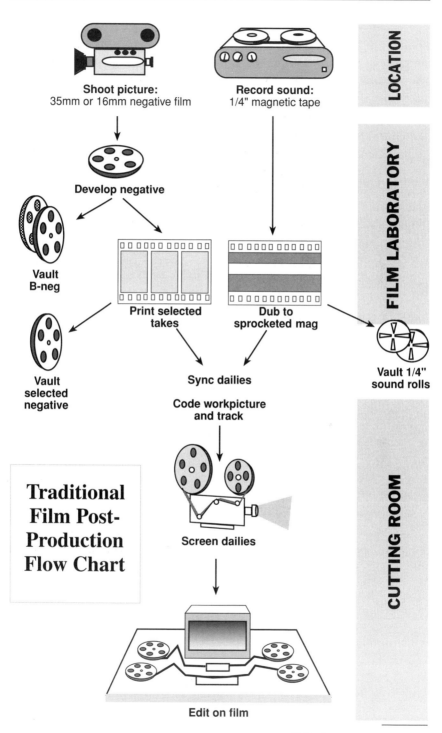

Shoot picture:
35mm or 16mm negative film

Record sound:
1/4" magnetic tape

LOCATION

Develop negative

Vault B-neg

Vault selected negative

Print selected takes

Dub to sprocketed mag

FILM LABORATORY

Vault 1/4" sound rolls

Sync dailies

Code workpicture and track

Traditional Film Post-Production Flow Chart

Screen dailies

CUTTING ROOM

Edit on film

▶▶ Nonlinear flow charts: p. 196

T H E   V I D E O   S T O R Y

Professionally, videotape projects often start with a film shoot. If not originating on Betacam SP (or Digital Betacam), they are often shot on 35mm or 16mm film. These film projects will finish on tape; so although they go through the *3:2 pulldown* in *telecine*, they do not track film information. The telecine process gets the film into the video domain, where it will remain indefinitely.

Telecine records the film onto a high-quality, usually digital, raw stock (Digital Betacam, or D2, D3, D5, DCT, *etc.*) with one or more formats of lower quality video recorded simultaneously during the session (sometimes referred to as a "simo"). This 3/4" or DV tape is recorded with the same timecode as the master Betacam tape, and a visible timecode window is generally "burned" into the lower portion of the frame.

The Denecke, Inc. Dcode TS-1 timecode "Smart" Slate

The audio, recorded on location using a Nagra 1/4" audio recorder or a Digital Audio Tape ("DAT") recorder, is synced up by the telecine operator (known as a "colorist"). To facilitate the somewhat slow process of syncing the 1/4" audio and the negative, many

productions utilize special *Smart Slates*. These are clap sticks that have a timecode display embedded in them. Timecode is fed to both the Smart Slate and audio tape and the colorist can use these numbers to quickly and electronically sync the dailies in telecine.

▶▶ Online and offline: p. 79

For music videos, or projects with playback audio, the 1/4" music tape is synced to the source picture, and laid down on the tape. The 1/4" tape's playback timecode is also burned into a separate window on the simultaneous 3/4" tape and occasionally placed into the user-bits of the videotape (in telecine if not during the production shoot).

Telecine records tapes until they have no more than an hour of source material on them, and often less. Once completed, the master tapes are stored in a special vault by the video post-production facility for use later.

A telecine

The 3/4" tapes are taken to an offline bay and screened. Sometimes, during screening, a frame-grab printer is used to take "snapshots" of the tape at desired frames and print them out.

Detail of a telecine "gate."

Since the 3/4" tape has a visible window containing the timecode, rolls of these images can be built that act as a select log of sorts.

Using a keyboard, the editor types timecodes for cueing, or can mark edit locations "on the fly" (while the tape is playing), simply by pressing "mark in" or "mark out" buttons at the appropriate moment. Using the keyboard edit controller, the editor can mark ins and outs of shots, preview

A Calaway editing keyboard. The basic design and functionality were pioneered by CMX in the 1970s.

photo courtesy of Cintel International, Ltd.

A reasonably typical linear editing set-up: monitors along the top (just out of frame here), the BetacamSP decks along the bottom, an edit control keyboard and monitor front and center. A small audio mixer sits on the desk surface, along with tape stock.

them, and record them onto a separate 3/4" tape called the "record." Edit controllers are manufactured by Axial, CMX and the Grass Valley Group, and others, all of which offer a range of systems that vary in price and functionality. Typical linear offline controllers include the CMX 3600, the GVG 341 or 351, and the Sony BVE 910.

Linear edit systems consist of a keyboard controller, display monitors, and some kind of central computer interfaced to necessary machine controllers that run the peripheral tape transports.

Since it takes time for a videotape deck to move a

Grass Valley Group's System 51. Typical of decades of linear and timecode-intensive editing

tape from a dead stop to 30 fps, you need a kind of running head start before watching a tape. "Pre-roll" is the distance a tape is backed up before the in-point of an edit in order to be moving at the correct speed at the edit point.

For a shot to be recorded properly, both the record tape and the source tape have to be moving together at the right speed. Consequently, editing systems must calculate an interlock relationship from the appropriate edit point, then pre-roll both machines a certain distance back, still synchronized up, then roll forward until they are at speed. Once at speed, the recording will begin at the chosen in-point. If one or both machines aren't moving at the correct speed, or fail to reach the edit point in time, the interlock relationship may be lost, and the edit will be aborted automatically. An editor wants sufficient pre-roll time, without making it so long that time is wasted. Five seconds is usually enough.

Slowly, linearly, an edit decision list (EDL) is built. This list is a chronological history of the editorial decisions made. As each edit is recorded, an entry is placed in the computer memory. If a shot is modified, often both the old and new versions of that event remain in the list.

Monitors in an edit bay generally present the "record" tape large, centrally, and in color. This monitor will either double for "previews," or there will be a separate monitor adjacent to it. Secondary source monitors are located peripherally. These smaller black and white displays are provided for each source videotape deck. This way an editor can see where all the tapes are located.

There is often an *effects switcher* in the bay for doing simple effects and keys (superimposing images). To create dissolves in offline, the outgoing and incoming shots must be on different tapes. If both shots are on the same reel of source material, as they often are, a *B-roll* must be created of the desired shot. This reel, named for the original source tape but usually with an appended "B," will be fed into the effects switcher along with the original source reel, and the switcher

Detail of an online effects switcher.

will superimpose the two as the effect dictates. Because of the technical properties of video, it is important that both images be synchronized not just in time but in framing and color. A peripheral device called a *time base*

▶▶ The EDL: p. 92   ▶▶ Special effects: p. 233

corrector (TBC) is used to balance both horizontal and vertical signals against visual timing anomalies like "jitter." TBCs are also used to synchronize multiple video decks' signals. In general, visual effects cannot be done in analog without a TBC on each videotape machine.

A "key" is a special superimposition of two or more video images. Keying is a complicated area of video editing. To simplify: one image, called the "matte," is a video element with holes in it through which other images can be seen. Sometimes, the key does not involve a matte with "cut" holes in it, but rather a background of one of the primary colors (red, green, or blue) that can be replaced by a keyed image. This *chroma key* is an older type of video keying. It is commonly seen in news and weather reports.

The offline editor edits and re-edits certain events, working in somewhat random ways, and will create numerous entries in the *edit decision list* (EDL). Some of these events will be current, many will be redundant, new shots will be added, and others removed; the list is not usually orderly nor is it easy to follow. Re-organizing the events of an EDL can be laborious and difficult; consequently, most EDLs are "dirty."

Once the offline session is completed, the "dirty" EDL should be "cleaned" in order to maximize the efficiency of the (expensive) online session. There are programs that will clean EDLs—the original for this was ISC's "409." Another significant program that cleans up EDLs is called "Trace," also developed by ISC. With Trace, the edited master tape can be moved from the record deck to a source deck, if desired. Now, with the edited tape as a source, new edits can be made from it to yet *another* record tape. In an effort to cull material from larger numbers of source tapes to fewer ones, this process—the making of "select" or "pull" reels—is common in video editing. However, the EDL will not automatically be able to follow where a shot originated after it has been copied from source to master a number of times. "Trace," using reel identification numbers and timecodes, will compile a single EDL that can re-create the edited master from the ORIGINAL locations of each source shot.

The EDL is copied from the editing system onto a floppy disk, formerly 5¼" but 3½" floppies are now generally the rule. Whatever the size, this EDL must not only be in the proper format (CMX, GVG, Sony, *etc.*) but the disk format itself must be appropriate for the online system (Mac, DOS, RT-11, *etc.*). Together with a printout (also called a "hard copy") of the EDL, the information is brought to online for re-creation using the master elements. Where offline might cost anywhere from very little to a hundred dollars an hour, online begins at a couple hundred per hour, but with digital equipment and full effects, it might cost $1,000+ per hour.

Once the master tape is finalized (and the EDL is *locked*), before, during, or after the online, the audio might be "sweetened" in a multitrack audio session. *Sweetening* is the post-production of audio for videotape masters (also called "mix-to-pix"). Traditionally, produc-tion tracks are laid down onto one or two of perhaps 24 tracks of a 2" audio tape machine. Although not quite as flexible, the 24 tracks of an audiotape are like 24 separate dubbers in a film sound department. Using SMPTE timecode to correlate sync between picture and sound, ADR, effects, and music can be edited to the 2" tape for mixing and eventual layback to the master videotape. In the mid-'80s, nonlinear audio technology was introduced, called digital audio workstations (DAWs). These subsumed the multitrack mixes into the digital domain.

The Modern Video Story

The primary change to the classical video paradigm involves the use of digital offline systems, most commonly those from Avid, replacing the linear CMX and GVG systems of the earlier decades, and High Definition video in the form of 1080i or 24P (another way of saying 1080p/24fps) as the master format in telecine.

Once shot material is telecined to HighDef, and simos are made in any number of lower resolution formats, a telecine log is input into the offline system, and dailies are batch digitized into the computer.

The EDL from the offline system is taken to a HighDef online ses-sion, effects are laid in, and color correction is completed tape-to-tape. The audio is sweetened and mixed (to 5.1 channels) and laid back to the HiDef master, ultimately for delivery to the client.

R T - 1 1

THE GHOST OF (EDIT) COMPUTING PAST

When CMX designed its first linear editing systems back in the early 1970s, they needed a computer to act as the brain for the edit controller, but ideally something other than the usual behemoths that took up a whole room. They found it in a "mini-computer" offered by Digital Equipment Corporation (DEC), called the PDP-11.

But this was in an era where most computing was accomplished in "batch" jobs, where you'd feed in a stack of punch cards and then come back the next day—clearly not an ideal way to, say, preview an edit. Thankfully, the PDP-11 also featured an operating system that allowed a user to interact with it using a keyboard, and in "real time," thereby making it an ideal choice for a linear edit controller. That operating system was called RT-11—**R**eal **T**ime computing on the PDP-**11**.

Since a computer's operating system determines the precise method used to format its storage media, the CMX system (and later, ISC/Grass Valley's) stored EDLs on disks that used the PDP-11's native format. Over the years, the original 8" and 5¼" floppies were phased out in favor of a smaller 3½" size, and the PDP-11 was abandoned completely; but the legacy RT-11 format itself still lives on for reasons of compatibility.

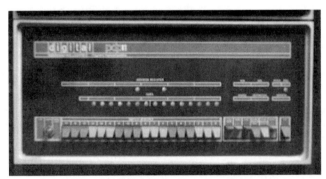

The front panel of a PDP-11 central processing unit.

The PDP-11 had been so popular and earned such a dedicated following in the computer industry that, amazingly, a few are still in use today.

HISTORY AND MOVING PICTURES

In all likelihood, from the point of view only possible from 50 years in the future, students of media will recognize about 100 years dominated by film—in acquisition and distribution—followed by 50 or more years where digital media ruled . . . and there will be this historically curious time where the two transitioned. The *Transition Years*, as they may be called, began with the invention of the videotape recorder (1957) and escalated with the standardization of timecode (1971) and then began to wind down in the '90s, as the economics of digital media changed dramatically. By the year 2000 analog video was on the wane, and the final years of sprocketed film as an acquisition, post-production and distribution medium were at hand.

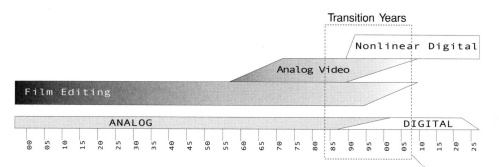

Today's professionals lived through the awkward Transition Years. Those Years were wraught with public confusion, gawky competing formats, dozens of intermediary technologies, and hybrid fits and starts.

From an historical perspective, it can be interesting to analyze the types of editing systems developed through these Transition Years. And while the systems themselves may have become obsolete quickly, the information they provide people interested in streamlining editing system

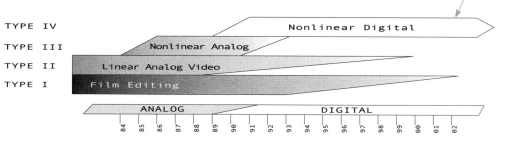

▶▶ Types of editing systems: p. 317

interface and functionality can be invaluable. Understanding details of the evolution of systems (and the reactions of the marketplace and big business) to those changes can be useful in evaluating how new technologies may affect people and industry. There are many parallels, for example, in how the film industry competed with the developing television industry, with each using technology to deliver enhanced consumer experience, and current changes in those now-mature industries facing the Internet.

Movies evolved from nickelodeons (individual silent movie viewers with images a few inches in size, not unlike a *RealMedia* viewer in 1999) to room-size projections over the course of decades. While movie sound was first introduced in 1927, it still took the decade of the 1930s to see it grow in popularity. Television was demonstrated in 1935, and made its big public debut at the 1939 World's Fair. The film industry, not standing still, saw the arrival of television and responded with color. 1939 was *The Wizard of Oz* and *Gone With The Wind*. More color worked its way into films over the 1940s, but not industry-wide until the new color processes of the 1950s. By 1955 the black-and-white NTSC television broadcast standard was revised to include a chrominance signal. But the adoption of color television sets for the public still took more than a decade. This back-and-forth of invention to capture the public imagination continued, through failures like Sensaround, and 3D, to the 1970s and 1980s with widescreen film projection formats followed by higher-resolutions and frame-rate (*e.g.*, IMAX, Showscan) films to this day. HDTV is the latest in the push of television as a film-competitor.

Not only are there parallels between the film *vs* television competitions (circa 1930–1950) and television *vs* Internet (circa 2000–2020). Even in relatively recent years, criticisms of new digital nonlinear systems (1989–1993) sound remarkably familiar to debates about usability and opportunity, as pundits evaluate and predict the future of broadband video on the Web.

The history and experiences in the development of nonlinear editing are not only quaint reminders of the speed of technological change, but may also provide one of the most appropriate benchmarks for understanding the Internet in the present era.

A   B R I E F   H I S T O R Y

Of Electronic Editing

Film editing has hardly changed since the upright Moviola was introduced in the '20s. Even after the flatbed was developed, although film editors often chose one "system" or the other, both involved many mechanical similarities and identical media.

Through the '40s and '50s, television was born and grew. TV was a strange and arguably inferior medium than the theatrical presentation of film, but it was a growing trend and unmistakably pervasive. In its early years, programming was LIVE, broadcast directly from studios in New York to viewers across the country. It was not edited, and thus didn't involve the editing community—it was just this other "thing."

But since Los Angeles time is three hours earlier than New York time, and LIVE television is seen simultaneously all over the country, there was no way for the early networks to provide ideal scheduling of shows for everyone. They wanted a way to delay West Coast broadcasts. But how do you delay a LIVE broadcast?

1956

↙ The delay problem is solved. A company in California—Ampex—invents an electronic recorder for broadcast television; the first videotape recorder (VTR) utilizes a 2" wide roll of magnetic tape and is called the VR-1000.

To properly set the stage, we must first recognize that of the three networks, CBS was clearly the leader. Not only did they have the top-rated shows, but also had a powerful research division called CBS Labs. Although a branch of NBC was actively doing research in mechanisms for using the new videotape, CBS was in the unique position of having both

November 30, 1956:

The historic first videotaped broadcast, "Douglas Edwards and the News," played back three hours after the fact for CBS's west coast audience. Within six months, both NBC and ABC were broadcasting from videotape as well.

photo courtesy of Ampex Corporation

film and video (electronic) production companies in CBS Television: an electronic facility in New York and both electronic and film facilities in Los Angeles. At the LA film studio (in Studio City), they shot shows like "Gunsmoke." CBS Television was in a position to know both worlds well and recognize the advantages and drawbacks of each.

The director of engineering at CBS was Joe Flaherty, a visionary in broadcast electronics and considered by many the greatest single force in the advancement of broadcast editing technology. If Flaherty had a mission, it was to move television production from film to videotape. Film shoots tend to shoot too many takes to cover themselves; you can never be quite sure that you got the shot you wanted. Flaherty believed that with tape, by seeing the material immediately, you could save shoot time, location days, actors' time, *etc.* . . . in other words, save money.

1957

⅄ All three networks now own the new videotape recorders and begin experimenting with ways to edit the videotape.

Originally, editing videotape was like editing film or audiotape—with a razor blade and adhesive tape. Unfortunately, unlike audio, video must be cut between frames, and there is no visual guide as to where a frame boundary is located. Editors used a chemical mixture containing extremely fine iron powder that, when applied to videotape, made the 1/200-inch space between frames—called the "guard band"—somewhat visible. A "clean" edit between frames was difficult; cut transitions frequently lost picture signal and broke up. Videotape editing was extremely hit or miss. In the following years, an expensive (over $1,200) cutting block, called a "Smith Splicer," was developed; it allowed the editor to see the guard band through a microscope. Still, there was an offset between where audio and video events occurred on the tape; even sync-cuts involved careful and slow measurements of picture and audio.

Engineers at NBC developed their own in-house methods for editing videotape that they called the "edit sync guide" or ESG. This was a track with an electronic beep every second, followed by a male or female voice announc-

The Smith Splicer (1957)

ing each minute and second. Coding the videotape and a kinescope (film copy) of the videotape with identical ESG codes enabled television to be edited via traditional film methods. Later, the ESG was used to help physically conform the master videotape, much in the way negative cutters use workprint to conform negative. NBC's methods were so successful that other networks often brought projects to them for editing.

It is somewhat ironic that in the late '50s, video editing was using film as its working media; by the '80s, film editing was beginning to use video.

1962

⋏ The first commercially available electronic editing system, the Editec, is developed by Ampex. ⋏ Time clocks are recorded onto magnetic tape; applications in editing are investigated.

The Editec involved the selective recording of one videotape (source) onto a target videotape (master). Since there was still no visible way to tell where you were on your tape, electronic pulse tones were recorded along the length of the tapes at desired edit points. An electronic edit meant a cleaner splice and the audio/video differential in cut point was rendered moot. But previewing or exactly re-creating an edit was impossible; you simply had to try it again. The problems were (1) the frame accuracy of the point the edit started recording on the master tape, and (2) how to synchronize the source tape and master tapes while rolling. These problems were solved very slowly over the following years.

In the mid-1960s, Dick Hill, a CBS technician in Los Angeles, happened upon a piece of Defense Department technology that allowed electronic recording of a time clock on magnetic tape. When the military

photo courtesy of Ampex Corporation

The Ampex Editec (1958)

was testing missiles, they had devices monitoring from various locations around the test site: some at the launch point, some thousands of miles away. Launch and missile data were recorded on magnetic tape, and along with that information was a time-of-day clock, which allowed them to track and relate events occurring at different data recording locations.

Dick Hill felt this technology might solve the pulse tone problems in editing and began sending reports to CBS headquarters in New York. Adrian Ettlinger, a CBS engineer, was sent to investigate Hill's reports. Ettlinger reported back to CBS that this was a good thing, and that they should encourage the development of what was to become known as TIMECODE.

1 9 6 7

⅄ EECO (The Electronic Engineering Company), in Orange County, California, begins manufacturing the first timecode equipment. ⅄ IBM creates the first floppy disk.

With timecode available, CBS Labs began developing an editing system to utilize these numbers. Ettlinger, with other engineers, first experimented with a system of Sony recorders that could perform a continuous play of an edited sequence by accessing a number of duplicated source tapes—the origin of "look-ahead" previews. But the 2" tape machines were unwieldy and too many would have been required to achieve any useful kind of nonlinearity. Later they began using computers to control various kinds of newly-developed 1/2" videotape machines, with mixed degrees of success.

1 9 6 8 – 6 9

⅄ NBC's "Laugh In" pioneers the use of videotape editing as something more than an extension of live switching. The show has 400–450 physical edits where most other shows have between 40–100. ⅄ The first version of UNIX is invented for the PDP-11 computer at Bell Labs. ⅄ ARPAnet is launched.

By 1969, none of CBS Labs' tape format experiments had proven particularly appealing, so they abandoned tape and began experimenting with magnetic disk platters for storing the analog video information. They ran into some mechanical problems involving the disk platters and turned to the Memorex Corporation for assistance. Memorex was intrigued with exploring the technical feasibility of recording video on disk media for random access.

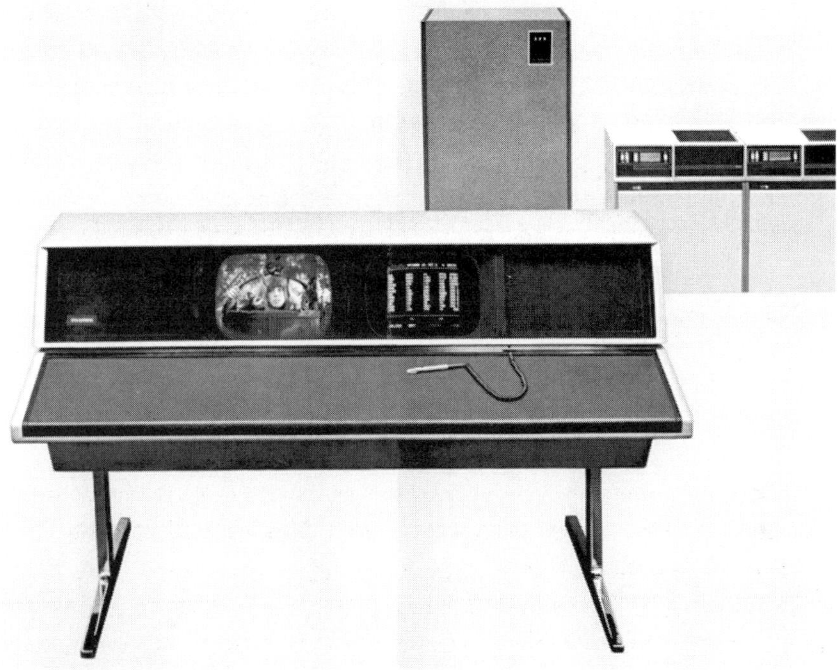

The CMX 600, as shown in the original CMX product brochure (1971).

1970

⅄ In January, CBS Labs and Memorex create a joint venture called CMX to make editing systems. They build the CMX 200 and CMX 600.

The business plan for the CMX (which stood for **C**BS, **M**emorex, e**X**perimental) venture was drawn up by Adrian Ettlinger and Bill Connelly (both from CBS), and Bill Butler and Ken Taylor (both from Memorex). Butler became the company's general manager. Their first product had two parts. The first part, the CMX 600, was a computer with a stack of removable disk platters, each holding about 5 minutes worth of material. The platters looked like a horizontal bread slicer—their cost: $30,000. With six disk drives working in union, the 600 could locate any frame in under a second from about 30 minutes of mediocre (half-resolution) black-and-white dailies. This part was a nonlinear editor, using what would become SMPTE timecode. It did not produce a master videotape, but a punched paper tape encoded with a list of timecodes for re-creating the edits on a broadcast-quality machine. The second part of the system was the CMX 200, a linear tape assembler that would take as input the 600's list and build a master 2" broadcast videotape.

1971

⅄ The CMX 600/200 combination costs $500,000 and is patented by CBS; its inventor is Adrian Ettlinger. In April it is demonstrated at the CBS stockholders' meeting.

The first commercial application of a nonlinear editing system, the CMX 600, is begun. CBS Television used the system on a movie-for-television called *Sand Castles*.

1972

⅄ After a number of years of investigation, the Society of Motion Picture and Television Engineers (SMPTE) adapts the EECO timecodes and standardizes their industry use.
⅄ Hewlett-Packard releases the HP-35, the first handheld calculator, making the slide rule obsolete virtually overnight.

By this time, five CMX 600s had been built—besides placements at CBS and Teletronics in New York, one was located at Consolidated Film Industries (CFI) in Los Angeles. Head editor there, Arthur Schneider (formerly with NBC) and Dick Hill (formerly at EECO) worked with CMX programmers, among them Jim Adams and Dave Bargen, to develop and beta test improvements to the systems. Problems in the CMX 200 led to the CMX 300. Where the 200 could only take input from a 600's printed papertape, the 300 allowed for a keyboard terminal to interface directly with the assembly editor. CFI suggested that there was a wonderful product in the linear online system, the 300. Both for practical reasons (perhaps due to its high cost, poor image quality, high maintenance requirement, and limited storage of source) and for social ones (it radically changed all accepted work rules and production concepts), the 600 did not catch on; however the linear editing products that CMX developed did.

At the National Association of Broadcasters (NAB) show, the CMX 300 was demonstrated. With guidance from users like CFI, the CMX 300 continued to evolve; by 1978 it had become the 340.

1973

⅄ Sony introduces 3/4" tape. ⅄ The CMX 50 is debuted at NAB. ⅄ The XEROX Alto is developed: the first computer with a graphical display, mouse, menus and icons.

Sony introduced the first 3/4" machine (and a standard for 3/4" tape called U-matic) with hopes of seeing it adopted for home television recording; but it was too expensive and too large for consumers, and ultimately did not take off. On the other hand, because it was more cost effective than previous editing formats, 3/4" videotape was picked up by the broadcast and industrial markets.

The keyboard controller for the original CMX 50 (1973).

At the NAB show CMX introduced the CMX 50, the first commercially available 3/4" edit controller—because of the inexpensive U-matic tape machines, it was considered the first viable "offline" system. The product used a standardized Edit Decision List (EDL), look-ahead previews and assemblies, and a style of editing that continued to be dominant for the next 20–25 years. With the commercial success of the 300 and the 50, CMX wanted to expand; however, Memorex was financially strained, and CBS would not contribute more than 50% of the company's financing. The decision was made to sell the upwardly mobile company.

1974

⚐ Orrox purchases CMX and names Bill Orr as president. CMX's manager of product development, Dave Bargen, leaves the company. ⚐ After two years of development, two separate corporations—MCA and N.V. Philips—announce that they are working together to standardize formats for their new invention, videodiscs.

1975–76

⚐ The MCA/Philips agreement produces a format for optical video discs known as the LaserVision standard. Around the same time, a number of companies, among them Sony, Philips, and Hitachi, announce their research into laser optical audio discs.

Due to popularity of the systems used by existing CMX clients like CFI and Vidtronics, Dave Bargen began writing software programs that would increase the functionality of the CMX systems. First he developed "409" (an EDL cleaning program), then "Trace" (another special list management tool), and finally "Wizard" (which became "Super Edit"). Eventually, Bargen marketed these products to other CMX clients.

1 9 7 7 – 7 8

⅄ The CBS-Sony system is created. ⅄ Laserdisc players are first sold. ⅄ A film from young director George Lucas dominates at the box office. "Star Wars" has what will become the largest ticket sales in the history of movies; the trilogy of films will gross over 4 billion dollars in ticket, cassette, and product rights over the next ten years. ⅄ IBM researchers build the first relational database.

Because Joe Flaherty still believed that television should be done on videotape, CBS Labs continued to develop systems. Late in 1977, CBS began an advanced form of their CMX 600 project, this time using a new type of 1/2" videotape, called "Beta," that was being developed by Sony. This new format unfortunately required expensive tape decks. CBS's new editing system used modern computers, and the interface, like the 600's, was through a lightpen. Adrian Ettlinger was moved from consulting peripherally on the project to being the software product manager. By 1978 the CBS-Sony system, as it was called, was in use at CBS. But both companies decided not to pursue the marketing and manufacturing of the product, and it remained in-house at CBS.

In 1977, the first LaserVision videodisc players were sold in the educational market. MCA teamed up with Pioneer to form a venture, Universal Pioneer Corporation (UPC) to mass-produce videodisc players. Magnovox introduced their competing disc format, "MagnoVision," utilizing completely different technologies to play videodiscs.

At the New York Institute of Technology (NYIT), Alvy Ray Smith and Ed Catmull invented the "integral alpha channel."

1 9 7 8 – 7 9

⅄ New video formats deluge the market. ⅄ ISC is formed. ⅄ Philips demonstrates the first audio "compact" discs. ⅄ Bell and Howell acquires Telemation. ⅄ CMX invests in DBS; begins developing new kinds of editing systems. ⅄ Lucas and Coppola investigate video applications to filmmaking. ⅄ Hayes ships the first 300-baud modem.

A host of new video formats began to inundate the consumer market: Sony's **Beta** format on 1/2" videotape; Panasonic also had a 1/2" tape format called **VHS**; Magnovox's MagnoVision videodiscs; MCA and partner IBM's 5-year-old entity DiscoVision Associates (DVA)'s 12" videodisc format, LaserVision.

Dave Bargen, formerly of CMX, worked with a former chief engineer of Vidtronics, Jack Calaway, who had built a somewhat more flexible machine controller interface than the previously best-known CMX I^2. Bargen used a DEC computer, Calaway's hardware, and his own CMX-like software to create a new editing system. He formed the Interactive Systems Company (ISC).

In mid-1979, CMX/Orrox's penetration into the equipment market was beginning to plateau (over 90% of all broadcast editing in 1978 was on CMX equipment). CMX/Orrox saw the Direct Broadcast Satellite (DBS) business as the next boom industry. In a bold move, they began to invest heavily in DBS.

At the same time, development began on the newest CMX products, the 3400 and 3400 Plus. The plan was to move videotape editors smoothly from somewhat difficult and number-intensive editing (in the 340) to a modern editing system; the 3400 Plus would have a database management system (DBMS) and soft-function keys, and would implement a new computer feature—a windowed graphical interface.

Also in 1979, George Lucas began investigating ways of improving the filmmaking process. His friend and mentor Francis Ford Coppola had been active in using video technologies to help in production. Coppola and Lucas, like their friend at CBS, Joe Flaherty, understood the cost savings that could be achieved on location by shooting in video instead of film—by being able to view immediately the material you had shot. But since neither felt video looked as good as film for production, it was generally understood that the video would only be a tool in the film process.

Coppola had pioneered the use of a video camera alongside the film camera on shoots—the video was fed to and recorded in a customized trailer that sat at his locations, called the "Silver Bullet." With his video specialist Clark Higgins, Coppola is considered the first to use video assist in film productions, in particular on 1982's *One From the Heart*.

Lucas was more interested in facilitating post-production by using computers and video technology. He hired an Ed Catmull, an expert from NYIT, an institution known as a leader in academic computer applications, particularly in computer graphics and animation. Catmull moved to Lucasfilm in California where he began investigating the post-production process and planning what type of technologies could be used.

Bell and Howell, a Chicago-based film equipment company established more than a half-century earlier as a manufacturer of 35mm film printers, purchased an equipment distributor/manufacturer called Telemation. Bell and Howell's Video division (consisting mostly of tape duplication) hired Jim Adams for Telemation, and in doing so, acquired his previously developed editing system, the Mach-1.

1 9 8 0 – 8 1

⋏ German equipment manufacturer Bosch-Fernseh purchases 50% of Telemation from Bell and Howell in 1980; by 1981, it acquires the remaining 50% (and the Mach-1) from Bell and Howell, and calls the new company Fernseh, Inc. ⋏ Early in 1980, George Lucas creates his Computer Division. Late in 1981, Lucas releases his third blockbuster, "Raiders of the Lost Ark," directed by friend Steven Spielberg. ⋏ Adrian Ettlinger begins work on his ED-80 editing system.

At Lucasfilm, Ed Catmull presented a plan for the development of three products: a picture editing tool, a sound editing tool, and a high-resolution graphics workstation. The proposed development cost of this three-project plan was about 10 million dollars. It got the go-ahead after the success of Lucas' new film, *The Empire Strikes Back*.

By the end of 1980, Catmull hired three computer experts. The picture-editor (which became the EditDroid) was led by NYIT friend Ralph Guggenheim, also an NYU film school graduate. The sound-editor (which became the SoundDroid) was led by Andy Moorer, a digital audio pioneer who had been with CCRMA (the Center for Computer Research in Music and Acoustics at Stanford University). The graphics-project (which became the Pixar) was led by Alvy Ray Smith, another NYIT friend and graphics expert from, among other places, Xerox PARC (their research division). By late 1981 the Lucasfilm Computer Division was actively developing these products.

But on what kind of workstation would they be run? At the time, there was no such industry as "desktop computers," and a number of companies were competing against the giant IBM for the market of smaller high-powered computer workstations. The host computer had to be relatively small, fairly inexpensive, and of high enough horsepower to run simultaneously all the things these systems had to do. Lucasfilm investigated all options, tried a few different hosts, and finally chose a SUN computer (with Motorola's 68000 processor) costing around $25,000.

As for what video media would be the source for the EditDroid, the decision was made in Catmull's '80 report. At the time there were only a handful of potential formats: 1/2" Betamax, 1/2" VHS, LaserVision and MagnoVision 12" videodiscs. Catmull, and later Guggenheim, saw laserdiscs as the defining characteristic of the EditDroid project—they were convinced the price of laserdiscs would drop, that a consumer adoption of discs was just around the corner, and that discs were the only way to achieve any kind of real nonlinearity.

Lucasfilm began talks with leaders at DiscoVision Associates about the possibilities of recordable LaserVision discs. The EditDroid, at the time called the *EdDroid*, was a nonlinear, computer-controlled, disc-based editor. The system would control tape (because you had to control 3/4" for emergency linear editing); however, the thrust of the development was on a new "style" of editing, manipulating the laserdiscs, and the idea that the human interface was going to be significantly more "user friendly" than computers or video editing systems.

The point-and-click computer style, mouse-based control, bitmapped and icon-based graphics were unseen in the consumer world, and only beginning to take the academic computer world by storm. The Computer Division's products all utilized these features.

Throughout the early 1980s, numerous individuals and organizations came through Lucasfilm's Computer Division to see what was going on. Among them, an English entrepreneur named Ron Barker, who had been working outside of Boston, and who was interested in developing a revolutionary new editing system of his own. Also visiting the team was Adrian Ettlinger, who had much earlier left CBS Labs and had also been developing a new editing system of his own.

Ettlinger's system was based on a smaller computer, the Z-80, and was originally called the ED-80; it was named for the host computer and for the year in which his project began.

1982

▲ The Montage Computer Company is formed and work begins on the Montage Picture Processor. ▲ The Optical Disc Corporation is formed. ▲ The Compact Disc debuts.

Ron Barker teamed up with engineer Chester Schuler and formed a new company outside of Boston, The Montage Computer Company. They began development of a tape-based nonlinear editor, using the 1/2" Beta format videotapes as their source media. They called the system the Montage Picture Processor® because in many ways it was like a word processor.

DiscoVision Associates (DVA), the venture between MCA and IBM, was shut down and most parts were sold to former-partner Pioneer Electronics. DVA remained as a small company holding the rights to all the laserdisc-related patents. Members of the DVA development team joined forces to start a company to develop a recordable laserdisc, initially for the editing market. They called the company the Optical Disc Corporation (ODC).

Lucasfilm hired Robert Doris to take over the management of the Computer Division. For the next 18 months there was a great deal of development activity both at Lucasfilm (outside of San Francisco) and at the Montage Computer Company (outside of Boston).

1983

⅄ A prototype of the Ediflex is built. ⅄ BHP spins off as an independent company. The TouchVision is announced. ⅄ Lucasfilm and Convergence create a joint-venture for the Droid products. ⅄ Lotus 1-2-3 is released, the first "killer app" for PCs.

During its time at Bell and Howell, the Mach 1 interested members of the R&D division, called Bell and Howell Professional (BHP). They were interested in potential applications of video technology for film. Bruce Rady, with the rest of the R&D team at BHP, began working on a new video-for-film editing system project which they called the Envision.

When Bell and Howell was taken over in a leveraged buyout, they began to analyze ways to divide up and sell off the company. Management chose to drop all things "video" related, and the Envision project was dismantled. The members of BHP were encouraged to purchase their division from the Bell and Howell parent company, which they did; Bruce Rady bought the rights to the Envision project for developing on his own time.

Rady began designing his own hardware and software, in particular an extremely low-cost VITC reader, for use on his system—its name was then changed to TouchVision.

CMX/Orrox decided to close its DBS manufacturing division. Changes in microprocessor technology along with other factors left them unable to recoup their huge investments. At the NAB show CMX/Orrox debuted their 3400 and 3400 Plus to crowds; following the show, CMX began layoffs and eventually stopped development of the 3400 Plus. Its designer, Rob Lay, moved to Lucasfilm's computer research department to work with the assembled team on the editing project. CMX's slow-down in the editing equipment market allowed other companies to enter the arena and offer new products. ISC began to erode CMX's high-end equipment line, and new smaller companies formed to move on lower-end systems.

Ettlinger, now with partner Bill Hoggan, completed a prototype for a new editing system with specific applications for the television market in

Los Angeles. He worked with a number of editors and designed a script-based editing system that would control 1/2" VHS videotapes as sources. The system, no longer the ED-80, was called the *Ediflex*.

Lucasfilm's Computer Division, under Robert Doris, begin looking for an equipment manufacturer to license and sell the Droid products they were developing. Their investigation turned up seven companies, among them CMX, ISC, Grass Valley, and Convergence. Lucasfilm considered the ideal partner to be a privately-held company (like itself), one that was relatively profitable, and one that was ready to sustain a joint venture. In September '83, Lucasfilm Ltd. and the Convergence Corp. signed a joint-venture agreement for the Droid products.

1984

⅄ The Montage and the EditDroid debut at NAB. ⅄ ISC merges with the Grass Valley Group. ⅄ Apple introduces the Macintosh personal computer. ⅄ ODC begins shipping laserdisc recording systems in the Fall. ⅄ The Laser Edit facility opens in Hollywood.

In late '83 and into '84 both the Montage Computer Company and Lucasfilm had been teasing the market with advertising and promotions about their new film editing systems. For the EditDroid, Convergence was beginning to manufacture the necessary hardware, and Lucasfilm was continuing the development.

NAB 1984 was the debut of these two editing systems, to a great deal of interest. Although both presented prototypes, neither the EditDroid nor the Montage was market-ready. The EditDroid had to deal with the high costs of making laserdiscs and also the software difficulties associated with releasing a product running on the UNIX operating system.

The original EditDroid product nameplate (1984)

The Montage had to deal with the seventeen 1/2" consumer Beta machines that the editor controlled. Betamaxes had no provision for machine control, and so each machine on each system had to be specially fit with a custom machine-control card. (The cost of 17 professional videotape machines would have been prohibitive.)

By the summer, Lucasfilm and Convergence were still hammering out the nature of their joint agreement. Lucas decided to spin off all parts of the Computer Division; Convergence, still running its own independent business, signed the spin-off papers officially creating a new company with Lucasfilm. Negotiations continued for more than six months.

Guggenheim, the leader of the editing project moved to the graphics project, and Robert Lay moved into his position.

ISC, after having grown tremendously since its inception, was merged with Grass Valley, an almost 20-year-old company known for its video production switchers, graphics and routing equipment.

Apple released the first desktop personal computer to use bitmapped and icon-based graphics and a Motorola 68000 processor at under $5,000—the Macintosh. Advances in memory and processor power began to dramatically affect the "computer-consciousness" of the public, with powerful computers that were easy and affordable.

By spring, a new post-production facility was formed in Burbank, called Laser Edit, Inc.; it owned its own in-house system called the Spectra-Ace (developed by Spectra Image exclusively for them). Laser Edit was the beta site for the prototype of an optical disc recorder made by the ODC, and developed over the previous two years. The Spectra system was similar to conventional linear videotape editing systems except that dailies were delivered on ODC laserdiscs and used in specially-developed two-headed laserdisc players. Although disc-based, the system was linear and proprietary at the Laser Edit facility. President Bill Breshears initially marketed the system to multicamera television productions.

1 9 8 5

⅄ Interscope purchases the Montage Computer Company. ⅄ Cinedco is formed to manufacture the Ediflex. ⅄ The CMX 6000 project begins at CMX. ⅄ Film-originated television programming begins testing the new systems. ⅄ AOL is founded.

In February, the Montage Company sought a new finance partner; Interscope Communications (which had just purchased Panavision from Warner Communications) bought the company from founder Ron Barker and brought in new management. The company was again on solid ground.

At the NAB show both the Montage and the EditDroid were shown again, this time considerably more stable than as prototypes the previous year. A prototype for the SoundDroid was also displayed by The Droid Works (the new Lucasfilm and Convergence company). The cost of the

Montage and the EditDroid were comparable, well over $200,000 each, and both already had placed a few "beta" systems in the field.

Also at the show was a prototype of Bruce Rady's film-style editing system, the TouchVision. As a developer from the film equipment business, Rady was relatively unfamiliar with the video editor's world. But in extensive interaction with film editors, Rady learned that 1) film editors didn't really want computerized equipment, and 2) every editor described what he or she did differently, so coming up with a grand new simplified model of editing would be problematic. Basing the TouchVision on a flatbed, his system essentially allowed editors independent control of the source decks, instead of disguising the multiple decks necessary for achieving nonlinearity. By being able to lock and unlock various decks, editors were allowed to work as if using an electronic (videotape) multi-plate flatbed. The TouchVision prototype controlled three 3/4" Sony decks for source and was target-priced around $100,000.

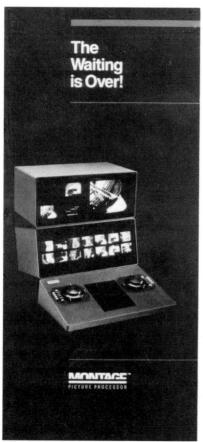

The Montage brochure cover (1985)

In May, Milton Forman and Adrian Ettlinger formed Cinedco to manufacture and rent Ettlinger's Ediflex system. Ediflex was not at the NAB show, because as part of a cautious marketing plan, it was only rentable in Los Angeles, for around $2,500 per week. By approaching the film market in a way it understood—system rental—the Ediflex, too, was being tested for use in the post-production of the 1985-86 episodic television season. The script-based system originally controlled eight JVC-400 VHS decks with its central computer, and used a light-pen to interact with the screen. Also for rental-only was Laser Edit's Spectra System, in Burbank.

CMX began to develop a new nonlinear system specifically for film editors. Seeing nonlinear as the direction all editing would go, CMX developer Robert Duffy became project leader on a completely disc-based system that will become the CMX 6000, a modern incarnation of CMX's earlier 600, and bearing little resemblance to their existing linear products.

Lorimar, more than any other studio, experimented with electronic post under Gary Chandler, Lorimar's vice president of post-production for both television and theatrical projects. Lorimar's motion picture, *Power*, directed by Sidney Lumet, was the first feature film to be edited electronically, using the Montage Picture Processor.

Pacific Video, the largest facility in Los Angeles in 1985, and a pioneer in new technology, began testing the new electronic systems—the EditDroid, the Montage, and the Ediflex—for use on film-originated television programming. President Emory Cohen created a film environment in a video post facility in an effort to streamline the new electronic process for traditionally film-oriented productions: he coined the term "the Electronic Lab."

1986

⅄ Television and some films begin using the new nonlinear systems. ⅄ The Montage Computer Company is closed and re-formed. ⅄ BHP's TouchVison and CMX's 6000 debuts at the SMPTE show.

With sales slow, and development costs high, heads of both Montage and The Droid Works are troubled. In spite of great excitement over their products, the president and principal owner of Interscope Communications, Ted Fields, announced in March that the Montage Computer Corporation would be liquidated. Similarly, Droid Works Chairman of the Board Doug Johnson announced the departure of TDW president Robert Doris. The demise of the Montage was ironic in light of the activity on the roughly 29 systems in the field: Alan Alda had switched from editing *Sweet Liberty* on film to the Montage; director Susan Seidelman was planning on using the system for *Making Mr. Right*, and rumors from London said Stanley Kubrick was planning to use the system for *Full Metal Jacket*.

NAB '86 did not see the Montage; however, improved EditDroids and SoundDroids were demonstrated on newer and less expensive SUN/3 computers. By May, the Montage technology had been purchased at auction for $700,000 by a New York businessman, Simon Haberman.

By June, The Droid Works' board had appointed a new president to the company, (the first president of CMX) William Butler.

In August a new company was formed from the remains of the Montage Computer Corporation. Called the Montage Group, Ltd., the new company announced a new smaller Montage II system that would cost $150,000 less than their original system, and promised that it would be shown in prototype at NAB '87.

Ediflex boomed in the television rental market. By August, Cinedco announced that all 21 Ediflex systems were in use and that they were turning away business. Lorimar-Telepictures, under vice-president of post-production technical services Chuck Silver, pioneered the use of the Ediflex, as well as other videotape-based systems, for use in network television. Universal was also experimenting with electronic post under vice president of post-production Jim Watters: Universal's Alan Alda-directed feature *Sweet Liberty*, cut first on film, went through additional cuts on the Montage, following the system's success on Lumet's *Power*.

At the SMPTE show, CMX introduced its CMX 6000—controlling only laserdiscs and no videotape. It was extremely fast and was the first system with a "virtual master," the ability to roll a "simulated" version of the edited master in all directions and at all speeds, without recording to tape.

Up until this point in time, nonlinear video systems could not simply *play* a cut; like rendering graphics a decade later, they had to devise clever ways to preview edits. But laserdiscs could do what hard disks could do better: cue quickly from bits here and there with no recording at all. The 6000 was the first nonlinear system to let editors grab a cut and roll at will, an experience more like the yet-to-be developed digital editing systems than previously seen. While the practicality of laserdiscs was

CMX 6000 advertisement (circa Winter 1986)

BHP advertisement (circa Winter 1986)

often in question, the experience was not.

Also at the show was BHP, Inc., the manufacturer of Bruce Rady's Touch-Vision system, now controlling 9 or 12 VHS tape machines instead of 3/4" tape, and featuring a unique touchscreen for control. Notably absent from the show was The Droid Works' SoundDroid. Two other products of interest shown were Kodak's prototype of their new film numbering system, Data-kode, and Cinesound International's Lokbox—for interlocking a video-tape machine with a film synchronizer. Both would aid in the electronic post-production of film.

1987

⅄ The Droid Works is closed. ⅄ Work begins on the E-PIX, the EMC and the Avid/1.
⅄ A decade of research and politics begins on what will eventually become DTV.

By the end of January, Lucasfilm announced the closing of The Droid Works, and that no more Droids would be sold. No SoundDroids were ever manufactured beyond the proto-type, and only 17 EditDroids were made. The post-production division of Lucasfilm, Sprocket Systems, absorbed the remaining Droids and continued in-house develop-ment of the EditDroid with a new devel-opment team.

NAB '87 saw the new Montage Group demonstrating their Montage II in prototype.

A software designer on staff at Adcom Electronics, a Toronto-based equipment distributor and manufacturer, wrote a pro-gram that would help some filmmakers edit their film project on videotape. Shawn

Carnahan, the designer, created a database for establishing a relationship between film edge numbers and timecode, and Adcom sold the product as "Transform LM." Soon, Carnahan began to add machine control and video switching to the product. He called the system the E-PIX.

Desktop computers continued to become more powerful, memory continued to increase in density, and applications became more sophisticated.

Bill Ferster, president of West End Film in Washington DC, had grown increasingly weary in the videotape assemblies of his company demo reels each year. He designed a PC-based editing system that accommodated his needs, storing images digitally on hard disk, and playing them back at 15fps. He decided to sell his seven-year-old animation/graphics company.

At around the same time, William Warner, a marketing manager at Apollo computers outside of Boston, had grown frustrated with the time and cost of the online he endured for the preparation of corporate projects. He, too, began developing an editing system; this one on the Apollo workstation—a high-powered professional system designed for computer graphics applications.

1988

▲ The Montage wins an Oscar. ▲ E-PIX debuts at NAB ; the EMC debuts at SMPTE.

The Montage Picture Processor won the 1987 Scientific and Engineering Achievement Academy Award® for its use in cutting films such as *Full Metal Jacket*, *Power*, *Sweet Liberty*, and *Making Mr. Right.*

By the NAB show in Las Vegas, two new systems were demonstrated in prototype. Adcom debuted its disc-based system, the E-PIX, after eight months of development. Carnahan had designed the system after examining the problems of the major systems in 1987; he did not want a huge bank of peripheral equipment, and ODC disc mastering was still somewhat expensive and difficult to come by. He wanted a stand-alone system that was not tailored to any single market, but generalized to many high-end markets.

William Warner arrived to the NAB show with no intention to market or sell his system, only to get input. He set up two of his prototypes in a hotel room; the system was called the Avid/1 and ran on the Apollo computer with a fixed hard disk, played video at 15fps, and had no audio capabilities. Still, interest

in the digital system was high, and
Warner absorbed the information given
him: the system had to run at 30fps,
video image had to be better quality, it
needed audio tracks, and storage quantity was a big concern.

After NAB, Warner left Apollo
and created Avid Technology with
venture capital from the Boston investment community. Apple approached
Avid, and by the end of the year
the prototype was ported to a Macintosh computer, video was moved to
magneto-optical discs for storage, and
playback was at 30fps.

In July, with $6 million from the
sale of his West End Film, Bill Ferster

Don't Be Fooled by Smoke and Mirrors!

E·Pix
Hybrid Editing System

E-Pix advertisement, Fall 1989

financed a new company to manufacture and market his digital nonlinear
editing system running on a personal computer. Ferster called it the Editing
Machines Corporation, and the system was called the EMC2; he had moved
the storage media from hard disks to magneto-optical drives. The EMC2
had its debut at the SMPTE show in New York. The system price was under
$30,000.

1 9 8 9

⚊ In January, EMC has shipped the first digital systems. ⚊ The Avid/1 debuts, and the
company begins taking orders.

ACADEMY AWARD® LEVELS OF TECHNICAL ACHIEVEMENT

There are three levels of merit associated with the Academy
of Motion Picture Arts and Sciences (AMPAS) awarding:

1. **Technical Achievement** (certificate)—for contribution to the
   progress of the industry;
2. **Scientific and Engineering Award** (plaque)—for inventions
   with definite influence on the advancement of the industry;
3. **Academy Award of Merit** (statue)—the highest degree of
   award for definite influence on the advancement to the
   industry, in particular, for changing the way the industry
   works.

At the NAB show, Avid Technology began to take orders based on their prototype of the Avid/1, now with full editing capabilities, although no systems had yet been manufactured. At the SMPTE show, Avid demonstrated a second channel of audio; plans were to ship the beta systems by year's end. By December 29th, the first Avids were shipped.

1990

⋏ LucasArts shows a new editDROID. ⋏ Steady improvements are made in all digital systems. ⋏ A researcher in Europe invents HTML and HTTP.

February, the EMC2 demonstrated 30fps digital video, and began using the C-Cube compression technology in order to increase resolution.

Early in the year, Lucasfilm unveiled a new company structure and a new "editDROID." LucasArts Editing Systems is the manufacturer and developer of the improved system; it was released after a three-year hiatus for re-design; 18 new systems had been built and 12 more were under construction at a plant located on Lucas' Skywalker Ranch. The system was announced as only rentable for its first year.

At the SMPTE show, the Montage Group demonstrated a digital-tape hybrid system, called the Montage II-H. This hybrid would lead to a fully digital system, expected to be shown in prototype at the next NAB show.

In London, Paul Bamborough, a founding partner of Solid State Logic (SSL), the professional audio console company, began work on a new system. He had been considering developing new editing systems since the early '80s, but had felt the EditDroid did about as much as could be done with existing technology. By the late '80s, Bamborough left SSL and began work on a new graphical system, primarily for film work, that would use compressed digital video. His intention was to make a system that was both simple to use and more playful than traditional editing systems. He joined with some friends of his who had experience in engineering and system architecture, and formed O.L.E. Ltd., a partnership to manufacture a new system, to be called the "Lightworks."

Tim Berners-Lee, a scientist working at CERN (the European Laboratory for Nuclear Research), invented Hypertext Markup Language (HTML). HTML allows programmers to embed links from one text document "page" to other pages, so readers need only point to a word and click on it to move around and dig deeper into desired information. The tool was designed to streamline record-keeping at CERN. Berners-Lee also launched the

Avid brochure cover (Summer, 1990)

Universal Resource Locator (URL) address for a page, and placed "www" in the header of this address, coining the term "wide world web."

A programmer at SuperMac designed a new software editing package to help demonstrate their new video digitizing cards for the Macintosh. Randy Ubillos calls his program "ReelTime." By the end of the year, SuperMac has sold the software to Adobe, who renames it "Premiere."

1991

⊿ OLE debuts the prototype of the Lightworks. ⊿ Apple releases QuickTime. ⊿ Chyron (and CMX) are purchased by Pesa.

Early in the year, a small group who left Grass Valley Group in 1990 gathered to form a company to make a new kind of video computer. Calling themselves "ImMIX"—a variation on the Greek word for a gathering of friends—they were acquired by Carleton and incorporated by February.

At NAB, there were eight familiar nonlinear editing systems, and a handful of new ones: the Montage III was shown in prototype as expected—a kind of an upgraded Montage II with DVI-based digital video source material; editDROID was demonstrated, after having been used on significant portions of Oliver Stone's *The Doors*; the Ediflex II was demonstrated with Panasonic write-once discs instead of the familiar VHS decks. E-PIX was demonstrated, as was the TouchVision. TouchVision, Inc., Rady's new company, demonstrated a prototype of its DVI-based digital system, called D/Vision. The TouchVision had just finished its work on the Madonna documentary, *Truth or Dare*. Chyron/CMX showed their familiar CMX 6000 after having cut a Bernardo Bertolucci film, *The Sheltering Sky*, and a Paul McCartney concert film during the past year; there was a fair amount of activity at both the EMC2 and Avid booths, both showing prototypes of upgraded systems—more editing features and better resolution images. Prototypes were demonstrated for the Lightworks system and D/FX's Video F/X.

In May, Apple Computer Company announced a new format—called "QuickTime," for running digital video in many Macintosh applications. With a grant from Apple, the American Film Institute opened their new AFI-Apple Computer Center, dedicated to investigating new technologies for filmmaking and professional applications of the new QuickTime format. Adobe releases v1.0 of *Premiere*.

1992

⅄ Avid introduces its OMF Interchange format. ⅄ Hollywood begins to adopt the digital nonlinear systems in place of the original analog ones. Image quality for offline editing reaches an acceptable trade-off between size and resolution. ⅄ NCSA Mosaic, the first Web browser, is invented.

As expected, NAB was rich with digital nonlinear systems. A release version of the much-anticipated Lightworks system was shown, as were a wide array of products from Avid Technology. CMX privately announced a low-cost digital system based on the 6000, to be called the CMX Cinema. Montage again showed a prototype of their Montage III. All the principal equipment manufacturers demonstrated upgraded systems, with the single exception of the editDROID from LucasArts.

The issue of digital image quality finally began to be seen as the red herring it was—diverting spectators from the serious issues of editing system interface, storage, and price. After a number of years in which image quality was the issue, all systems demonstrated "reasonable" images at comparable compression rates.

Encore Video, in Los Angeles, made the first significant strides in the use of the Avid for episodic television. Paramount Pictures began the first significant use of the E-Pix for television and, finally, theatrical film work by the end of the year. The Post Group, also in Los Angeles, became the principal Lightworks installation, giving the system some of its first serious trials on television programming.

By the end of the year, it was clear that 1992 had been a watershed year for nonlinear editing, and digital nonlinear systems in particular. Avid had placed over 1000 systems worldwide over the previous few years, successfully introducing the concept of nonlinear editing to the video and broadcast communities that had long been unaware of the available products. Now, too, there were viable competitive systems to the Avid being evaluated. Television was now taking the digital offline systems more seriously—and theatrical features, the long-time holdout, moved one more step toward the acceptance of these systems as the future of theatrical post.

The first shot of the Internet revolution was fired, when undergraduate Marc Andreesen developed a new computer program called NCSA Mosaic. (NCSA is the National Center for Supercomputer Applications, located at the University of Illinois.) Mosaic made it easier to access files written in HTML such that a "web page" could deliver more than text. It gave a human interface to the Internet, and opened it up to a new degree of access for the non-technical user. Andreseen gave away Mosaic to anyone interested.

1993

⅄ Avid Technology IPO. ⅄ Montage III and CMX Cinema, digital versions of established systems, are beta-tested. ⅄ J&R Film Company takes over the editDROID inventory from LucasArts and begins shifting from Moviolas to digital film systems. ⅄ "Online" nonlinear systems are debuted. ⅄ Unprecedented numbers of feature films move to digital nonlinear systems.

January saw beta shipments of CMX's digital Cinema, the only digital departure from the windowed and mouse-driven environments of other systems. Both Montage III and CMX Cinema were anxiously awaited for their national release at NAB. Both were introduced more slowly than expected.

Avid Technology, riding high as the leading digital nonlinear vendor, made an initial public offering (IPO) in February, and also announced its purchase of DiVA's VideoShop—a Macintosh application software product that was a low cost desktop video system—and the principal competitor to Adobe's Premiere. Avid also began shipping first versions of Airplay, NewsCutter and MediaRecorder products. They entered the corporate market with their Media Suite Pro.

NAB was for the first time filled with software companies, offering a whole range of editing products. New arrivals to the scene included Adobe and SuperMac.

Ediflex announced the impending arrival of their Ediflex Digital, one more in the growing list of established nonlinear systems moving into the digital domain; although the system was not shown at NAB, it was demonstrated throughout the spring and initially released that summer.

J&R Film Company, the owner of Moviola, began the slow phase-out of their Moviola rental business by acquiring digital nonlinear systems for film rental—a dramatic step that was precipitated by other Hollywood establishment film rental houses (like Christy's) doing the same. It was a landmark move in an industry where the Moviola had been the bedrock of traditional editing equipment for almost 100 years.

By April, J&R took over the ownership and support of the editDROID from LucasArts, thus ending the Lucasfilm connection to the product, began more than a decade earlier. Lucas, in turn, announced a plan of future development with Avid Technology.

The buzzwords of the year involved "online nonlinear systems," with the prototype of ImMIX's VideoCube, Avid's 4000 and 8000 series, and an early prototype of Lightworks' Heavyworks. There were also many new products targeted towards the industrial and corporate "online" desktop: Matrox Studio, Data Translations' Media 100, and the German FAST Electronics Video Machine. Many manufacturers began to unbundle their products into available software and boards, for end-users to use with their personal computers.

The year was a key for theatrical features moving towards electronic film; after well-watched trials for Oliver Stone (on the Lightworks) and Martha Coolidge (on the Avid), those two systems made the move into feature films and began steady work as accepted tools for Hollywood.

The first experimental search engine was created: Webcrawler was the Ph.D. project for University of Washington graduate student Brian Pinkerton; at about the same time, Architext Software was born from the interests of six recent grads from Stanford: Mark Van Haren, Ryan McIntyre, Ben Lutch, Joe Kraus, Graham Spencer and Martin Reinfried.

1994

⅄ Avid Technology increases massive market penetration and differentiation of products. ⅄ CMX Cinema project is ended. ⅄ Lightworks' Heavyworks is the first viable digital nonlinear system used for traditional multicamera television programs. ⅄ Cinedco, owner of the Ediflex, closes its doors. ⅄ Many manufacturers begin steering their video products into digital and nonlinear domains while existing nonlinear manufacturers push the image resolution limits with better and better images—further blurring the online/offline distinctions. ⅄ EMC is purchased by Dynatech. ⅄ Softimage is purchased by Microsoft. ⅄ Netscape Communications is created.

After five years in business, Avid reported gross sales of $203 million dollars and profits of $4 million. Their R&D budget exceeded the gross sales of every other nonlinear manufacturer to date. At NAB they made the first public demonstrations of their OMF (Open Media Framework) Interchange. By December, Avid had purchased Digidesign, a leading digital audio company.

Montage settled into slow adoption of their digital Montage III, but continued to aggressively pursue enforcement of their patents concerning digital images in editing systems. Adcom phased out their E-PIX editing system, and replaced it with a high-end digital "online" nonlinear product called Night Suite, boasting D1-quality video output. Sony demonstrated a prototype of their first nonlinear product, called the Destiny Editing System. One of the key designers of the Destiny system was Robert Duffy, the engineering designer of the CMX 6000.

In April, Jim Clark (former Chairman of Silicon Graphics) joined Marc Andreesen (inventor of NCSA Mosaic, the first web browser) to form Netscape. By the end of the year, the first commercial Netscape browser shipped.

Yahoo! was created in April 1994 for the personal use of Stanford Ph.D. students David Filo and Jerry Yang. Yahoo! grew in usefulness and popularity until it stood on its own by year's end. Also by December, Architect Software was funded by venture capital, and AOL purchased Webcrawler.

1995

⅄ Tektronix purchases Lightworks. ⅄ Avid purchases Parallax Software Group, developer of high-end paint systems; and Elastic Reality, a leading developer of image special effects software. A prototype of an AvidDroid is shown at NAB. Avid licenses Ediflex's Script Mimic technology for the next five years. ⅄ Inventors of the Avid and

Lightworks products each win 1994 Scientific and Engineering Academy Awards. ⅄ All major networks have broadcast programs directly from nonlinear systems. ⅄ Sony stops development of its nonlinear Destiny product. ⅄ According to an NAB survey, 38% of broadcasters use nonlinear equipment. ⅄ Netscape IPO. ⅄ Search engine companies become newsworthy; Amazon.com opens for business.

In March, Avid acquired Parallax Software, developer of high-end paint and compositing systems Matador (2D paint and animation) and Advance (compositing and image processing). At the same time, Avid also acquired Elastic Reality, a software developer known for their advanced morphing and image restoration products. At NAB, Avid showed versions of products that run on Macs, PCs, and introduced new Silicon Graphics configurations. They also debuted an SGI Onyx-based online system that incorporates their products. According to company literature, Avid employed 1,200 people. Their presence at the NAB show was pervasive.

Announced at the NAB show was the acquisition by video giant Tektronix (revenues of $1.3 billion in 1994) of Lightworks, at a cost of around $75 million. At the time, Lightworks employed 100 people and had approx. 1,000 high-end film cutting systems in the worldwide market.

A new company arrived at NAB made up of 11 senior staff members from NewTek (makers of the Video Toaster) who called themselves Play, Inc. The hip team demonstrated a prototype of a PC-based, D1 video bus (rather than PCI) post-production system called Trinity.

Following NAB, Avid announced a patent infringement suit against Data Translations, designer of the Media 100 and a rising competitor to some of Avid's products. The suit concerned proprietary audio and video manipulation methods internal to the Macintosh.

Ironically, in a year with more than 120 hard disk-based nonlinear-type products, there were more video tape formats introduced than at any previous NAB. The buzz at the show was on MPEG compression, networking systems, and in dedicated broadcast news nonlinear systems.

The year showed a marked increase in development on the PC as a video platform: with the introduction of Microsoft's Windows 95, plus the creation of the Open Digital Media "Language" (OpenDML) consortium, on top of the continued development of PCI-bus plug-and-play video boards. Windows-based products had a surge in popularity.

Scitex, a market leader in the prepress and publishing industries, purchased two video companies: Abekas (well known for its 3D devices) and ImMIX. The businesses were incorporated into a new Scitex Digital Video division.

The D/Vision team scrambled to prepare their online nonlinear system using the yet-to-be released Matrox hardware. The product was expected to release at NAB '96, but the hardware board was taking longer than expected to become available. Rady bets his company on this new high-end product based on their software. A failure to meet the release date would be cataclysmic.

The NAB show itself was in many ways a metaphor for the state of the industry. The giant convention center in Las Vegas had an annex in 1991 (and for the previous few years) dedicated to high definition video; by 1992 that was superceded with something called Multimedia World. By 1995 it was as if Multimedia World was beginning to reach adolescence; the convention was split and it was not unlike the "adults" in the main floor, and the "kids" in multimedia world. It perfectly represented the transition from the broadcast realm to the computer realm; from the establishment to the revolution. It was not uncommon to see midlife broadcasters wandering lost and confused among the multimedia booths; and to see youth drunk with their new-found power, but not quite sure what to do with it. More than anything, NAB exemplified the still-enormous knowledge gap between the two worlds, and the few participants who could speak both languages. The corporate giants of the establishment having failed to truly invent the future were forced to grab it by brute force, buying and merging with the new technologies and placing them, like badges, on their corporate jackets.

With the move of Yahoo! from Stanford to Netscape's computers early in the year, the opening for e-commerce business of Amazon.com midyear, followed by the August Netscape IPO, 1995 was the year the Web first truly reached the consumer's mindspace. Webcrawler was sold to AOL in April. Architext Software changed its name to Excite.com, and had its IPO. WebTV was founded in late summer, to deliver the experience of the Internet through regular television sets. For the most part it appeared broadcasters were unaware of the future implications of the Web; they were way too focused on the rise of digital, and the coming of HDTV.

The first all-digitally created/animated full-length feature film, *Toy Story*, was released in November. The 77-minute film from Pixar (the former Lucasfilm Computer Division group, now in the hands of Apple pioneer Steve Jobs), required 800,000 hours of number-crunching by 117 Sun SPARC 20s.

1996

⅄ Netscape Navigator 2.0 ships. ⅄ Discreet Logic releases FIRE ⅄ Macromedia releases Final Cut. ⅄ Supercomputer company Cray merges with Silicon Graphics. ⅄ 4MC begins strategic purchasing and aggregation of Hollywood post-production facilities.

Now that digital nonlinear offline had settled into reality, interest shifted to nonlinear online, now three years more mature: first with ImMIX and their Video Cube Lite, D/Vision's onLINE, and Discreet Logic entering the arena with FIRE, a 4:4:4 system that fit into its high-end graphic products (Flint, Flame and Inferno). On the D/Vision front, the Matrox Digisuite boards, on which the D/Vision Online was predicated, were behind schedule, and capital in the extended company was running short. onLINE v1 was shown in prototype at NAB using the Matrox Illuminator Pro boards. By the end of the year, founder Bruce Rady left his company, and Paul Reilly took over as president of a totally downsized D/Vision. Reilly set a target of getting the product stable by the next NAB.

LaserPacific, an early nonlinear proponent (a post-production facility) in Hollywood continued with early MPEG and digital video disk research, begun the prior year, and moved strongly into HDTV development.

Macromedia introduces Final Cut (Spring 1996), and then sells it to Apple 14 months later.

The most valued products at NAB '96 included the Avid Media Composer, Sony's Digital Betacam, Discreet Logic's Flame, the Media 100 and Evertz KeyLog telecine log system. Adobe announced at the show that they intended to port their successful Mac-based After Effects to Windows and NT. A number of companies, from FAST to Avid Technology, found the year to be one of transition: to determine how to leverage their present position and products in a rapidly changing field.

The OpenTV version 1.0 began shipping from an alliance of Thompson Multimedia and Sun Microsystems. The product, like WebTV, was designed to offer interactivity to television viewers through dedicated set-top boxes, although unlike WebTV, was not based on an Internet backbone.

Avid Technology finished the year with $429M in revenues, ending a six-year run of more than doubling the prior year's revenues every year (beginning in 1990 with $7.4M in sales), and began a slowdown in the meteoric expansion of the industry leader. Data Translation, long positioned as the price-performance competitor to the Avid, changed its name to match its key product: Media 100.

Braveheart won Mel Gibson a best picture Oscar, the first feature film to win the award that was cut digitally (Steve Rosenblum cut it on the Lightworks the prior year). But the Best Editing Oscar eluded Rosenblum.

1997

⅄ DVD issues rise on the post industry radar. ⅄ Best Picture and Best Editing Oscars for a film edited on a Macintosh. ⅄ Discreet Logic purchases D/Vision and Iluminaire products. ⅄ QuickTime 3.0 is released. ⅄ IEEE 1394 products begin entering the marketplace. ⅄ It is the Windows NT "year." ⅄ Amazon.com IPO. ⅄ WebTV is purchased by Microsoft.

For the first time in television history, viewer share of the "big 3" networks—once greater than 90% of all TV viewers—dropped below 50% due to competition from cable.

Windows NT was everywhere at NAB '97. PCs were inexpensive and ubiquitous and video products were increasing market potential by moving to the platform. By NAB it was clear that Windows could be accepted as a professional video platform. Also on the rise were high-bandwidth formats, like Apple's FireWire, and 10/100Mbit Ethernet on the lower end; HIPPI, FibreChannel, and other gigabit formats on the higher end.

Also at NAB, D/Vision showed onLINE 3.0, a watershed version of the product, with significant real-time capabilities. Following the show, Discreet Logic purchased D/Vision for about $21 million. Discreet also purchased NT-based software company Denim, and their key product: Illuminaire, with compositing and paint tools. Over the next year, D/Vision would become Discreet's edit*, Illuminaire compositing software would become effects*, and their paint tools would become paint*.

Avid had consecutive quarters in the black, and prevailed in the patent infringement suit brought by Media 100. *The English Patient* won the Oscar for Best Picture, edited by Walter Murch on the Avid Film Composer; it was also the first time in history that a film edited electronically won Best Editing. Where films had dabbled with nonlinear equipment over previous years, this year there was an explosion of film activity, and resistance to the digital equipment was virtually eradicated.

1998

⅄ Avid Technology wins an Academy Award of Merit — the highest technical achievement award given. In August, Avid purchases Softimage from Microsoft. ⅄ MPEG-2 nonlinear editing systems debut. ⅄ Premiere 5.0 is released. ⅄ Discreet Logic acquired by Autodesk. ⅄ Television networks begin taking sides on HDTV formats. ⅄ Former industry staple CMX folds.

Early in the year, C-Cube announced the first MPEG-2 editing demonstration, something formerly considered too complex to accomplish. By year end, FAST debuted 601—"the world's first MPEG-2-based editing system."

Apple purchased the entire video operation from Macromedia (itself focusing on successful Internet tools like Dreamweaver and Flash), including their release of Final Cut. Apple continued in its work in both professional and consumer video.

After two years of leading-edge work in HDTV, post facility pioneer LaserPacific entered into a technology agreement

Avid advertisement for Illusion (circa 1997), a high-end set of tools with sights set at Discreet's Flame and Quantel's Henry.

with Sony to develop a 24P post-production system. The work led to the landmark posting of *Chicago Hope* in the fall, the first episode of a regularly scheduled primetime network series to be finished in HD. LaserPacific had been the first post-production facility in Hollywood to develop real-world procedures for nonlinear editing 14 years earlier.

OpenTV was spun off as an independent venture to develop interactive television experiences. The company focused on trials in Europe as it had been over the prior year.

Avid released v7.0 of the industry standard Media Composer and debuted the Symphony v1—an online nonlinear workstation. Avid also announced AAF—*Advanced Authoring Format*—a new file format set to replace their earlier work in OMF.

NAB 1998 introduced HDTV to the mainstream; where in earlier years the products were mostly in prototype and technology demonstrations, now post-facilities and networks were finally shopping for products to meet with FCC broadcasting regulations. Affordable desktop systems for professionals were commonplace, with the maturing of numerous products coinciding with the continued drop in hard disk prices per gigabyte. Adobe Premiere 5.0 was released, but was rapidly followed by v5.1—a landmark version, that crossed over from a mostly-amateur consumer product to a production-ready tool.

In November, the feature "film" *The Last Broadcast* made cinema history. The 90-minute mystery was shot on digital video cameras, edited with Adobe Premiere, released to select theaters via *satellite,* and projected digitally. Although numerous corporations donated goods and services to the project, production and post-production were reported to have cost $900.

Late in the year, acclaimed software developer Autodesk—well known in the PC market as the leader in architectural and design tools—purchased Discreet Logic. In December, Accom—manufacturer of the Axial online controller—purchased Scitex's digital video products (formerly Abekas and ImMIX). The ImMIX VideoCube nonlinear online system was repackaged as the Sphere line of products.

1999

▲ Final Cut Pro 1.0 is released. ▲ Pinnacle acquires Truevision. ▲ Radius changes its name to Digital Origin. ▲ Personal hard disk video recorders debut at NAB: TiVo and ReplayTV. ▲ OpenTV IPO. ▲ DVD encryption hacked.

IEEE 1394 "FireWire" products became widely available to consumers, in the form of MiniDV digital camcorders and Apple's new G3 computers. Final Cut Pro 1.0 was released at NAB, a new generation of desktop high-quality editing, specifically designed to take advantage both of the new DV formats and FireWire-fitted Macs. Final Cut Pro was developed by a team led by Randy Ubillos, the inventor of Adobe Premiere, and in many ways was a next-generation iteration on editing, built with the experiences of the successful Premiere under consideration, but without the baggage that comes with updating existing products. Apple debuted the higher powered G4 computers before year end.

Media 100 purchased Terran Interactive, maker of the unique product Media Cleaner Pro (MCP). MCP is a facile tool that allows video from numerous formats to be converted and prepped for online streaming, a key functionality in the nexus of nonlinear editing tools and Internet video. The purchase was a coup for Media 100.

Avid, due to both political and technological issues, appeared to abandon long-term development for the Macintosh platform. An outcry from the company's significant installed-base of (Mac) users resulted in spin doctoring and reconsideration of their plan. After Q1, Avid stock prices tumbled and a management shake-out left users wondering about long-term viability. Could Avid *really* be the new CMX—once holding the majority market share of editing systems, and gone a decade later?

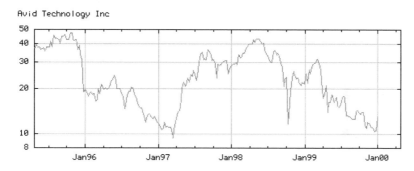

Avid Technology Inc

Web fever reached a critical pitch by midyear, as venture capital poured into Internet start-ups in unprecedented quantities. The Industry held its breath in Q4 to see how the holiday commerce would meet expectations.

DVD market penetration increased, as PCs shipped with DVD drives rather than CD-ROMs. DiVX, a proprietary DVD pay-per-view concept, died by mid-year. As was bound to happen, the DVD encryption algorithms that gave movie studios comfort with the format was hacked. The Content Scrambling System (CSS) was undone by a Linux-based DeCSS and posted to the Internet.

In May, the long-awaited prequel to *Star Wars* was released. Lucasfilm's *Episode I: The Phantom Menace* also became the first major studio release exhibited electronically, that is, using high-definition *video* projection. The test in a handful of specially-outfitted theaters utilized playback from a customized 360GB RAID-3 array configuration, as well as new DLP digital projection technology. Less widely known was that the simultaneous

release of Miramax's *An Ideal Husband* in a few theaters also played back digitally (from D5 videotape). With similar experimental exhibitions held later in the year for Disney's *Tarzan* and Pixar's *Toy Story 2*, the year would go down in history as the dawn of eCinema (or Digital Cinema) and the beginning of the end of celluloid. The summer theatrical hit was *The Blair Witch Project*, shot mostly on consumer-level videotape. After an overwhelming response at Sundance (and through marketing on the Internet), it was acquired by Artisan Entertainment and released as a theatrical feature.

Personal digital video MPEG recorders reached consumers, in the form of TiVo and Replay, two competing products that transparently use hard disks (rather than videotape) to record, timeshift, and manipulate television programs. The products are effectively nonlinear consumer "VCRs" with DVD-like quality. In June, Panasonic backed ReplayTV, and in September, Sony backed TiVo.

4MC in Hollywood wound down its ambitious facility round-up, having over the prior years purchased Los Angeles post-production staples such as Digital Magic, Encore Video, Anderson Video, Hollywood Digital, Todd-AO (itself handling more than 60% of Hollywood sound work), Pacific Ocean Post, and others. The aggregation of facilities for the first time on this scale impacted the post industry in LA, creating a corporate monolith among numerous boutique and rapidly evolving facilities. The long-term impact on post-production was unclear.

Big questions came to light: will consumers watch video on computers, surf the 'net on their televisions, will nonlinearization of television be competitive with net-streaming video on demand?

2 0 0 0

⅄ The first TV advertisement for editing directed at consumers runs during the Superbowl. ⅄ Montage is sold to Pinnacle Systems. ⅄ A new team debuts a new Lightworks. ⅄ The year's mergermania involves "solutions" for Internet streaming and HD postproduction. Video aggregating and streaming websites begin to proliferate.

During the Superbowl, Apple ran commercials for its iMovie software running on the iMac; this was the first network advertisement in history for *editing*—post-production crossed over with the new millennium from industry technology to home product.

Avid restructured executive management, introducing new President David Krall and acting-CEO Bill Flaherty. At NAB, Avid showed a prototype for a system that managed interactive tools with webstreaming video, ePublisher, expected to ship before year-end. Avid also announced Media Composer (NT) v10 and Symphony v3.

Media 100 jockeyed ahead with strategic positioning in the Internet future through its partnerships with Digital Origin, Canon, Beatnik and Digital Fridge—to produce *icanstream.com*, one of a dozen new Internet video sites to go live in the prior months. Others included market leaders *AtomFilm*, *iFilm*, and newbies *pop* (backed by Dreamworks), *icast* (backed by CMGI), *eveo*, *nibblebox* and *cameraplanet*, to name a few.

After Tektronix failed to integrate the Lightworks product family into their core market, the products were sold to a new company, Lightworks Nonlinear Editing, led by former Lightworks aficionado Mark Pounds. The new team, made up of many of the people responsible for the system's original success, debuted a prototype of the editing system, now rebuilt on Windows NT, with considerable effects capabilities and their familiar ergonomic controller.

Final Cut Pro was demonstrated playing back (and editing) uncompressed high-definition video on a Macintosh G4. Through an announced relationship with Pinnacle Systems and Matrox, the Apple software tool scaled from prosumers to broadcasters. Pinnacle also acquired software company Puffin Designs, and their popular Commotion 3.0 graphics product. The stock deal was worth roughly $13 million.

Hollywood's 'Quantum Project' heads for the Internet

LA Times Business section announcing the first feature film shot for and distributed on the Internet (May 2000).

Streaming Internet video, MPEG-2, and tools for HDTV were the mantras of NAB, with a quieter omnipresence of DVDs and online nonlinear systems; the convergence of webmania and the leading edge of video technologies fueling deep contemplation and outright scrambling in the establishment of NAB. While HDTV (DTV) was a forefront issue for broadcasters, video editing on computers was growing in feasibility and popularity for individuals.

The Eidos Judgement was billed as "video content creation software," and was shown in prototype at NAB. Eidos Technologies LTD was a subsidiary of Eidos PLC, best known for it's interactive division's CD-ROM publishing: the Laura Croft-Tomb Raider series of games. Aside from the product's innovative timeline interface, another aspect of Judgement which was noteworthy involved the product's database—built on an SQL server, it was implemented more like an enterprise-level Internet database than a traditional editing system. By July, Eidos announced they were closing the Eidos Technologies subsidiary.

Liberty Media completed its acquisition of Four Media Corporation (4MC), which had itself been purchasing a series of the largest post-production facilities in Hollywood.

Meanwhile, with the May release of *Quantum Project,* the Internet took its baby steps as a legitimate long-form distribution medium. The 32-minute, $3 million film was designed for the Internet and was distributed by SightSound.com for $3.95 (purchase, not rental). According to reviewers, the landmark film was thin on plot and heavy on music-video*ness.* At 90MB, the download took 4–6 hours via dial-up modem, although merely 10–15 minutes via broadband.

Also in May, LEGO (the kids' toy company) announced an alliance with Steven Spielberg to launch LEGO® Studios, a new product that allows kids to make videos, and do it *nonlinearly.* The product included LEGO-based character and set elements (in the familiar LEGO style), plus custom Pinnacle editing software that runs on a PC. Along with a mini digital video camera (with USB port), kids can shoot easily, edit and distribute home movies. The product was priced at $180 and targeted for an early 2001 release.

Chapter 3

Fundamentals

There's an old woodworker who makes furniture. He finds the wood, carefully cuts and assembles the pieces, and then sands and polishes. Sands and polishes. Sands and polishes.

People come by his studio and tell him how beautiful it is. He thanks them, and keeps on sanding and polishing. Sanding and polishing.

One day someone asks him: *"When are you going to be done?"* He replies: *"Done? Oh, it isn't ever done. Someone just comes and takes it away."*

—Anonymous

OFFLINE AND ONLINE

What is the difference between *offline* and *online*? Does nonlinear editing make these terms obsolete?

The words themselves have to do only with the disparity between element creation at *delivery* quality, versus any (image or audio) quality below that. Furthermore, there is no absolute benchmark for what "online" quality *is*, as it is relative to what is required for delivery. Thus:

ONLINE is the production of any aspect of the completed product—whether an entire production, or simply an *element* to go into a finished production. Either way, the inherent image or audio quality is that which is required for delivery of the project. Specific online image qualities vary depending on the needs of the product's recipient. Projects finished for television broadcast need "broadcast quality" signals and "high" quality pictures. But if a lesser quality is acceptable for delivery, as for personal use or many non-broadcast uses, what is considered "online quality" may be quite different.

OFFLINE is the creation of elements at any image or audio resolution less than online, and tends to be the domain for the creative "decision-making" stage of the project. "Offline quality" might only be a tiny bit less quality or it might have considerably less quality. Since quality has many factors (frame rate, image/pixel resolution, color resolution, aspect ratio, *etc.*), any of these might be compromised to decrease the cost of the offline edit session, where all that is needed is any quality deemed good enough to make creative *decisions*.

Background

There was a time when online and offline *quality* coincided with specialized online and offline *tasks*, *equipment*, and *people*. Because of the high costs of online editing equipment, offline evolved for the more time-consuming construction of the edit—narrative, story, "horizontal" structural decisions—the paring down of enormous volumes of source material into smaller sequences. It was a cost-effective way for editors (along with direc-

04:12:22:08

SONY

tors and producers) to make their initial creative decisions, and revisions later.

Scenario 1: The Bad Ol' Days

The offline editor goes to the offline bay with a stack of tapes; these are actually lower-quality dubs of the original master tapes, since the decks required to play the master tapes are too expensive. (Besides, the master tapes are irreplaceable, and she can't afford to risk damaging them from all the constant shuttling back and forth.)

She spends a few weeks cutting the material down to 30 minutes. No point in putting in titles or effects, as they need to be done by artists with special skills, and they need to be at online quality, so they will be done separately and later.

When she is done, she hands an online editor her EDL. He is in a very expensive edit bay, tying up some expensive video decks for a day or two. He carefully recreates her edit with the high-quality master tapes, doing a little color correction as he goes. This person is particularly skilled at watching video levels and tweaking knobs. He is also excellent at compositing effects, and integrates elements from other sources (*e.g.*, graphics bays) into this videotape. He builds the final show for final delivery.

Observation: Clear task division based on the economics of equipment. There are online and offline "tasks," "rooms," "machines" and *people*. She is an offline editor; he is an online editor. To be honest, she isn't even remotely interested in "online."

Scenario 2: Here and Now

The "picture" editor goes to the edit bay with the master videotapes and digitizes a low-resolution version into her editing system to do the offline edit. She edits the dozens of hours of video into 30 minutes over the course of a few weeks. In another bay down the hall, an artist is focusing on building some graphics and effects. Again, he starts at low resolution while he roughs out the look and movement.

Meanwhile, when the editor is done, she "locks" the cut. Then, she re-captures the needed material from the master tapes, this time at *online* resolution. Her workstation, while designed for storytelling, is resolution-independent, and capable of working at a range of image qualities. She may not be an expert in the details of video signals, but this only means that her associate comes to the bay to supervise the high-resolution recapture, and help her tweak a few video levels during the process.

Down the hall, the effects editor gets everyone's approval on some compositing, then re-renders at high-resolution.

In another bay, connected via a network to these two bays, an "online" editor works—his job is primarily integration. Like publishing or DVD premastering, this individual gathers up the "media assets" and integrates them into the final product. Pulling in the final online video from the storyteller-editor, tweaking a few audio levels, inserting the final effects from the graphics department, and maybe even handling the titles here himself, he preps the final project for delivery to the client.

Observation: Many workstations can do both offline and online. Division of labor is by skills (music, story, effects, audio, titles) but not necessarily by rooms and quality level.

The goal of *offline* (linear or nonlinear), is to select and assemble the best performances while seamlessly joining pictures and sounds together. The video effects that the editor is concerned with might be simple ones (cuts, wipes, and dissolves), or more elaborate ones (DVE moves, layering, *etc.*). The product of the offline session is *the creative decisions themselves*, whether represented by a file or printout of the edit decision list (EDL), or a machine-readable "sequence."

Except for frame accuracy, everything is allowed to be compromised in order to minimize expense and maximize efficiency. Compromises might include image quality, visual effects, and the lack of final elements such as music, sound effects, and titles.

The goal of *online* is the efficient re-creation of the offline editing decisions, at the quality of the finished product. Online equipment, linear or nonlinear, tends to be expensive (almost by definition, more expensive than offline equipment).

Sometimes, for efficiency's sake, there are tasks that are best done once, at online quality, as opposed to being done inexpensively first and then re-created later. Effects building is often done only once, at high quality. Same with titles. And audio. Even the picture edit itself, if highly effects-intensive (a commercial, perhaps), might make economic sense to be executed once, in "online."

▶▶ Online/offline theory: p. 378

Hence, online and offline editing both involve different kinds of attention to detail. Whereas in offline, the focus is on the *content*: story, structure, pacing, *etc.*, online editing focuses on more *technical* attributes like video and audio quality.

Is Film Editing "Offline" or "Online"?

In the film world, the closest parallel to online/offline is editing workprint and cutting negative. The film editor traditionally made creative decisions and built an edited reel from positive duplicates of the original negative. The reason for this has little to do with cost, but rather the fact that you don't want to handle your priceless yet fragile negative—film editing being a sort of destructive-then-constructive process. Besides, you need a positive image to view material.

Once "offline-types" of creative decisions are finalized on workprint, that film is given to a negative cutter, who performs the rough equivalent of "online." The negative cutter does not have a paper list (like an EDL), but rather the cut duplicate print from the negative. Using that duplicate print as a guide, the negative cutter reads the numbers off the edges of the film to locate and assemble the original negative. It is a slow and dangerous process. Any mistake in the negative cutting can be disastrous. Negative cutters, like online editors, are not concerned with creative decisions about the edit, only in the faithful re-creation of the editor's decisions.

Since nonlinear systems themselves do not create a cut film workprint, but do create instructions for cutting the original negative, the only way to inform the negative cutter of what to assemble is a printed list—a film-version of an EDL—called a "negative assembly list" or "cut list." This list has entries for each shot's in-points and out-points, in a format very similar to the video-style EDL. Videotape dubs of the final offline cut are also commonly given to the negative cutter to provide a visual reference. A special electronic controller (*e.g.*, the LokBox) allows that videotape to be viewed in sync with the film negative as the latter is being pieced together.

T I M E C O D E

In the early 1970s, videotape editing changed from control track editing to using timecode. An 8-digit timecode defined and electronically identified each video frame with a unique "code" number broken down into:

HOURS : MINUTES : SECONDS : FRAMES

01:00:00:00

Timecode made videotape editing more efficient than it had been before. In addition to speeding up the editing process, it was considerably more "frame accurate" than the earlier and cruder control track editing. It also allowed any edit to be previewed and repeated.

Later, computer-controlled editing systems were introduced that could read timecodes, identify edit points, and store lists of these edit decisions.

"Drop Frame" vs "Non-Drop Frame" Timecode

For our current discussion, videotape plays at 30 frames per second (30fps). Videotape timecodes, therefore, count from frame :00 to frame :29 before rolling over to the next second—

Unfortunately, for a variety of electronic and technical reasons, videotape doesn't really run at precisely 30fps; in reality it runs at 29.97 frames per second. So, even though timecode will accurately identify every single video frame with a unique number, it isn't precisely measuring REAL TIME.

Say you've edited an infomercial, and began recording it on a piece of timecoded videotape, starting at 01:00:00:00 (called "one hour, straight up"*). If the show ends exactly at 01:29:00:00 you might be led to believe

* Video people never start counting from time zero, but rather start at one hour. This allows machines to pre-roll (back up) ahead of the first edit, without getting confused—or running into a spot on the tape with no timecode.

that your show was precisely 29 minutes long. **THIS IS NOT CORRECT.**
Since your videotape is actually playing slightly slower (0.1% slower,
to be exact), your actual program duration is almost **two full seconds
longer**!

The people who work with videotape often want timecode to do two
things: (1) to uniquely identify each frame *and* (2) to give accurate indica-
tions of running time (duration). Clearly, regular old timecode doesn't do
the latter very well.

REGULAR OLD TIMECODE that has a single number for every frame,
that counts from frame :00 to frame :29 and then rolls over—but is temporally inaccurate (by 0.1%)—is called **NON-DROP FRAME TIMECODE
(NDF)**—because it never drops any numbers while it is counting.

DFTC
01;03;59;25
01;03;59;26
01;03;59;27
01;03;59;28
01;03;59;29
01;04;00;02
01;04;00;03
01;04;00;04
01;04;00;05
01;04;00;06

The only way to make timecode keep anywhere
close to REAL TIME is to leave out certain numbers.
If you skip some numbers (remember that this doesn't
affect the video pictures at all; it is only a numbering
scheme), your calculations can be extremely close to the
actual elapsed time of a segment.

Timecode that skips certain timecode numbers is
called **DROP FRAME TIMECODE (DF)**. The way in
which it skips is very precise:

**Drop the :00 and :01 frame number every minute,
except for every 10th minute.**

This way, source and record times DO reflect real time, and thus can
be used to determine length. (To calculate the length using timecodes, subtract the "in" timecode from the "out" timecode. It can be difficult; you
might want to use a special calculator. Editing systems do this automatically.) For many reasons, source material (dailies) tend to be transferred
from film to videotape using NON-DROP FRAME TIMECODE—this way,
every frame has a number that is one greater than the preceding frame, and
the algorithms that convert timecodes into film key (or code) numbers are
"safer." In actuality, it makes no difference. Record times for almost all
broadcast television, however, are in DROP FRAME because this way you
can easily see if your program is running long or short—you know how
long it will play on the air.

▶▶ EDLs: p. 92   ▶▶ Tape formatting: p. 366

Remember—

☞ **You MUST know whether you are using DF or NDF timecode. It makes a *big* difference.**

☞ *Almost* **all nonlinear editing systems deal equally well with Drop or Non-Drop Frame timecode.**

FACTOID, Part I • The logic behind Drop and Non-Drop is similar to what we follow in our calendars for leap years. We pretend that a year is 365 days long. In reality, a year is 365.24 days long. Because of this, we drop a day (February 29th) three out of every four years to "keep in sync" with real (astronomical) time. Like with videotape, dropping a day out of our 366-day year prevents a "cumulative temporal error" in our calendar. If we didn't correct for it, eventually (in less than a thousand years), we would be celebrating Christmas in the middle of the heat waves of summer.

In this sense, our calendar is DROP-FRAME . . .

What About Audio?

When ¼" audio tape is synced with film and transferred to videotape (in a telecine session), it is also being slowed down 0.1% to keep it in sync with the picture. The speed change is virtually imperceptible, but if you ever want to resync the audio from a videotape back to the film (either the cut film or even the original film dailies), it must be sped back up—this is called **resolving**.

With nonlinear editing systems, final cuts are created electronically, and lists for cutting film negative or print can be provided. Assuming your nonlinear editing system did not change the timings of your edited sequence, the videotape you have created in offline can often be SYNCED to the film cut.

But it will not sync directly (remember, the videotape runs a tiny bit slow—even if it is "accurate"). The audio portion of the offline videotape master is often **"resolved to mag"** to bring it up to the correct speed and record it onto sprocketed MAGnetic film.

FACTOID, Part II • . . . just as "dropping" a day from the calendar doesn't leave each of us with 24 hours mysteriously missing from our lives, "dropping" a frame **number** from the timecode count doesn't affect the actual CONTENT of video material in any way. (In addition, this is NOT the same as dropping frames when streaming web video.)

TYPES OF TIMECODE

VITC, LTC AND BEYOND

There are two ways you can record timecode on a piece of videotape. Running lengthwise along the videotape, as an audio-type signal; or vertically, in each frame, as a video-type signal.

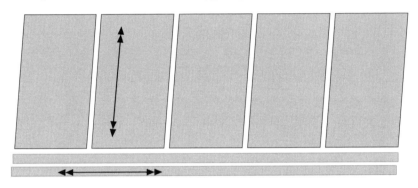

An audio-type encoded signal can be placed on one of the audio channels of the tape (running along the bottom of a tape) or in a special separate channel (running along the top). This timecode, since it runs LONGITUDINALLY (*i.e.,* lengthwise) along the tape, is called Longitudinal Timecode (or LTC). LTC might be called *audio timecode*, if it is recorded in one of the tape's audio tracks, or *address track timecode* (if it is recorded on the dedicated address track).

On the good side, LTC is inexpensive to generate, record and read. The encoders and decoders are affordable and are often the choice for budget-conscious facilities and projects.

On the down side, LTC is notoriously prone to getting confused.* Because the timecode is running longitudinally, it can only be read by the videotape machine when the tape is moving within a narrow speed range. When you slow it down or stop it, the timecode can't be read. It is very much like a regular audio signal—as you slow down an audio tape, the words on the tape get more and more difficult to understand. When the tape stops, you hear nothing. Because of this, LTC is not the perfect way to edit. The constant slowing down, stopping, changing direction, stopping, jogging some more, then making an edit can cause the machine to "read" timecode a number of frames from where the tape is actually parked.

---

*Really good decks—like BetaSP, DigiBeta, etc.—are actually much better reading LTC than older formats, like 3/4".

It is because of LTC (and the now-truly prehistoric *control track* editing that preceded it) that editors first became concerned about editing systems that were FRAME ACCURATE. In many cases, LTC is not.

If you want really accurate timecode, it must be able to be read while a tape is not "real-time" playing. Here is an *interlaced* video frame—on videotape, and then scanning and refreshing on your monitor:

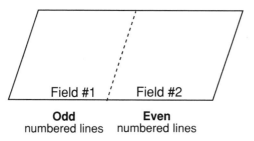 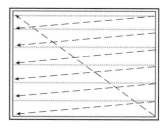

Odd   **Even**
numbered lines   numbered lines

There are 525 discrete lines of video, interlaced (while playing) from two fields on a length of tape. First, Field #1 fills the odd lines on the monitor, then 1/60th of a second later, imperceptible to the eye, Field #2 drops

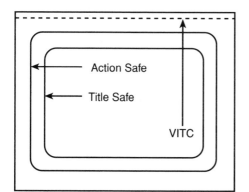

into the even lines. The electron beam in the monitor scans the screen from upper left to lower right, then when it hits the end of the field, seeing no video picture (but rather a "blank" signal), turns off the beam, zips back up to the top and starts again. The first 21 lines make up this blank space between adjacent frames—where *vertical blanking*

takes place. Although most monitors don't actually show you all 525 lines, you can usually see at least 400 of them (in overscan). Once the picture is on your televi-

Actual video frame, with title and action safe zone. Notice the raw VITC (like little dashes) showing at the top of the screen, outside the safe zones; and burn in windows of timecode, key numbers, and a time-of-day clock along the bottom.

sion at home, you could conceivably see even fewer. This is why there is a TV SAFE zone in the middle of professional monitors that gives you an idea of what TV viewers will see at home.

Since you have more lines of video on videotape than anyone sees, engineers use the blank area for inserting data. Each line has a number, 1 through 525; video engineers often encode the timecode into the picture on line 12 or 14 (or both, just to be safe; lines 16 and 18 are also used). To the untrained eye, a line of timecode in the picture is just a bouncy black and white bar along the top of the screen. But this timecode can be decoded and read by tape machines and, if desired, converted into a human-readable number (by a *timecode reader*) that can be superimposed over the images on a monitor, or recorded into other tapes.

Timecode inserted into each picture frame is called *Vertical Interval Timecode* (or VITC); "vertical" because it resides in what's known as the "vertical interval" between fields. This might appear confusing because although the timecode is roughly vertical on the tape, it is horizontal on the monitor.

In practice, though, VITC is usually used in conjunction with LTC. Being part of the video signal, VITC cannot be read accurately in fast shuttle or rewind speeds, something LTC has no problem with at all.

DV Timecode

The purpose of timecode is to accurately identify every frame on a video medium, and be able to access it when required. It is important to recognize that there are timecode issues relating to its functionality and differences relating simply to how they are encoded on tapes. The SMPTE standardized timecode formats—VITC and LTC—are *somewhat* obsolete in the digital domain. VITC and LTC are analog signals—in picture or in audio channels, and for a digital tape to use them, it pretty much needs to convert the analog timecode to digital, then back to analog for "reading." While SMPTE timecode is **critical** for compatibility between generations of professional equipment, and for a handful of certain functions, digital formats of timecode can perform with the same precision as SMPTE if used properly.

Today, in addition to *Hi8 Timecode*, and its sister *Hi8 Rewritable Consumer Timecode* (RCTC), is the newer generation of *DV Timecode*. Each of these variations record the frame timecodes as digital information on the tapes, and provide that data to decks and computers as the data is decoded.

While the various consumer-types of timecode are neither LTC, VITC, nor standardized by the SMPTE/EBU organizations that formalize these conventions, they perform the identical function of SMPTE timecodes and with comparable accuracy. If need be, these digital timecode formats can be converted to SMPTE-standardized signals for the rare occasion when they are required. Perhaps the most important reason for a DV set-up to include SMPTE timecodes is not for functionality *per se*, but backwards compatibility with the large installed base of established professional facilities and projects.

The DV format does allow for SMPTE timecode on the tape. While this feature is built into the professional DV variants (*e.g.,* DVCPRO) consumer MiniDV (*aka* DVCAM) does not.

With regard to accuracy, when DV is captured by computer, the video and the timecode are sent separately, and can easily arrive at the computer at different times. Consequently, there will often be an *offset* between the displayed timecode in the editing system and the actual code of a given frame. While most editing systems offer a setting that will adjust the timecode with an offset, the potential frame error is often within the margin of accuracy for many consumers.

BLACK

There is a difference between a videotape that is "blank" and one that is "black." A new videotape, just purchased and never before used, is called *raw stock*. If you played it in a VTR, what you would see on your screen would be "static," familiarly called "snow." This is a **blank** tape. It has no video on it, no timecode, no nothing.

Black is a special video signal. A picture of absolute blackness must be electronically generated and recorded on videotape for you to see, well . . . black . . . on the monitor.

When you take a professional videotape with video or whatever on it, and then magnetically erase it (called *de-gaussing*) the tape is returned to a blank state, but not black. A tape used, and then de-gaussed, loses some video quality—slightly more each time it is erased. Professionals rarely re-use a tape for critical applications more than once.

Tapes have a number of "tracks" on them, designed for recording differnt kinds of signals. There is a video track, two or four audio tracks, a timecode track and a control track (a metronome-like pulse which is essentially invisible to the user).

▶▶ Tape formatting: p. 370

Timecode is recorded, or *striped*, on a piece of videotape to prepare it for use before most editing sessions. In general, black video signal, timecode and control track are recorded at the same time. (Remember that just because a tape has timecode recorded on it does not mean that you can see a visible burn-in window of that timecode number on the screen. Generating a visible timecode window is an option you have when playing or recording a tape; a window is rarely if ever burned in to black tapes.)

Most offline editing, linear or non-linear, must record onto a **black and coded tape**. The computer controllers use that timecode to determine record points, to synchronize the source and record tapes, and to search for specific locations. When these edits take place in offline, they are only recording audio or video over the black, but are not altering the timecode track. This is called **insert editing** (it has nothing to do with the nonlinear "insert.")

If you are editing to all the channels of a videotape (recording to the timecode track *and* control track as well) you are **assemble editing**. Online and offline rarely assemble edit; however, telecine sessions often do.

When you assemble edit, you generally record timecode as each edit is made. If this timecode is based on the previous shot's timecode (continuing it where the previous shot left off), it is called *jam syncing*, and you will end up with a series of edits with continuous timecode running on the tape. A tape made this way is indistinguishable from one that was blacked and coded first, then edited using video/audio inserts, except that after the assemble edit session, the timecode will only extend as far as the last shot. Only *very accurate* timecode generating and editing equipment can assemble edit and pick up *exactly* where the previous timecode left off.

If you try to assemble edit on a machine that is not capable of jam syncing (since the timecode track will get disrupted at the out-points), the numbers on the tape will not be continuous. When the timecode track is broken in this way, editing can be difficult. You cannot do an insert edit across a break in timecode; computer edit controllers can easily become confused when the timecode jumps for no apparent reason or is missing altogether, and edits and cues may be aborted.

For offline editing, 3/4" videotapes are pre-striped with black and code. **Continuous and ascending timecode** on source tapes is required for the smooth function of many, but not all, nonlinear systems, including

▶▶ Videotape assemblies: p. 257

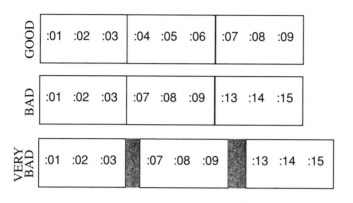

Continuous and ascending timecode.

Discontinuous and ascending timecode (with **continuous** control track).

Discontinuous timecode and (broken) control track.

consumer DV systems. Systems that continuously monitor the timecode on the tape *can* handle discontinuous timecode, as long as it is unbroken—meaning as long as there are no places where there is NO timecode in the middle of the tape. Editing systems with a log calculating the relationship between the videotape and film must be alerted where discontinuities in timecode affect changes in the relationship to the film. Many systems controlling videotape (whether editing video, digitizing source for a nonlinear session or outputting to tape) will abort recording if they find a timecode anomaly.

Desktop nonlinear systems can generally capture DV with *or* without timecode; breaks in source tape timecode, however, can cause system errors, and can also make it impossible to recapture a piece of material at a later date. When shooting DV, the camera is effectively jam-syncing new shot's timecode based on the previously recorded shot. For this to work, you must begin recording on top of the tail of the previous shot (if you do nothing after ending a recording, the camera does this for you). Make sure you leave room at the end of shots for this overlap. MiniDV tapes with "IC memory" streamline this process, with new shots automatically beginning at the end of previous ones, but more importantly, if you shuttle the tape between shots, the camera can always find last frame via "end search" functions.

When printing to tape, a process that reasonably replaced auto-assemblies, desktop systems often *assemble edit* to tapes, so blacking DV tapes beforehand is of little consequence. It is always a good idea to record leader before and after each assembly session.

T H E   E D L

Edit Decision Lists

Prior to 1972, there was no standardized way to repeat the edits of an editing session. Before there was timecode, and before computers kept track of timecode, you could not automatically re-edit old work, or perform an auto-assembly.

CMX pioneered the development of the Edit Decision List, or EDL. Although currently there are other formats for the EDL, in particular those from the Grass Valley Group, Sony and Accom, the information contained in the EDL may differ slightly in style, but is basically the same in substance.

In the CMX-type of EDL, there are 8 primary columns and 2 ancillary columns. At the top of the page is a header (the title), and below that from left to right, are the following columns: *event number, source reel ID, edit mode, transition type* (sometimes followed by 1 or 2 other columns, *wipe code* and *transition rate*), *source in, source out, record in* and *record out*:

```
01  02  V    C            02:28:32:03  02:28:49:00  01;00;00;00  01;00;16;27
02  02  A1   C            02:28:32:03  02:28:49:25  01;00;00;00  01;00;17;22
03  04  V    D  45        04:20:37:20  04:20:41:10  01;00;16;27  01;00;20;17
03  04  A1   C            04:20:38:15  04:20:41:10  01;00;17;22  01;00;20;17
04  12  VA1  W  094 60    12:06:00:28  12:06:07:28  01;00;20;17  01;00;27;17
05  BL  VA1  C            00:00:00:00  00:00:01:00  01;00;27;17  01;00;28;17
```

EDL Components

• HEADER: At the head of every EDL is the list's TITLE, and whether or not the record times are in Drop Frame (DF) or Non-Drop Frame (NDF) timecode. *Source timecodes are specified as NDF by colons, or DF by semicolons.*

DROP FRAME TIMECODE   01;00;00;00
NON-DROP FRAME TIMECODE   01:00:00:00

Sometimes source timecode-type will vary from tape to tape within an edit session. Some kinds of systems will place a special comment in the list whenever source timecode has switched. For example, it might say "FCM: Drop Frame," meaning that the Frame Code Mode (FCM) had changed, and the new source timecodes are now in Drop Frame.

◀◀ Timecode: p. 83   ▶▶ Printouts: p. 260

• The EVENT NUMBER is simply an identifying counter, beginning at 1 and increasing with each edit. While each edit is given its own number, notice that a single edit can take up two lines in the EDL, in the case of dissolves, wipes, and "split edits." The event number is important in uniquely identifying each edit for eventual list cleaning or assembly—especially since offline editing is usually not done in sequential order.

A **note** in the list is an *unnumbered* comment line, generally associated with the preceding event.

• SOURCE REEL ID is the name of the videotape on which a particular shot was originally located. Every videotape that is recorded in telecine is numbered, and all duplicated reels ("dupes") of that master source tape get identical numbers. For editing, additional copies of a source reel will have identical source reel IDs *with an appended letter*; for example, reel 004 and 004B are identical material. This kind of "B-roll" is usually created for use in effects such as dissolves. In general, but not always, the "hours" portion of the source timecode is equivalent to the reel number (*i.e.* videotape reel #4 begins at 04:00:00:00—four hours straight up). If the source for an edit is video black, the source ID appears as "BLK," "BL" or sometimes just "L."

SOURCE REEL FACTOID • Nonlinear editing systems will often let you name reels pretty much whatever you want (e.g., "Joe's Birthday Tape #234!"); however, if you ever expect to take your EDL to a **linear** online session, the reel names must conform to that editing system's naming restrictions, namely: a maximum of 6 alphanumeric characters (or 5 to be safe). Talk to your local online editor for specifics.

Other non-VTR sources, like electronically-generated color bars or other switcher "cross-points" (for doing special video effects) are called auxiliary sources, and are often labeled "AUX," "AX" or just plain "X."

• EDIT MODE denotes whether the edit takes place in video only, audio #1 only, audio #2 only, or any combination of these. The letter "B" was originally used to signify a "Both" cut, where both picture and sound cut at the same point, but the introduction of a second channel of audio made this obsolete. "V" is video, "A1" or "1" is audio #1, and "A2" or "2" is audio #2. (Also note that many digital VTRs have *four* channels of audio.)

▶▶ Nomenclature of scenes, takes and reels: p. 211

• TRANSITION TYPE describes whether the edit is a "C" for cut, "D" (dissolve), "W" (wipe), or "K" (key). Transitions other than CUTs will be followed by a transition duration, in frames. WIPEs and some kinds of KEYs will have an additional code following the letter, indicating what kind of effect is being triggered. Unfortunately, the wipe codes from the various switcher manufacturers have not been standardized.

• SOURCE IN and SOURCE OUT are the first two columns of timecodes. These numbers describe the timecode of the first and last* frame of the shot, as it is played from the source videotape.

• RECORD IN and RECORD OUT are the last two columns of timecodes. These describe where the source shot is to be recorded on the master videotape. Note that, due to the linear nature of videotape, a change in duration to any but the final shot in a sequence may require ALL following record in and record out points to be shifted earlier or later. Source in and source out points for all these shots do not change. A shift in record times is called a "ripple" because of the ripple effect even a small change makes in all following edits.

Since nonlinear systems need only manipulate numbers in the computer's memory rather than images on a piece of videotape, all nonlinear systems can "ripple" effortlessly as a matter of design.

The EDL is traditionally the most important output of a nonlinear editing system. The reason is that these systems were originally used only as *OFFLINE* systems, and the creative decisions made in offline were then later conveyed to *ONLINE* for re-creation of the edits using the master source tapes. As systems' storage capacities and computing horsepower improved (and hence, so did the quality of the digitized video), they eventually commandeered more and more actual online *finishing* work, and the importance of the EDL began to wane.

---

*Actually, the established convention in video editing is that the "out" timecode is that of the frame **following** the last one used (i.e., it is not inclusive). This means that the next event's **in-point** is the same number as the previous one's **out-point**.

T H E   E V O L U T I O N   O F   T H E   E D L

DIGITAL FILE EXCHANGE MEETS THE JAPANESE LUNCH BOX

The original "edit decision list" (EDL) invented by CMX, and further developed by other manufacturers of linear edit gear, was intended to convey instructions on where to find all the shots for an edited sequence, and how to execute various effects in order to replicate a finished online master from an offline list. But these effects were quite simple by today's standards: just wipes and dissolves of various durations, nothing fancy at all. No DVE moves. No layering capability worth mentioning. So, for offline edit decisions that required often-times extraordinarily complicated written or verbal instructions for the online re-building process, a plain-vanilla EDL was (and continues to be) insufficient ...

The ability to readily exchange digital video and audio (and the editorial decisions made about them) directly from one type of digital workstation to another has long been a holy grail for digital artisans: the picture editors, sound editors and mixers, graphics/effects artists and others who need to collaborate on a particular project. And while it's always possible to output the result of your work to a linear tape format (whether analog or digital) to give to the person next in the creative chain, this imposes potentially serious limitations in efficiency, quality and flexibility. Clearly, the ideal workflow would allow you to pass along the raw digital source media that you already took the time to capture, as well as all the editing and layering decisions you've made about them along the way. If this weren't desirable enough, the sharing of networked storage between different types of workstations (*e.g.*, FibreChannel) makes such a common "language" even more compelling. And if you can convey all the information you need to in just one self-contained digital file, so much the better.

OMF: a case study

One of the first serious attempts to bridge the "digital divide" separating the workstations of various manufacturers was the Open Media

Framework Interchange (OMF or OMFI, depending on who you ask), a file format spearheaded by Avid Technology with other strategic partners, in the early-mid '90s. It conveniently took advantage of something called *Bento*, a digital file "container" format originally developed by Apple Computer.

Contents of a modern-day "Bento" lunch box

But the choice of Bento was noteworthy in another sense, as the name is derived from an elaborate lunch box dating back to ancient Japan. A "bento box" features a number of discrete, separate compartments that can each be used to hold a different type of food. The engineers at Apple (and the folks behind OMF) obviously recognized a good analogy when they heard one; this was just the right concept for embedding different types of digital data into one self-contained file, which could then be closed up and conveniently moved from one digital workstation to another.

Just like the lunch box, each of the compartments in a Bento *file* could hold discrete "helpings" of various kinds of data—some digitized raw video footage here; digital audio there; source reel names, timecode and KeyKode information over there; each type as needed, *a la carte*. Yet OMF went beyond the mere ability to transfer digital media; it also provided the means to convey the all-important "recipe": instructions that told the destination workstation just how to manipulate those raw picture and sound "ingredients" in order to re-create a previously-edited sequence. This recipe capability made the scheme unique, an heir-apparent to the rapidly-antiquated text EDL, and well suited to the needs of those who wanted to be able to collaborate digitally.

Imagine a timeline with complicated video layers, some superimpositions, picture-in-picture effects with a motion path, some of those with internal fades and traveling mattes—and that's just for the video alone. When it comes time to online all these creative decisions, an ideal workflow would utilize the equivalent of an "EDL on steroids" to retain all those exactingly detailed relationships, and rebuild all the effects automatically, though at a higher "finishing" quality.

Transferring the digital media and all editing/layering decisions directly from one system to another using such a scheme has several important advantages:

- It eliminates the need to output from one type of system, and then re-digitize a tape back into another system, with a potential loss of quality.
- Provides for potentially faster-than-real-time transfers.
- Allows for the inclusion of "handles"—adjacent extra material that isn't needed in the current version of the cut, but which might still be useful to the next person in the chain (finessing of the soundtrack by the sound department is one frequent example).

Even by the time an official attempt was made to unveil an updated EDL standard (SMPTE 258M, in 1993), it was already too late—the dizzying array of capabilities being developed for even lowly nonlinear systems quickly outpaced the ability of a mere ASCII text file to keep up (let alone have any chance of carrying the digitized *media* files also needed for the project).

And while Bento itself was ultimately abandoned by Apple, as eventually was the original OMF format by Avid and friends, the underlying concept of a modular "digital lunch box" is still germane to future standards that are being developed to fulfill similar needs.*

As digital nonlinear online finishing grows in popularity, and the power of desktop video systems continues to increase, the desirability for bento-like interchange formats and facile digital media exchanges increases commensurately. While for the time being, it might still be most expedient to do both the offline and finishing of any given project using the same "family" of nonlinear systems—for the sake of a broader and more streamlined digital workflow—one would hope that various (and otherwise competing) manufacturers will see the ultimate value of a truce, at least just long enough to share a little conceptual "nourishment" from the same box of lunch.

---

* In 1998, the successor to OMF was announced: AAF—the Advanced Authoring Format. This evolution was precipitated by both technological and political factors.

F I L M   E D G E   N U M B E R S

R unning along the side of all 35mm film are some pretty small numbers, updating each foot. Actually, all major types of film have some
kind of edge numbering. Edge numbers that are pre-printed on negative
stock are also referred to as *key numbers.* When edge numbers are stamped
onto a positive print of the original negative ("workprint"), and identical
numbers are also applied to sprocketed magnetic audio stock to correspond to synchronized tracks, they are called *code numbers,* or sometimes
rubber numbers.

Both types of edge numbers are human-readable characters that are
essential in the smooth process through post-production. Here is a typical
edge number on a small piece of film:

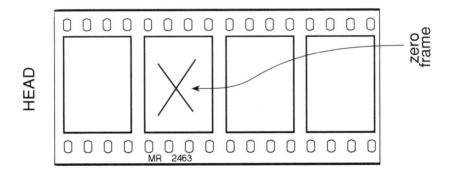

The first part of an edge number is a set of numbers and letters called
the *prefix,* which is constant throughout each original roll of film. For code
numbers, the prefix also indicates on which reel a shot originated. Following the prefix is a *footage count.* This number increments every foot
throughout the entire roll of film.

These numbers allow you to identify any single frame on any roll.
Frames without numbers (actually the majority) are identified by counting how far they are located from the closest preceding edge number. This
way, every frame can be distinguished by a unique number: the prefix, the
footage count, plus the additional number of frames from the location of
the actual edge number—for example, MR 2463 +11 is the ID of a frame
11 frames after the key frame MR 2463.

◀◀ The Film Story: p. 24

Because the numbers are spaced a foot apart (and there are 16 frames per foot), counting frames from the key number to the frame you want to identify can be tedious and error-prone. Even if you didn't make mistakes, a method of counting still has to be consistent among all the people using the film. You could count from frame 0 to frame 15, then roll over to the next foot (known as zero-counting), but if the next person calls the first frame "1" and counts to frame "16," you'd both be in trouble. Neither is incorrect, but the method used must obviously be standardized. The most common numbering scheme is the 0-to-15 count ("X" marks zero).

| TAIL | +4 | +3 | +2 | +1 | X | +15 | +14 | +13 | +12 | +11 | +10 | +9 | +8 | +7 | +6 | +5 | +4 | HEAD |

You also must determine from which frame you will start counting. If edge numbers were perfectly placed on the film with relation to the frames, it would be simple to determine the "0" frame (sometimes called a *key frame* or *index frame*). Officially, the key frame is the frame in which the key number ENDS. However, in the real world, edge numbers often end on a frame line, making things even more complicated. Everyone using the film *must* determine key frames in the same way.

Tracking key numbers and code numbers is a significant part of the film post-production process. Especially with the electronic post of film using video (nonlinear, in particular), the proper entry of these numbers into relational databases is the critical factor that allows film to be conformed from video offlines. The greatest single "frame accuracy" issue in electronic systems has usually been the human error factor involved with the reading and typing of these numbers into computers.

K E Y K O D E

MACHINE READABLE FILM EDGE NUMBERS

Kodak spent a number of years developing a better kind of film edge number for their film stocks. In the early '90s they introduced KEYKODE®  numbers, which are human- *and machine-readable* edge numbers. The applications of KeyKode have not yet been exhausted. But for electronic nonlinear systems they significantly reduce the human errors involved in the logging of film numbers into relational databases for the eventual reconforming of the negative or workprint.

How KeyKode Works

KeyKode offers a number of significant advantages over earlier methods of numbering (remember? a single tiny number for each foot of film):

• Along with the alphanumeric edge numbers are machine-readable bar codes. A bar code scanner can easily read and input edge numbers into computer logs and databases without the necessity of human reading and typing.

• Edge numbers have also been made easier to read by eye, spaced better, and given an additional digit in the prefix, which minimizes the possibility of duplicate prefixes on different rolls of film.

• A special dot has been inscribed following each number, denoting the Zero-Frame. This should standardize the reading of edge numbers and eliminate discrepancies created from different counting schemes.

• An added mid-foot edge number has been included, printed smaller than the regular footage count, which should decrease problems associated with identifying short (less than one-foot) shots.

• There are a number of marks, checks and symbols located on the stock that make verification of frame lines, negative matching, and offsets much simpler. There are also film stock identification codes, manufacturer identification codes, and various product and emulsion codes.

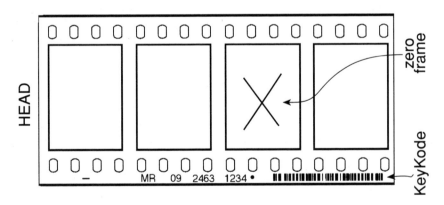

Since 1995 KeyKode has become the industry-wide standard in film numbering, and ancillary products are available in many film and video post-production settings. Bar code readers are now included in most telecines and anywhere that the reading of edge numbers needs to be done. Telecines fitted with KeyKode readers can begin the logging process automatically. Accuracy of editing systems, speed in telecine, and efficiency in the preparation to edit are all significantly increased with the adoption of this technology.

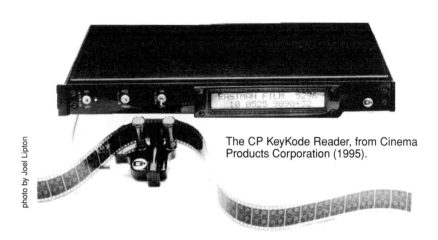

photo by Joel Lipton

The CP KeyKode Reader, from Cinema Products Corporation (1995).

24P

NOTICE! The following pages concerning *telecine* are about the idiosyncratic gyrations involved with transferring 24fps film to 30fps video. The HDTV standard has created a hybrid format, 1080p/24fps—known in the industry as **24P**. This video format runs at 24fps and thus has a 1:1 transfer relationship between a broadcast video and a celluloid print. Ultimately, it will render the 3:2 process moot. It will remove the need for telecine, for questions of frame accuracy, for telecine logs, for drop and non-drop timecode (its new timecode counts from 0 to 23 then rolls over). When 24P technology is combined with electronic distribution (see Chapter 5), the medium can be digitally projected in theaters. If you're lucky enough to be on a production with 24P technology, skip ahead to page 120 (*and collect $200*).

▶▶ Electronic cinema: p. 312

T   E   L   E   C   I   N   E

Getting the Film into the Video World

Here is a piece of film, 35mm, 4 perf, shown actual size:

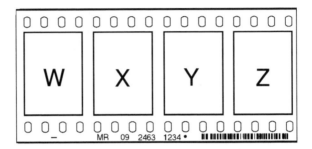

As you may recall, most film is played at **24 frames per second.** Another way of saying *24 frames per second* is that each single frame is 1/24th of a second long.

If a single film frame is 1/24 of a second long, then two frames are 2/24ths (1/12) and three frames are 3/24ths (1/8), and so on:

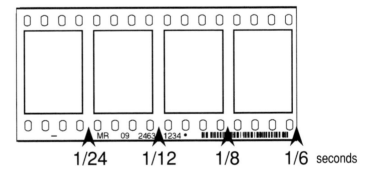

If you continue with this, you can see that the time it takes four film frames to pass is 1/6th of a second. Exactly. Remember this for later.

AUDIO, unlike picture, does not have "frames" and thus does not move at 24fps—it just keeps on moving all the time, smoothly and constantly. While your picture jumps from frame to frame, the audio (usually on 1/4" magnetic tape) plays smoothly along.

VIDEOTAPE is also different from film. It doesn't look the same. It isn't recorded the same. It doesn't play the same. One-inch (1") videotape—although one continuous strip—is usually diagrammed something like this:

Notice that each videotape frame is slanted. This is because, unlike most 1/4" audio tape, videotape runs across the play/record heads of a VTR at an angle (actually, it's wrapped around a cylindrical head, but ignore that right now):

Also, you may remember that videotape records and plays at **30 frames per second**, so that if we do the time thing again we would see:

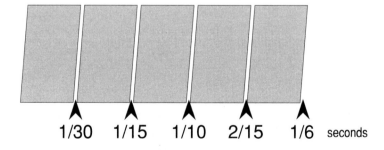

1/30   1/15   1/10   2/15   1/6 seconds

—where the first frame is presented for 1/30th of a second, two frames play in 2/30ths (1/5) . . . and so on. Unlike film, on video five frames play in 1/6th of a second. And if this wasn't strange enough, every single video frame, although uniquely identified by a timecode number, is ACTUALLY made up of two nearly identical VIDEO FIELDS (known quaintly as "field

1" and "field 2"). So let's do the videotape diagram again, this time with fields drawn in, and timecode labeling each frame:

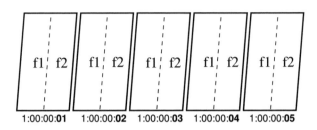

1:00:00:**01**   1:00:00:**02**   1:00:00:**03**   1:00:00:**04**   1:00:00:**05**

Every frame has its own timecode: either encoded longitudinally, in an audio track on the tape (called *Longitudinal Timecode*, or LTC), or encoded in the picture portion of the tape, vertically in the space ("interval") between field 1 and field 2 of each frame (called *Vertical Interval Timecode*, or VITC). Or in both.

When videotape is recorded—for instance, when you go out with your VHS camcorder—the same image is recorded into *both* field 1 and field 2 of every frame.

FACTOID • The image in field 1 is only half the video image—but not the top half, nor the left half, but the odd half. A video image is made up of 525 lines of video per frame; field 1 contains the ODD numbered lines (1, 3, 5 ... 525) while field 2 contains the EVEN numbered lines (2, 4, 6 ... 524).

When the tape is playing, each full video frame is the result of interweaving (called "interlacing") of the two fields. When you inspect a freeze-frame on videotape, you are only seeing one of the two fields—or HALF the resolution of the moving image.

A TELECINE is a machine that transfers FILM to VIDEOTAPE. No simple task. You put in your film (either positive print or original negative), load up a record videotape (in a format of your choosing), and get to it. But how? Film runs at 24fps and videotape records at 30fps.

If you transfer (copy) one film frame into each video frame, look what happens:

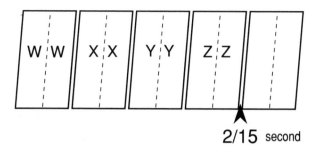

2/15 second

We took **4 FILM FRAMES**—that should go by in 4/24th of a second, and copied them into **4 VIDEO FRAMES**—which play by in 4/30ths of a second—FASTER than film! In fact, this kind of direct transfer will speed up the action on film by **20%** and, as you can imagine, it throws the audio on the 1/4" tape completely out of sync with the picture.

This kind of transfer is called a **2:2 transfer**, because every film frame is copied into 2 video fields, repeatedly (they could have called it a 2:2:2:2:2:2... transfer, because that's exactly what's happening).

If you want your videotape copy to play at the same speed as the original film, you have to take every FOUR film frames and SOMEHOW stretch them into FIVE video frames.

A telecine transfer does this by performing what's called a 3:2 transfer (or **"pulldown"**). Rather than copy each film frame twice—once into each of the two video fields—a 3:2 pulldown transfer will make every other film frame *a little bit longer...* it will copy every other frame into one extra field:

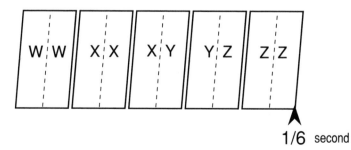

1/6 second

The key here is that it is okay for certain video frames to be made up of two fields with *totally different pictures* in them. Technically speak-

◀◀ 3:2 vs 2:3 pulldown name conventions: p. xvi

ing, some film frames have been transferred to 1/30th of a second-types of video frames (2 fields long), and others have been transferred to 1/20th of a second-types of video frames (3 fields long). This is a 3:2 pulldown (3:2:3:2:3:2…)

Now 4 *film frames* occupy the space of 5 *video frames*—all playing in 1/6th of a second—faithfully remaining in sync with the 1/4" audio tape.

To make matters slightly more confusing, it turns out that there are actually FOUR different "ways" that the 3:2 transfer can take place:

• Starting on field 1, alternating 2 fields, then 3...2...3...2...3...
• Starting on field 1, alternating 3 fields, then 2...3...2...3...2...
• Starting on field 2, alternating 2 fields, then 3...2...3...2...3...
• Starting on field 2, alternating 3 fields, then 2...3...2...3...2...

They ALL produce the same end result. So what's the difference between them? In some ways, nothing, at least when you are editing videotape. BUT, if you ever want to convert your videotape (timecode) edit decisions back to FILM, the kind of 3:2 pulldown is **extremely important**.

On every videotape created from a 3:2 pulldown, regardless of the KIND of 3:2 pulldown that was done, the 4 transferred film frames are converted to one of FOUR TYPES. See if you can follow this:

TYPE A: 2-field "frame" beginning on a field 1
TYPE B: 3-field "frame" beginning on a field 1
TYPE C: 2-field "frame" beginning on a field 2
TYPE D: 3-field "frame" beginning on a field 2

To demonstrate this, let's look at all four of the possible 3:2 pulldowns. Below each transferred film frame is its "3:2 frame-type":

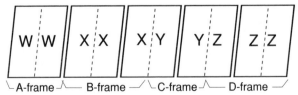

2 Starting on field 1, alternating 3 fields, then 2...3...2...3...2

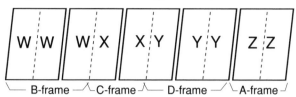

3 Starting on field 2, alternating 2 fields, then 3...2...3...2...3

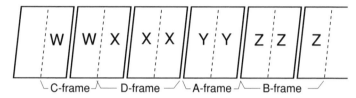

4 Starting on field 2, alternating 3 fields, then 2...3...2...3...2

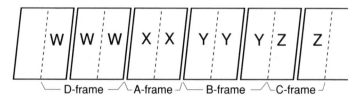

It should be clear from these diagrams that no matter what type of 3:2 pulldown is performed, only four kinds of transfer frames are created.

All editing systems that deal with both film and tape will *NEED TO KNOW THE EXACT RELATIONSHIP BETWEEN THE ORIGINAL FILM AND THE VIDEOTAPE.* This relationship is always entered into some kind of database (or "**log**") as an "interlock" point between film key numbers (or code numbers) and a videotape timecode. In addition, the 3:2 type IS REQUIRED.

Film frame "W" for example, may have originated as key number MR 2462+15. On videotape, after a telecine of type 2, it will have a timecode reference of 1:00:00:01, and will be a **B-frame**. If, however, it is a telecine of type 1, the same frame will have the same references, but will be an **A-frame**.

CUTTING FILM USING VIDEO

FRAME ACCURACY OF "MATCHBACK"

Any piece of film can be transferred to videotape for editing with little difficulty; 3:2 pulldowns are standard operating procedure for all telecine bays—that's exactly what a telecine machine is designed to do.

Once your film is on videotape, you do not need to go through all these frame and field gymnastics, if you don't want. Film is edited on videotape all the time. Linearly. You can always take a 3/4" tape with a burned-in window of timecode and offline to your heart's content; the online of this edit session can air on television, can be dubbed to other tapes for distribution, whatever.

The issues associated with cutting film on tape do not pertain to projects finishing **only** on videotape. As long as you never want to go BACK to having cut film, there are no concerns of "frame accuracy," or "slipping sync with audio," or any of those things that everyone talks about when discussing electronic film editing.

But, to edit on videotape or some video media and then to CUT NEGATIVE OR POSITIVE PRINT—that is where the 3:2 pulldown and issues of frame accuracy come into play.

Let's say you take a 3/4" dub out of telecine: you've transferred your film and now want to edit on tape (linear or nonlinear, it makes no difference). Here's your videotape:

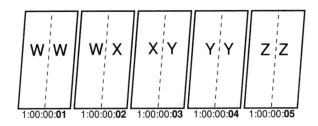

You can see the 3:2 pulldown frames clearly here. Every film frame transferred to this tape can be classified either an A, B, C or D-type frame. This transfer is of type 2, as the first frame is a B-frame (a 3-field frame beginning on field #1). Remember, it doesn't matter what kind of transfer you do as long as you know what kind of 3:2 frame you *began* with.

Now, let's say you decide to cut between tape frame :01 and :02:

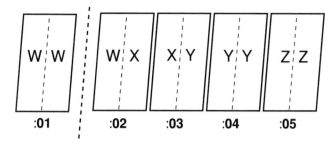

The deceptively simple question is, "Where is this location on the film? Is it between frame W and frame X?" The answer is...sort of. It certainly looks like it could be. Realize that all videotape, when parked in "freeze frame" can only show you ONE FIELD. *Which field?* Either one, it depends where you stop. So when you park your videotape, and choose to edit AFTER frame :01 (in this case, in the middle of an A-frame), you still have one more field of film frame W after this point...hmmm...okay...so maybe cutting your film between frame W and X isn't the right spot. But what else could it be? Cutting between :01 and :02, and cutting between W and X both give you a 1-frame shot...

Okay, we'll skip this one and try making an edit on our videotape between :02 and :03.

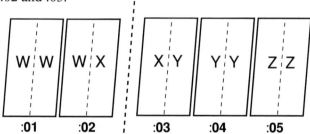

Not better...in fact, considerably worse. Now where should you cut your film to match this edit decision? Video frames :02 and :03 have split film frame X! "When I made my decision to cut out on video frame :02, did I mean to *INCLUDE* or *EXCLUDE* the image in film frame X?"

Since all videotape only shows you one field at a time, we might be able to say "If you were looking at field #1 of both frame :02 and frame :03 when you made the decision to edit here, then you *were* deciding to cut between W and X, *NOT* X and Y." But with videotape editing, the EDL will not take into account which field you were looking at when you made your editing decision.

To make matters smoother, when many offline editing systems DIGITIZE video they save space by only digitizing each FIELD #1 (yes, this gives you 1/2 the resolution of the image). So unlike interleaved video, if you present each field #1 from the videotape, the editor would see this:

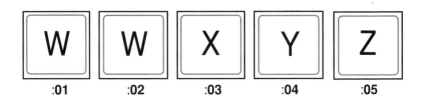

:01   :02   :03   :04   :05

The *five* video frames look like *four* film frames, with the W frame printed twice. Now, it is pretty easy to determine visually where to cut the film. *So what that you have two W frames*, it still has a direct visual relationship to the film (and, no, you can't edit between the two W's). Note: Some digital systems can also digitize at 24fps—meaning that they just ignore the repeated field. This saves disk space and reduces motion artifacts, as well as eliminating the need for any 3:2 calculations when producing the film cut.

NONLINEAR EDITING TIP • When you do an offline edit, and make editing decisions based on the visual information in only 1 of the 2 fields in a video frame, you run the (very real) risk that the video in the OTHER field is something you don't want to see—in particular, the beginning of another shot.

This is especially something to be careful of when using telecined or previously-edited material. These glitches (known as "flash fields") probably won't be discovered until online, where they can be difficult—and expensive—to fix. The best precaution is to avoid editing with the tail (or even head) frame of any particular shot.

But there are more than VISUAL issues involved with accurately cutting film from video media.

There are also very significant issues of TIMING.

For example, if I am editing a very special "flash cut" kind of sequence for a musical section of a film, and I make 24 one-frame edits—I have made 24 extremely careful edit decisions. I have chosen 24 special frames and timed them to the music.

▶▶ Head and tail frames: p. 216

But on my electronic editing system, each of the 24 frames, even if *each* one represents an *actual* and easily identifiable film frame, is still playing on my video media—where each frame plays at 1/30th of a second. That means that this sequence appears to play in 24/30th of a second.

But when I go to cut my film, and cut all 24 of these frames, it will take an entire second to play (24/24ths of a second). I won't be in sync with my film music track. (Actually, **no** digital system can be temporally accurate if all edits are 1-frame long. Some of the edited shots need to have at least 2 video frames in order to be adjustable.)

Many nonlinear editing systems will automatically alter the "film cut" to time out to the duration of the video cut. These systems use information about the 3:2 sequence to calculate film-cutting locations that will allow pictures to remain in sync with audio. *Even if you have carefully chosen* each of the 24 frames, by the time this sequence appears on film, it may only be 19 or 20 frames long in order to have the same duration.

The question arises: "What is *frame accuracy?*" Is it an exact frame-for-frame correspondence between the video editing media and the final film? Or is it a temporal exactness that makes 100 minutes on video become exactly 100 minutes when the final film is conformed?

There is no one answer. Each nonlinear editing system deals with these issues slightly differently.

Is there anything to be done? Just as film editors are precluded from making cuts that interfere with their negative cut margins,* nonlinear editors working at 30fps must make some small concessions if they safely want to match back to film, without undesired material showing up in the finished film cut. An edit may shift up to two film frames during this match back (statistically, this will occur in around 20% of all cuts). To protect yourself, the best workaround is to avoid cutting at extreme points in the source—the first and last usable frames of any given shot (near the clapboard, for instance, or near flashes, blinks, or just before the proverbial "dead" actor's eyes open). Similarly, where the "good" frames are nestled between bad ones, you may need to steer clear of potential problem areas.

In practice, though it may sound harsh, editing is a forgiving art. Even the most exacting and precise editor, when cutting narrative material, might

---

*When negative is spliced, a portion of the frame adjacent to the splice is used when the shots are "melted" together, thus obliterating that extra frame. This "negative cut margin" means that editors are prohibited from using certain frames in their cut.

not notice if you added or dropped a frame here and there in his or her cut to make it run the "correct" length. When about 20% of edits are affected, and of those, only a small percentage occur in places where the "cheat" might be noticeable, the reality is that it's generally not a catastrophic issue. Yes, once in a while an edit is precise and a frame added or subtracted would change everything, but this is still the exception and not the rule.

All of these things are based on the nature of the 3:2 pulldown.

The simple fact is that virtually ALL nonlinear editing systems are editing timecode, just like traditional videotape editing systems. They create timecode lists, and then "trace" those numbers back through a relational database to generate the film numbers. It is simply because there are 30 frames per second in video that there is *no* one-to-one relationship between the editing media numbers and the original film.

But this is not always the case.

The only way to have visual and temporal frame accuracy is to actually have a one-to-one relationship between the film and video media. One way to do this is to play and edit your videotape at 24fps—like the film. If you have special tape decks and an appropriate edit controller, you can do a 2:2 telecine transfer and edit *frame accurately.*

Digital media is the best choice, as it can digitize video at either 30fps or 24fps, *if* it has appropriate digitizing hardware and software. Not all digital systems can digitize only 24fps; more common is a digitizing process that records 30 frames each second, but only identifies 24 with unique numbers (exactly like a laserdisc with white flags). This latter method provides identical accuracy to a real 24fps digitization, but still requires more space and contains motion artifacts. *Remember that the accuracy is not just a function of the editing media, but rather HOW an editing system DEALS with numbers in that media.*

Determining Frame Accuracy Type

If you want to ascertain, for whatever reason, whether your system is editing "film frames" and then calculating video; or "video frames" and calculating film, there are a number of simple tests you can perform. These will give you some idea as to the scope and degree of accuracy a system possesses.

Test 1: Cut a list of short edits. How long is your video EDL? How long is your film cut list? Now, add a SINGLE frame to the head of the very first shot. Print out a new EDL and a new film cut list. What changes?

A 1- or 2-frame modification of a single edit on a 30fps-based system will result in a change in many film edit points throughout an edited sequence.

Test 2: Will an auto-assembly (either the online session or a simple print-to-video output from the nonlinear edit session) play in sync with the cut film? Are they the same length? (Don't forget that you still must resolve the video speed from 29.97fps to 30fps to make them sync.)

When most systems claim frame accuracy, they are not lying. Because videotape is NOT film, and because of the 30fps of video, there is no identical location to cut film when editing tape. All video systems try to make the best possible choices in the cutting of film—best visually and best temporally.

For many theatrical films, cutting negative directly from the nonlinear system is not an option; there will ALWAYS be a workprint cut. Because of this, any problems that might have been created can be easily fixed on film. Many editing systems allow for cutting negative DIRECTLY from the final cut on the editing system. Productions can utilize these features to save the huge costs associated with printing the negative. Most nonlinear systems will generate a negative cut list, and will edit for temporal frame accuracy—keeping the picture in sync with the sound. Because of the forgiving nature of editing, the changes in the cut made by timecode-based calculations are usually imperceptible.

But the only way to have both visual and temporal frame accuracy is to have the electronic system based on the same frame rate as your final product. Although *all* systems must use a mathematical algorithm somewhere in the video-film conversion, it is much more *forgiving* to convert film numbers (at 24fps) into a frame accurate EDL than it is to convert timecodes (at 30fps) into a frame accurate film list. This is solely due to the fact that the smallest unit of film (and therefore the smallest error) is 1 frame = 1/24th of a second; for video it is smaller: 1 field = 1/60th of a second.

PREPARATION FOR TELECINE
"THE FILM FINISH"

Telecine is simply the transfer of film—negative or positive—to videotape. For projects that are finishing on videotape, there are no special precautions to take. But for projects finishing on film (or both film and tape), there are some important specifications that must be met in the telecine. It is THIS process that will ultimately allow the film to be cut correctly.

Like printing dailies on film, the telecine session usually involves choosing selected takes to transfer to tape, not entire rolls. Every time you choose another take, you must stop the telecine machine, find where you want the take to begin on the film and the videotape, sync up the 1/4" audio, and then record.

Because of the need to sync the audio tape, you are stopping and starting recordings on every single selected take, even if they are already in the right order and consecutive.

To use this tape in editing electronically, the 3:2 sequence (3:2:3:2:3:2...) must be maintained through the entire tape, with no mistakes. This means that if the last take ended on an A-frame, the next take recorded MUST begin on a B-frame:

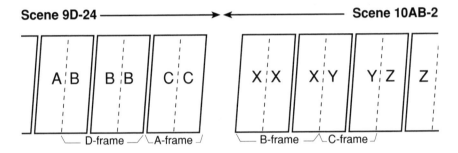

If this relationship is changed, the editing system may not able to determine the accurate position of the film relative to the videotape. There are a number of ways to make this process go smoothly in telecine:

1. Assemble negative or positive into entire reels of the exact length needed for the tape—then telecine the entire roll at once. If you don't stop the telecine recording, the 3:2 sequence cannot be accidently changed. Later, go back and record the sync audio from the 1/4" tapes.

2. Standardize telecine edits. Make every edit go in on an A-frame, for example. This means that the film edge numbers are not consecutive and MUST be logged as a new entry in the editing system's relational database. Often, this is of little consequence since transferring the select takes from the negative involves skipping whole areas of film, and the film numbers are already going to be discontinuous.

When telecine edits are standardized, it is easy to know that when timecode :00 is the beginning of an A-frame, every timecode multiple of 5 frames is also going to be an A-frame: :00, :05, :10, :15, :20, :25, :00…

AA	BB	BC	CD	DD	AA	BB	BC	CD	DD	AA	BB	BC	CD	DD
:00	:01	:02	:03	:04	:05	:06	:07	:08	:09	:10	:11	:12	:13	:14

3. The best method is to use KeyKode stock with appropriate readers. This automates the process, and can generate a machine-readable EDL-of-sorts, indicating the key number, timecode, and 3-2 type (among other data) at every edit. If this data list is converted to a format that an editing system can read—like Aaton or Evertz—jumps in 3-2 type can be managed.

Without the advantages of KeyKode, and short of using one of the other two methods, the telecine operator or assistant must check the tape at each edit to determine the 3:2 frame type for the last recorded frame. They would do this by actually viewing each frame of tape, field by field.

Another piece of equipment helpful for the telecine of material finishing on film is the Time Logic Controller, or TLC. This is a box that accurately counts film frames in telecine. This is one way it is used:

1. At the head of each film reel loaded onto the telecine machine, a single frame is scribed or punched to make it easily recognizable via picture. The punch is made at a key frame—a frame with a clear film edge number. This number is entered into the TLC.

Now, wherever you roll in your film, the TLC will show the film edge number of that frame on its LED display. This number can be noted or encoded into VITC on the videotape for retrieval later.

2. At each telecine edit, even when one or more film frames are going to be skipped in the transfer, the TLC continues to track the film frame numbers. The TLC displays film numbers accurately, regardless of what is being transferred.

3. At the end of the source film reel, the last key number is punched and checked against the TLC readout: they should be identical. This provides a good double-check for the accuracy of the film numbers for each edit; it confirms that there are no misprints of the film's edge numbers, and no mistakes in the videotape's recording.

If the last punched key number matches the TLC, you can be assured that ALL points in between are accurately encoded.

▶▶ Telecine trouble: p. 362

WHO CARES?

THE REAL LIMITATIONS OF "FRAME ACCURACY"

Although it seems reasonable to be concerned over the visual accuracy of electronic film systems, in actuality this may not be as important as most people would believe. For cutting film, there is really no way to be visually accurate on a 30fps-based system. So although people speak of visual accuracy, they tend to be more aware of temporal accuracy. As long as edited sequences are durationally accurate, pictures will sync to audio tracks built on film or tape and will play fine on television. And since no one will really notice the difference between an electronic cut and its film cut, should we really care about this type of frame accuracy?

It is a valid point. Most productions that are cutting film are doing one of two things. If they are television-based, the film images will be as identical to the cut images as the media will allow and they sync up to real time. So visual "inaccuracies" are irrelevant.

If the project is theatrical, like a feature film finishing on negative, serious productions are going to conform the lists to workprint before cutting negative. If there are any problems with the cut, and they would only be minor, they can be corrected on film. The video system is then only a tool to create a first cut, and future changes will be done on film. This kind of project is also undaunted by visual-accuracy issues.

Neither of these situations implies that the errors are not present. It only means that in the real world of editing, the errors are of little practical concern to the people using the equipment.

In instances where absolute frame accuracy is imperative: where workprint changes are being done on the system, or where negative is being cut directly from the system, only 24fps paradigms—found with *some* digital systems—can achieve the essential 1:1 relationship between the editing media and the film. **True 24fps digitizing is the best choice** from both a visual and temporal accuracy standpoint.

THE BIG QUESTIONS:

"WHY DOES THE US HAVE 24FPS FILM AND 30FPS VIDEO?"
"WHY IS PAL 25FPS?"

For years, longer than there has been video, film has been recorded and projected at 24fps—in all nations. When individual frames are shown in succession 15 or more times per second, they begin to fuse together and give the perception of motion. Clearly, the faster you display frames, the smoother the illusion. For a variety of perceptual and economic reasons, modern film makers decided upon 24fps as a good speed.

When video was pioneered, it too was intended to play at around 24fps. However, at the time, you needed some kind of clock (or constant pulse) to synchronize frames to play at a constant speed. The easiest location of a natural synchronizing ("sync") signal was the alternating current (A.C.) in an electric socket. In the US, A.C. power is 60Hz. This provided the basis for our video sync. Although video has 30 frames per second, it is really 60 fields per second—60 anything per second is 60Hz. A perfect source for sync for the early black and white televisions.

In Europe, as in many nations, A.C. power is at 50Hz. This would mean that a 25fps video signal is more natural (50 fields per second). It also means that film productions shot at 24fps can be transferred to video-tape at 25fps and the change would be so small that no one would notice (a 24fps to 25fps speed-up is about a 4% change in run time; 24fps to 30fps is a 20% increase).

"Why doesn't NTSC video really run at 30fps?"

It used to. The first black and white televisions, getting their sync from wall current, really played at 30.00 fps. The problems arose when Americans wanted to add color to the black and white signal. The addition of a chrominance signal to the luminance signal required the NTSC people to come up with a very slight modification in the sync signal (and thus the frame rate). To explain exactly why adding color to black and white necessitates this change would require a detailed explanation of video signal encoding and wave physics. That isn't going to happen here.

Anyway, with the development of color TV, the perfect 30fps signal was changed to the irregular 29.97 signal—a slowdown of 0.1%—and requiring the subsequent development of such things as drop-frame timecode and sync generators. We still refer to NTSC video as 30fps video because "30" is a more convenient number and is sufficient for many discussions.

TELECINE   LOGS

A combination of developments in the '90s made it possible to automate many parts of the logging procedure for nonlinear system projects finishing on film. The core of the automation is due to KeyKode—machine readable negative key numbers. These numbers are bar codes that run along the film that can be read by a barcode reader.

By employing a barcode reader, and taking the numeric output of the reader into other telecine devices, every edit made in telecine can be recorded, including the exact key number where each shot begins, the exact timecode of the tape frame where the shot is being recorded, and the 3:2 telecine frame type.

Prior to the evolution of these automated functions, telecine assistants were required to physically punch the negative at the head, read off the key number at the punch, whereby the telecine operator hand-typed the key number into a counting device (like a TLC), which displayed the calculated key number at any given frame. During the telecine edit session, a paper list was generated that showed the key numbers at each edit, along with the timecode. This list was taken to the nonlinear editing system along with the 3/4" videotape dubs made in the session, and hand-typed into the log of the nonlinear system. Regardless of how "accurate" the editing system was, there were numerous points in the process that allowed for human error: reading the key number in telecine, the telecine operator typing it into the system, the assistant mis-reading or mis-typing the number from the paper list into the nonlinear editing log. . . .

With KeyKode, the key numbers are automatically entered into and moved through the telecine process. In telecine, the edits are recorded along with any additional information the telecine operator or assistant chooses to type (usually the scene and take numbers, and sometimes a brief description of the shot). As well as managing all this data, the telecine "log" that is created also includes the audio timecodes from the 1/4" Nagra tape that is being synced to the negative (these numbers will be important after the editing, during the audio conforming or layback).

The telecine log is a special kind of Edit Decision List (EDL), but on this new kind of EDL the source column contains film key numbers, and the record column contains timecodes. There is no standardized format for the telecine log, and so the manufacturers of some telecine equipment have developed their own. There are three principal telecine log formats in use:

TLC offers *Flex*™
Evertz offers *KeyLog*™
Aaton offers *Keylink*™

These telecine logs, like EDLs, can be output directly onto disks; then the disks can be taken to any of a number of nonlinear editing systems and input directly into the system (although it should be noted that not every system can read every format of telecine log). Avids, for example, convert telecine logs into their own internal format. In this way, human error has been minimized, and the accuracy of the editing system for film cutting has been effectively increased.

Log Date: 07/13/93		LAW AND ORDER					Page 1 of 2	
EPISODE TITLE	69009		ACT	COMP REEL	901			
EPS #		MEDIA #						
Scene	Take	Time Code	Fr	Cam	Prefix	Key	+Fr	SNDR	Nagra TC
16	5	01:00:00:00	A	A1	KJ15 5796	5184	+15	1	09:58:36:11
16	6	01:01:06:25	A	A1	KJ15 5796	5289	+07	1	10:01:36:04
16A	2	01:02:17:05	A	A1	KJ15 5796	5452	+13	1	10:13:06:28
WT1002	1	01:10:09:05	A					1	12:53:27:10
9	1	01:10:37:05	A	A5	KJ06 5688	4965	+08	2	15:45:22:18
9	3	01:12:12:25	A	A5	KJ06 5688	5166	+14	2	15:51:27:04
19C	2	01:35:33:10	A	A8	KJ05 5742	5385	+15	3	19:59:21:28
19D	1	01:37:00:20	A	A8	KJ05 5742	5522	+07	3	20:03:58:06
19E	1	01:37:47:00	A	A8	KJ05 5742	5595	+15	3	20:28:14:03

list courtesy of The Post Group

◀◀ EDLs: p. 92   Avoiding telecine problems: p. 362 ▶▶

COMPONENT AND COMPOSITE VIDEO

PART I

B efore we discuss component and composite video, let's talk about television history. When TV first came out, it was showing images in black and white. The technical term for a video signal that is in black and white (*i.e.* monochrome) is a luminance signal.

A decade later, scientists figured out that they could broadcast and record in color. Problem was this: a lot of Americans already had black and white sets—and they weren't just going to toss those out. Our government decided that whatever kind of color signal was broadcast, the old black and white sets still had to display the images. So rather than design a really great color set with an ideal color video signal, the scientists had to compromise. Since black and white TVs could only read luminance broadcast signals, color signals had to be part luminance and part...something else. What they added (for everyone with color sets) was a chrominance signal to go with the luminance. A color set could combine both signals together and produce a color TV picture; an old set would just ignore the chrominance.

Thus, today we have inherited an electronic video image that is made up of these two kinds of signals: *luminance* and *chrominance*. Luminance is the brightness of the signal, from black, through grey, to white. Chrominance (or "chroma") is the color part of the signal, relating to the hue and saturation. You could think of it this way: luminance draws a picture and chrominance paints it.

PART II

If you recall fingerpainting in kindergarten, you might remember that you could create any color in the rainbow by mixing three primary colors. For paints and inks, these colors are cyan, magenta and yellow.

Light is a little different than fingerpaint. If you mix colors of light, they don't turn black (like the paints), but rather they turn white. For light, the "additive" primary colors are red, green, and blue.

PART III

The smallest unit of a video picture is called a *pixel*. You can see the pixels (or dots, roughly akin to pixels) if you push your face close enough to a TV screen. For a color TV, every phosphor dot is illuminated by one of three electron guns inside the set, and aimed (from behind) at the screen. One gun shoots at red phosphor dots; one at green dots; one at blue. It is through the careful mixing of the amounts of red, green and blue that you get a pixel of any other color.

Since you can create virtually ANY color from red, green and blue, it follows that you can also create all the shades of grey—from black to white. Sound familiar? That's what luminance is. This means that a luminance signal can be created from red, green, and blue; by definition, so can chrominance.

If you have the time (and money), you can record and play the red, green, and blue signals separately (but simultaneously). That's what a component video signal is. You are keeping the three colors as separate components.

COMPONENT VIDEO is a kind of video signal where the RGB color information,* either in analog or digital format, is maintained in separate channels, thus using separate cables and separate internal processing for each color.

This causes two things. On the good side, the red, green, and blue signals stay very pure and very clean. Your pictures look good and suffer little generation loss with each dub. On the bad side, it is expensive—you need special equipment that maintains the separate-ness of the colors. You need at least three cables of exactly the right length for the video signals. To use component video, you really need component tapes, component VCRs, component switchers... .

---

* This is neither the time nor the place to get too technical, but in fact, the RGB signals are carried as Y, Cr and Cb—but, for now, it amounts to the same thing.

COMPOSITE VIDEO is much, much simpler. Composite video takes the red, green and blue signals it "sees" and mixes them together into a single signal. Get it? It *composites* them. The method by which the RGB signal is compressed and encoded together is based on government-approved standards (like NTSC, PAL, or SECAM); in the US we use the NTSC standardized specifications for encoding the RGB signals—this is known as NTSC video.

TV receivers decode (and decompress) the NTSC video to play on monitors. On the good side, composite video is inexpensive and simple. The signal may be marginally more mushy, but it doesn't become a problem (or even noticeable), until you have re-recorded the signal many times. This encoding and decoding degrades the quality of the colors, and thus the accuracy of the chrominance and luminance information. Almost all professional equipment is composite.

Component video, because of the additional expense and trouble, is really only necessary where multiple generations of a videotape are necessary; for example, in special effects.

Actually, there is a third option. If you don't want the trouble and expense of three cables (RGB), and don't want to mush everything together in one cable, you can use two cables.* If you have two separate cables, you can separate the luminance (or "Y") in one cable, from the chrominance (or "C") in the other. This is what high-end consumer video does in the form of S-VHS and Hi-8. They are not quite component, but not as

*In practice, these "two" cables are generally bundled into a single multiple-wire component cable with a standard S-video connector.

mixed-up as composite. Technically, they are called Y/C video (or *pseudo-component*):

Prior to the debut of digital HDTV, broadcasts in all nations were exclusively composite video. The *method* by which the video is composited, however (NTSC, PAL, *etc.*), varies by nation.

Internally, cameras produce (and monitors require) red, green and blue. However, it's unwieldy to feed RGB signals from the multitude of sources through all the intermediate steps that lead to broadcast (*i.e.* recording, editing, effects... up to transmission and reception). So pictures are encoded into composite video at the camera and left that way until they get to your TV. Besides, the less they're decoded and re-encoded, the better they look.

THE TECHNICAL PART
[Only Read This If You Are Really Smart... Or Really Care]

To understand how luminance and chrominance can be related to red, green, and blue, first you need to know a little something about colors.

Lesson one:

Red + Green + Blue = White

This means that if you mix the three primary additive colors in the right proportions you will have *perfectly* white light.

Lesson two:

Scientists enjoy making 3-dimensional graphs for discussing colors—with red on one axis, green on another, and blue on the third. Every color imaginable is on this graph, somewhere, with three coordinates—a value of red, green and blue.

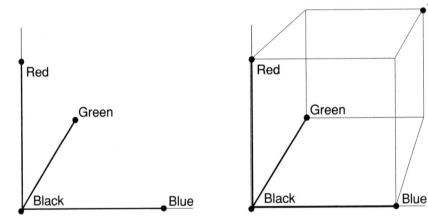

If you have trouble viewing the drawing in 3-dimensions, try visualizing the corners of a box.

In this box, black is at the zero-point for all axes. The corner opposite black—the corner where you are adding red, green and blue—is white.

Since there is a "corner" that is black and a "corner" that is white, we can easily draw a line between them. (See the figure at the top of the next page.) This line denotes a perfect gray scale (all the shades of grey from black to white). This line is the **luminance**. For no obvious reason, scientists call luminance "Y."

Chrominance, in one manner of viewing things, is everything else.

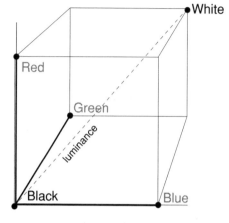

Let's review. The bottom left corner is black. One adjacent corner is red, one is green and one is blue. The far corner is white. But what are the other three corners? Answer:

Red + Blue = Magenta

Blue + Green = Cyan

Red + Green = Yellow

There are electronic scopes that analyze video signals. These scopes have rotated the RGB cube so you are looking down the "barrel," so to speak, of the luminance. Black and white are in line.

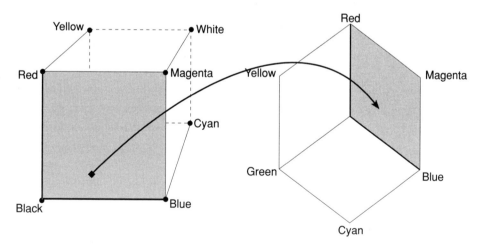

Now that we are talking about rotating the RGB cube, it's worth noting that scientists tend NOT to view the cube as an RGB cube. Like on the scope, they make the luminance (Y) one axis, and then they create two other axes at right angles to the luminance, called, in the US, "I" and "Q." YIQ components make up NTSC video signals. A more international standard (as in PAL broadcasting) uses "U" and "V" to create YUV "color space."

Chrominance = I and Q

"RGB" is used by engineers for television, some cameras, and computer CRT (cathode ray tube) monitors, and "YIQ" or "YUV" when working with broadcast signals.

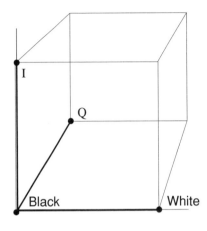

I

Q

Black   White

Deeper into the Technospeak

Truth? Engineers don't really say YIQ or YUV all that much. In the first place, when they graph "I" they tend to graph "negative I" (–I) to make the graph sit better. And those component video cables are not carrying R, G and B—they're not even carrying Y, U and V. They are carrying Y, red minus Y ("R–Y") and blue minus Y ("B–Y"); it really all amounts to the same thing. Artists find all of this too technical; they describe their color space as being made of HSB (hue, saturation, and brightness). In some sense, there is really no difference.

To make matters slightly more complicated, our RGB cube is not particularly cubic in shape. If you added equal amounts of red, green and blue, and showed it on a color TV, you still wouldn't get white (as you might expect), since the human "eye" varies in its sensitivity to different colors. In the real world of CRT monitors, the color formula for white is:

30% Red + 59% Green + 11% Blue = White

This formula is also how engineers derive luminance:

30% Red + 59% Green + 11% Blue = Luminance (Y)

So, although the reality of this "cube" is a little distorted, the fact remains that luminance and chrominance broadcast signals can be created from the colors red, green and blue. To record video in a camera, or play video on a monitor, you must manipulate RGB. To broadcast, you must convert the RGB to NTSC composite video. But to record and edit, you have choices: either component or composite can be used, each with its own costs and benefits.

▶▶ Digital video: p. 145   ▶▶ Color sampling: p. 165   ▶▶ Light and color: p. 180

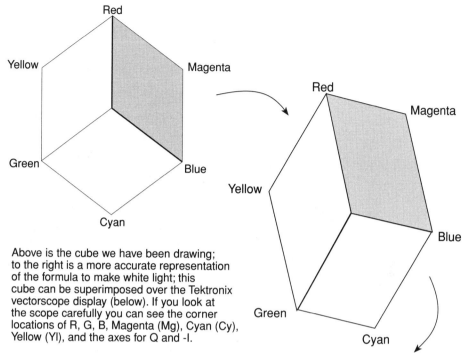

Above is the cube we have been drawing; to the right is a more accurate representation of the formula to make white light; this cube can be superimposed over the Tektronix vectorscope display (below). If you look at the scope carefully you can see the corner locations of R, G, B, Magenta (Mg), Cyan (Cy), Yellow (Yl), and the axes for Q and -I.

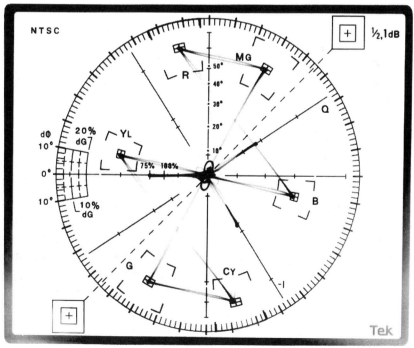

COMPONENT & COMPOSITE VIDEO FORMATS

You can't always tell from the name of a video product whether it is component, composite, or Y/C video. All video, even digital video, must be encoded as one of these types. Below is a list of common video formats:

	analog	digital
component	Betacam (1/2") BetaSP (1/2") MII (1/2")	D1 (19mm) D5 (19mm) Digital Betacam (1/2") DCT (19mm) DVCPro (6mm) DV and MiniDV (6mm) "4:2:2" "4:1:1"
Y/C	S-VHS (1/2") Hi-8 (1/2")	
composite	1" type C (1") 3/4" U-matic (3/4") 3/4" SP (3/4") 8mm (8mm) VHS (1/2")	D2 (19mm) D3 (19mm)

Although D1, D2 and D3 tapes are all 19mm formats, the actual tape used is different and the cassettes are not interchangeable. The same is true for the 1/2" formats.

When editing and manipulating video, it is the encoding and decoding between component and composite that can degrade the signal. Unfortunately, due to the costs of component video, many facilities use some component equipment, like D1 tape machines, and interface them through composite switchers. Realize that you will only receive the full benefits of component video when the entire video path is maintained as component.

19mm is approximately 3/4" in width; 8mm is approximately 1/3"

V I D E O T A P E   Q U A L I T Y

It is worth noting that although the Y/C formats theoretically could produce a better signal than the composite formats, the tape stock used for consumer videotape is of significantly lower magnetic oxide densities than those used for professional purposes. That oxide density, when combined with tape speed, defines the maximum *signal bandwidth* that can be recorded.

Also note that tape **width** (1/2", 19mm, 3/4", 1") has no direct correlation to videotape quality. The first videotapes were 2" wide; they are extinct today. For all practical purposes, it is only the *area* of tape covered by the record heads with each scan (for each field) combined with the density of oxide particles in that area, that corresponds to quality.

Oxide particles on videotape are in some ways like sand on sandpaper. Course sandpaper has few grains per inch, like inexpensive tapes. Very fine sand on high grade paper has thousands of grains per inch, like high density and professional tapes. The more density, the more detail.

When the videotape is moving slowly, the video fields are not very long and have a smaller area.

…but when the tape is moving more quickly, the same video field is going to be longer—and will therefore have more area. More area means more particles per field, which will result in better looking images.

Larger tape formats might mean better looking images; the real factor is area. These fields on a 1" tape are about the same size as the faster moving 1/2" tape format in the middle diagram.

A videotape may be manufactured via *metal particle* methods (MP) that produce a high quality distribution of metal oxide, or it might be made with *metal evaporated* (ME) methods, for lower qualities. A little mark indicating which method is used is often printed on tape labels.

COLOR SPACE

GAMUTS, DITHERING . . . AND EVEN THE KITCHEN SINK*

An explanation by way of analogy.

You're remodeling your kitchen, and picking out a new faucet for the sink. There are lots of styles to choose from, but they're all variations on just a couple of basic designs. Let us help you.

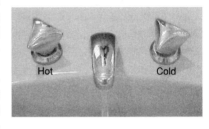

First, let's consider the old-fashioned two-knob job—where one knob controls the flow of hot water, and the second controls the flow of cold water. Traditional. Simple. Proven.

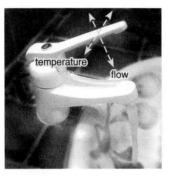

Then there's the newer style, where one fancy lever does it all: you move it up and down to control the flow, and pivot it left and right to control the temperature. Difficult choice, but let's say you choose option one: the two-knob system.

It's a few weeks later. The water's running and you're busy washing some dishes. You couldn't be happier with the flow, but realize it'd be nice to have the water a little warmer. What do you do? Well, there's no Temperature control, per se, which is what you actually need right now. With this style of faucet, your only option is to turn up the hot "component." The water gets warmer, all right, but now you're also using more water than before, and more than you need. So you turn down the cold "component" by about the same amount, and now it's hopefully the temperature you wanted, with roughly the same flow as before. Close enough. With this style of faucet, you have to guess a little, do a little trial and error, and make two adjustments (more Hot, less Cold) rather than the one adjustment you actually needed to make (higher Temperature, but with a constant Flow).

Take two: Let's imagine that you'd gone with the one-armed faucet instead. When you do the (above) washing-dishes test, this one-arm faucet makes the adjustments with ease. Now your task is to fill up a teapot with plain old hot tap water. Of course there's no hot water knob per se, so you lift the lever to get the right amount of water flow, then move it to the left to set the temperature. (If you're really an experienced kitchen pro, you might do all this in one move, but you'd still be making adjustments along two different "axes.")

Both types of faucets have the same plumbing "input"—a hot water pipe and a cold water pipe. The difference is the mechanism for controlling the amount of each. Because of their design, each has advantages and disadvantages depending on your fluid needs. We'll come back here in a moment.

---

*Please don't write to say these photos are bathroom—not kitchen—fixtures.

Applied to the world of video, "hot" and "cold" are pure *colors*, and the flow is the intensity of those colors. By mixing them, you create new colors. Each set of faucets represents a specific set of controls for mixing its colors and making *new* colors. The entire range of colors you can make with those particular faucets is the system's "**color space**." What we're doing is controlling the "flow" of red, green and blue light onto a point on a TV screen (but it could also represent the value of that point on an analog or digital videotape, inside a data file, or transmitted via a TV station or streaming Web server). For a specific color, the numbers representing the position of the various "knobs" might be different, depending on the color space being used to represent or transmit it.

Most people do not mix red, green and blue to make new colors. What

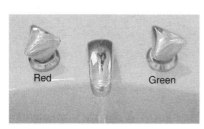

we do is choose the colors we like and ask a computer to deconstruct that choice into the quantities of red, green and blue that are required to encode it. If you only have faucets for red and green, it is impossible to make BLUE. It can't be done; blue is not in the color space of a red and green world. Likewise, there are many "normal" shades of colors that cannot be described in simple terms of red, green or blue; these shades are outside the RGB color space. Every color space is a different "language" or "dialect" for communicating information about color. When you graph those color spaces, the axes reflect this dialect.

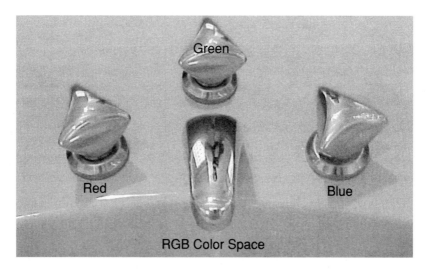

RGB Color Space

Where life gets complicated is that, while similar in many respects, the color space made by mixing red, green and blue is not the same is the one you get by mixing Luminance (Y), R–Y (C*r*) and B–Y (C*b*). And sometimes the controls you have for adjusting color are in one space, but the output you need is in another. For example, an editor often needs to get a certain output on a professional video deck, but the knobs that control the color are for set-up (black level), video (white level), chroma (color saturation) and hue.

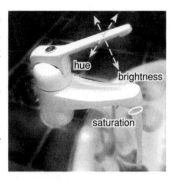

Timebase corrector controls (left) provide an editor with knobs for adjusting hue, chroma, etc.; the output from the video deck (right), however, offers component, composite and S-video—each within its own color space.

Or we might be tweaking the picture on a TV set, which receives a signal as YIQ, and displays it as RGB, but lets us adjust it with the more user-friendly *brightness*, *contrast*, *saturation* and *tint*. Or perhaps we're creating a graphic image on a computer screen, and using HSB (hue/saturation/brightness) or even a "crayon" color picker to choose values for an RGB-displayed image or a CMYK printout. All of these translations of one color space into another are riddled with potential problems.

When it's necessary for us to make the conversion of a video signal from one color space into another (like feeding the output of a component VTR into a capture card that only "understands" RGB), we can use a small converter box called a *video transcoder*. You may actually be doing that conversion already, with-

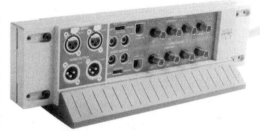

A typical breakout box (this, from Media 100 circa 1995), with balanced audio in/out on the left, component video in/out on the right, and composite in/out nestled between.

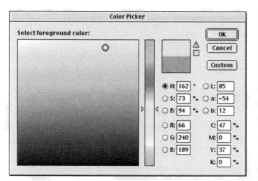

A typical format for a software color picker (this, from Photoshop), where colors picked on the left are represented on the right in different color spaces: H-S-B, R-G-B, L-a-b, C-M-Y-K.

out knowing it—say, using an editing system with a "break-out" connector box that has inputs for a number of formats (including analog component, composite, and S-VHS). In that case, the transcoding is done by circuitry in the capture card, or even in the breakout box itself (actually the ideal place, since there's less noise and interference there as compared to putting the converters inside the chaotic electrical environment of the computer itself).

Let's go back to the second faucet analogy (from page 130), and say that the lever isn't perfect: moving the temperature lever fully to the left gives us mostly hot water, but doesn't close the cold water valve completely; and moving it fully right gives us mostly cold water, but still lets a tiny bit of hot water through. If it were important for us to have purely hot or purely cold water, we'd have been better off using the old-fashioned hot-and-cold style faucet knobs, since they let us block off either supply completely just by not turning on that knob. Each style of faucet therefore gives us a slightly different range of temperature choices.

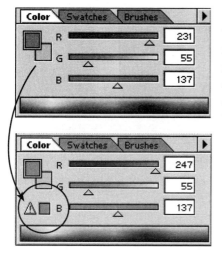

GAMUT WARNING

In the world of video and color, that range of choices available to us in a particular color space is called its **gamut**. RGB color space, for example, provides a greater gamut of colors than can be conveyed by the YIQ color space used in NTSC television. Graphic designers, for example, will find this to be a problem because they have a greater palette of colors available to them on their RGB computer screens and graphics applications than can actually be transmitted to the viewer's TV. That can be frus-

▶▶ Light and color: p. 180

trating—when, say, a bright vivid red either can't be broadcast as intended, or *smears* when it finally does arrive on some poor viewer's TV screen.

Many software applications are designed with those limitations in mind, and provide a "gamut warning" or "safe colors" mode, which either warns you when you use an "illegal" color, or better still, prevents you from using it in the first place. Generally speaking, it's important to understand that **different color spaces (or recording/transmission schemes) each have their own peculiar limitations.** *Since you can't change the basic limitations of any given medium, the best approach is to try to understand those constraints, and make any necessary decisions accordingly and deliberately—rather than having those limitations imposed automatically and by brute force, somewhere later in the signal chain.*

All right, hang onto your plumbing, 'cause it's back to the kitchen . . .

Let's now pretend that we're using the fancy single-lever faucet, but this time it's been redesigned so that the lever moves with little "clicks," as small discrete steps both up and down, and left and right. So instead of a continuous, smooth movement (analog), the lever only gives us a certain number of choices (digital) for the amount of flow, and for our choice of temperature.

In addition, there are two different brands of this single-levered faucet. Both allow a maximum flow of two gallons of water per minute (when the lever is fully up), but Brand A lets you make your flow adjustments in clicks of 1/10th of a gallon per minute, while Brand B only gives you larger increments of 1/4 gallon per minute. (And yes, Brand A, with more choices, is more expensive.)

In the digital world, forcing incremental "clicks" in the choices that we can use to control or represent something is known as **quantizing**. The more choices we have within a given range, the greater subtlety or detail we have available in our palette to represent finer shades of color, brightness, etc. For example, a professional D1 digital videotape recorder (costing hundreds of thousands of dollars) can record pictures with a tremendous variety of colors, and in great detail. A more affordable DV camcorder, on the other hand, may be able to capture colors spanning the same range of hue, brightness and color saturation, but has fewer shades of those colors it can "paint" onto the screen, and with a larger, more coarse "brush." Depending on the material you're trying to record, you may or may not see a noticable difference; in an outdoor scene, for example, you might see a roughness or "banding" in the gradations of color across the sky, or you might not notice much difference at all, depending on the subtleties of the image, and how picky you are.

Finally, suppose that the expensive style of modern "digital" faucet lever can be moved left to right in temperature increments of 1° F, from 40° to 120°. The cheapo version gives you the same overall range of temperatures, but can only be adjusted left to right in increments of 5° F (you get what you pay for!).

Now suppose that you want to fill the sink with water at precisely 78° F.

With the expensive faucet, you'd just move the lever to the 78° position and fill 'er up. But say you cheaped out with the other kind—and the nearest choices are either 75° or 80°, with nothing in between. Well, you could set the lever to 80° and pretend that it's a close enough approximation. Or you can get clever ...

Remember, the water doesn't actually have to be 78° as it leaves the tap—it just has to average out to that temperature by the time the sink has filled up. Let's say it takes 5 minutes to fill up the sink. You could run the faucet at 75° for two minutes, and then raise the temperature a notch to 80° for the final three minutes. (Or maybe alternate back and forth more often, but still in the same 2-to-3 ratio.) Finally, you stir it up so that any temperature variations in the water are evened out, and Eureka: you've got a sink filled with water at 78°!

This form of clever cheating, by alternating back and forth between coarse values in order to achieve the appearance of something more refined, is called **dithering**.

Dithering is a versatile and powerful technique in digital processing—it's the digital equivalent of the "halftone" method of printing. It is used for everything from simulating finer variations of brightness and color in a video or Web image, to improving the perceived fidelity of digital audio recordings.

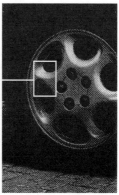

(Above) A newspaper halftone screen (enlarged at left) illustrates dithering in the printing world. (Below) Dithering in an image processed to severely restrict the number of colors, as might be necessary for GIF images on the Web or in low-resolution streaming video. From a distance, the arrangement of dots creates an illusion of a greater number of shades than there actually are.

CABLING

Cables are relatively new territory to film people, but common to the video and computer literate. Perhaps you have played with your home cable TV hookup — then you've seen coaxial video cable with a threaded connector; stereophiles may have seen RCA cable plugs; every Walkman has an RCA-mini headphone jack. These are all just different kinds of cable connectors on different kinds of cabling.

Cables are combinations of strands of wiring and appropriate shielding and insulation, usually wrapped up in a neat rubber-hose package. When new hardware devices become standardized, the variety of inputs those devices require often begin as a tangle of wires, and end up bundled into convenient single cables with an easy-to-use connector.

Wires come in a variety of **gauges** (thicknesses) which can impact how much electricity or data they can carry without overheating; and **materials** (copper is inexpensive and a good conductor; gold is more expensive, and a better conductor*). **Shielding** is a critical component to cables: high-quality cables tend to have more industrial degrees of material preventing the electromagnetic fields generated by cabling from interfering with things (other wires, recording devices, *etc.*) on the outside. Cheap cables are often cheap because they have minimal gauges and marginal shielding. This, along with the quality of the manufactured connector, will most dramatically impact price. Two cables that purport to do the same thing can have significantly different costs, simply due to the variations of these parameters.

Coaxial (or "coax"): Coaxial cable is a remarkable versatile material for moving video and audio signals in a highly-shielded set of wires. They tend to cost more than their unshielded alternatives, but are the standard for professional video (analog and digital) applications. Coax is made of a thin gauge center wire surrounded by insulation and a braid of shielding metal. The shielding minimizes electrical (AC) and radio (RF) interference. Inexpensive coax may have aluminum braids; expensive versions

---

*Some consumer connectors come with "gold" coatings which purport to improve conductivity; this is without merit, particularly as most gold coatings are superficial.

provide better shielding with copper braids. Coax delivers cable TV to millions of US homes.

SCSI (pronounced *scuzzy*): SCSI, or Small Computer System Interface cables, connect peripheral devices to a central computer. Technically, it is a family of parallel interface formats utilizing a cable consisting of many small wires, with limited shielding around each, but a big shield around the outside, resulting in a reasonably thick beefy cable. SCSI has a number of advantages over standard serial or parallel computer interfaces, including faster bandwidth and the ability to connect up to 7 devices to a computer (each with a unique SCSI ID). SCSI cables tend to have larger connectors to terminate the wiring, in one of several formats.

There are a number of SCSI variations, and many have unique connectors. SCSI-1 and SCSI-2 (4Mbps); Fast SCSI (10Mbps); Fast-Wide and Ultra SCSI (20Mbps); SCSI-3 and Ultra2 (40Mbps); Wide Ultra2 SCSI (80Mbps).

FireWire® (IEEE 1394): FireWire is a small bundle of wires that carry digital data with a bandwidth of 100, 200 or 400Mbps, most commonly at 100Mbps which is substantial for (25Mbps) DV. FireWire may connect cameras to computers, but it also is used for any high bandwidth connection, between mass storage devices and computers, and peer-to-peer between cameras and other post-production equipment.

Ethernet: Thicker than standard telephone twisted pair cabling, Ethernet is designed to carry digital information at 10Mbps (10base-T) or 100Mbps (100base-T), common for local networks moving all but the largest data. It is important to know that while most Ethernet cables are designed to work with hubs, some are "crossover cables" which have 4 wires swapped receiver to transmitter and vice versa, so computers can be connected peer-to-peer. If they are unlabeled, this can be challenging for those setting up networks.

Data Cables: Another typical bundle of wiring delivers computer data and control information between computers and external devices. Connectors range from RS-232 and RS-432 to USB and MIDI.

Optical Fiber: The fastest type of cable is one made of glass—fiber optic cable. These cables can manage bandwidths of a gigabit in magnitude, but actual bandwidth will vary depending on the protocol used (e.g., HIPPI, FibreChannel).

AC: While it probably goes without saying, power cables are pretty standard copper wiring with suitable insulation. AC cables, because they carry 60 Hz current, can cause 60 Hz "hum" in audio recordings when cables and devices are not well shielded.

Standard AC cables are bundles of 3 copper wires: positive and negative terminals and a ground. The primary difference in AC cables is the gauge of the copper wire in them; larger gauge will handle more current without overheating. Cheap thin wires can get hot and be a fire hazard.

CABLE CONNECTORS

While the science of cable connectors isn't necessarily the hottest technology topic, they are an everyday part of life for most professionals.

The first basic understanding you must have is the difference between MALE and FEMALE (if this perplexes you, you're on your own).

Next, recognize that cables may have different connectors on either end: effectively translating from, say 3 RCA plugs (video, left and right audio) into a single BNC; some people never quite get that a cable is anything but some kind of extension plug.

Third: some cables have the same connectors on both ends, but only work "one-way." Thus, even if it plugs in okay, it is possible you have the cable 'backwards.' Similarly, some cables, like FireWire or SCSI, may have slightly (or significantly) different variations on connectors. S-Video cables connectors may have 4 pins or 7 pins; FireWire may be 4-pin or 6-pin; there are a host of SCSI connectors. Here are some of the most common cables connectors and their basic functionality.

ANALOG AUDIO/VIDEO

RCA

These very common connectors can carry either video or audio. They are found on most consumer video and audio products. A signal travelling through this cable is "unbalanced"— i.e., somewhat prone to picking up extraneous noise and hum.

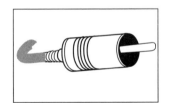

BNC

Convenient because of its twist-lock connector, BNC is a common type of video connector. It is found on most professional coaxial cables—for the video and timecode-in and -out of professional VTRs. Component video systems often use bundles of 4 video cables with BNC plugs—one each for red, green, blue, and sync signals. BNC stands for "bayonet Neill-Concelman" (named for the developers) and <u>not</u> "Bayonet Connector," or "British Naval Connector," as often stated.

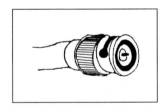

BNC connectors on coax, it should be noted, can be used for digital data.

XLR

These large 3-prong connectors are used for audio in most professional equipment, including VTRs. Unlike cables with RCA plugs, XLR connectors generally carry "balanced" audio signals. The connectors have a small latch that snaps into the jack to hold them in place; the release switch is on the top of the jack.

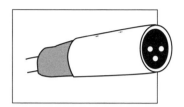

Phone

Phone connectors are like giant-sized RCA-minis. They are the kinds of cable connectors that early telephone operators used in making circuits; today they are used in patchbays, and in heavy-duty audio set-ups. They are also seen on the end of many headphones.

RCA-mini (*aka* 1/4" phone)

These compact connectors carry audio. They are found on most consumer audio products in the form of headphone plugs and jacks. Occasionally, VTRs and monitors will have RCA-mini jacks for audio-in and -out.

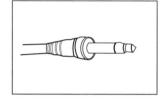

S-Video (mini-DIN 4 pin)

Although S-Video connectors share anatomy with Macintosh ADB connectors, they are significantly different in functionality. S-Video cables maintain the separation of luminance (Y) and chrominance (C) signals resulting in a far higher quality input/out than composite analog signals over the (above) video alternatives.

DIGITAL (DATA/AUDIO/VIDEO)

FireWire (4-pin and 6-pin)

There are two typical connectors for FireWire cables. The 4-pin is the most common variation (i.Link); Apple uses a special 6-pin connector with many of their products.

MIDI

MIDI Cables have MIDI connectors. No male/female variations.

RJ-45 (Ethernet, DSL)

While it looks a great deal like a standard phone jack connector (RJ11), this connector is wider and thicker and is used for the high-bandwidth local networking cables of Ethernet.

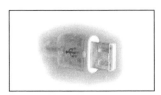

USB

This Universal Serial Bus connector standardizes what formerly had been a host of proprietary Macintosh and PC cabling. USB connects desktop computers to external devices (Zip drives, Scanners, etc.) as well as to a Mouse.

Computers move data in one of two ways: in *serial* or in *parallel*. The usual method is serial. There are two common protocols (like languages) for transmitting digital information serially. They are called RS-232 and RS-422.

While VTRs generally use RS-422 for machine control, most computer peripherals use the RS-232 protocol, especially in traditional editing systems utilizing computer-to-peripheral machine communication. RS-232 cables have a variety of sizes and shapes of connectors. All of them are composed of a number of "pins" (tiny metal wire endings), which carry information. RS-232 connectors often have 9-pin or 25-pin configurations (called D-9 or Din 9; D-25 or Din 25). These connectors are usually "D" shaped to prevent them from being connected upside down. They tend to have snaps or screws on either side to prevent them from accidently being moved once attached.

Computer cables, like all cables (for digital or analog video and audio), have length restrictions. Certain computer cables cannot run longer than 50 feet; for instance, video cables may need signal amplifiers if their lengths are too extended.

BALANCED and UNBALANCED AUDIO

Generally considered one of the important differences between "professional" and "consumer" audio tools is the presence of balanced or unbalanced audio signals as they are carried in cables between analog audio devices, such as microphones, cameras, mixers, and a host of other equipment.

A **balanced** line is characterized by having *three* wires making up the cable: two are isolated and heavily shielded, for carrying the audio signal, and the third is the ground. Balanced lines have vastly superior signal-to-noise ratios and tend to minimize introduced noise from AC lines and other audio cables.

Unbalanced audio is simpler: *two* wires in the line, one for the audio signal and one for the ground. While this configuration does provide for excellent audio, it can be unsuitable where there is considerable processing of audio signals back and forth across numerous lines, in multiple generations, and in the proximity of other audio signals and pieces of gear.

The familiar three-pronged XLR connector is the hallmark of a balanced audio line. These cables and connectors (and the circuitry that goes with them) are required for most professional audio, and are appropriately more expensive than their smaller cousins, the RCA cable and connector. Also, larger professional audio equipment can support the larger size XLR

. . . Just don't be swayed by labels printed on Japanese-manufactured gear: most native English speakers will use the more traditional adjectives "balanced" and "unbalanced."

connectors, whereas smaller (more consumer-like) devices choose out of necessity to limit the space allocated to the jacks.

Does it make a difference which you use? For simpler, "low-impact" post-production of audio, it probably doesn't make much noticeable difference. But analog audio signals are easily degraded as they travel in cables, move through equipment, get processed from analog to digital and back to analog, before finally residing on a master somewhere. The less degradation the better, and balanced lines aid in maintaining high-quality audio throughout post-production. Thus, audio coming from a microphone, a desktop mixer, or any other analog source is ideally balanced. It should be noted, however, that *digital* audio moving around in computers, over FireWire lines, and on and off DV tapes, is free from such analog degradations.

Higher costs of balanced lines and devices, though, must be weighed against the sensitivity of the listener and the likelihood that the differences will matter. For all intents an purposes, under ideal circumstances, neither a professional nor an amateur can *hear* a difference between a balanced and unbalanced line. In many situations, the costs of balanced audio outweigh the benefits, and the impact can be overstated.

That being said, there is **one other significant difference** to keep in mind. Professional equipment uses much *stronger* audio signals (called +4dBm, or "plus 4") than consumer equipment does (–10dBm, or "minus 10"). Simply put, this means that if you ever have to *interconnect* the two types of gear, you need to either amplify the audio signal (when going from consumer to professional gear), or reduce the signal (when going the other way). Neglecting to do so is the reason that BetaSP tapes and some NLEs often sound too loud or distorted when dubbed to

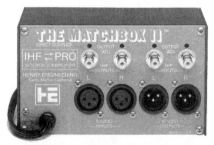

An example of a consumer/professional audio level interface device. RCA (unbalanced) connectors are on top; XLR (balanced) connectors are on the bottom.

a VHS deck. Audio mixing boards generally make a good intermediary between the two schemes; but when patching or "routing" one device *directly* to another, you might want to use a special adapter box that can do the proper conversion instead.

F   I   R   E   W   I   R   E
AKA IEEE 1394 OR I.LINK

In the mid-1990s, Apple Computer conceived of a new kind of low-cost digital interface: it would be scalable, flexible, easy-to-use and integrate the worlds of computers and video. The IEEE assembled a "working group"—a consortium of companies and experts—to develop an independent standard that would meet the demands of modern digital video and audio.

The result is IEEE 1394, called *FireWire®* (by Apple and most consumers), and branded as *i.LINK* (by Sony). "1394" is:

- a digital interface standard (which addresses the constant signal degradation that occurs in repeated digital-to-analog conversions);
- a physically small and thin cable;
- easy-to-use: no need for terminators, device IDs, screws;
- hot-swappable: unlike SCSI, devices can be added and removed while the computer is on and active;
- flexible: it can connect computers to devices, or skip the computer entirely and connect device to device (called "peer-to-peer");
- **fast: cables support data rates of 100, 200 and 400Mbps.**

At NAB 1999 the first devices with FireWire ports were released, notably prosumer digital camcorders (*e.g.,* Sony's TRV900; Cannon's XL-1) and personal computers suitable for managing digital video (*e.g.,* Apple's G3 line; Sony's VAIO).

One keystone to FireWire's adoption is that software in the form of nonlinear editing systems have FireWire support, meaning, in essence, if your digital camera is connected to your computer via FireWire, then your editing software can control your camera remotely. This delivers functionality previously seen only with expensive machine controllers, timecode

readers, and professional editing systems. But even better than that, it delivers true plug-and-play experiences with the transparency of a desktop laserprinter plugged into your PC: you plug in the simple cable, the software "sees" the camera (or video deck, or hard disk) and "runs" it as required. It is a truly revolutionary experience and it is available for consumers on middle- and even low-cost editing software.

A FireWire connection is not synonymous with digital video and it is not limited to consumer products. More and more professional products are emerging with IEEE 1394 ports; more "non-video" devices (like hard disks and RAIDs) are becoming available with FireWire protocols.

FireWire, like Ethernet, is simply a digital cable and format that is designed to move lots of data very fast. More and more computer products and professional video products will come with IEEE 1394 ports, making this standard the likely dominant force in the new era of democratized video.

An i.LINK port (IEEE 1394) on a Sony digital camcorder.

THE DIGITAL WORLD

AN INTRODUCTION TO DIGITAL AUDIO

This is an audio wave:

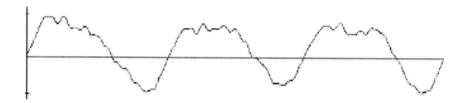

In many ways, it is just a more complex version of a simple sine wave, like this:

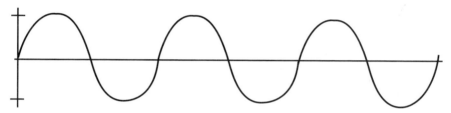

Audio waves, like ocean waves, go up and down through time. Exactly how quickly waves go up and down over time is defined as their **FREQUENCY**. The frequency of a wave is commonly called its **PITCH**.

A high-frequency wave might look like this:

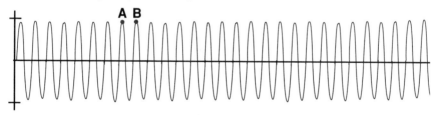

If you were listening to this wave it would be a high note. Notice how two points at the peak of adjacent waves (A and B, above) get closer together when the frequency goes up. On graphs like these, the passing time is measured along the horizontal line (or X-axis, if you're into this stuff), beginning at time zero on the left and continuing toward the right.

A low frequency note (perhaps made by a tuba), looks calmer:

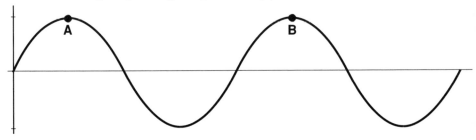

A wave is a series of repeating curves. One chunk of the wave, the smallest amount that repeats, is called one *cycle*, also known as a *period*:

The time it takes for one cycle—or the distance between any two corresponding points on consecutive cycles—is called the **WAVELENGTH**. On the waves illustrated here, the distance between two corresponding points (A and B) is

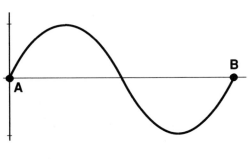

each wave's characteristic wavelength. If you examine the diagrams, you can see that a high frequency note has a *shorter* wavelength; a low frequency note has a *longer* wavelength.

Don't go on if this isn't clear.

The frequency of a note can be measured by its wavelength. You might say that a high note (with high frequency) does this (⌒⌣) a lot—many, many times a second. The wavelength of this high note, let's say, is 1/1000 of a second. Another way of looking at it is that the wave can do this (⌒⌣ —one complete cycle) 1,000 times per second.

The number of times a wave makes a complete cycle in one second is measured in units called "Hertz," which is abbreviated "Hz." Our high frequency sound, then, could be described as a 1000 Hz tone; our low frequency sound is a 10 Hz tone.

> FACTOID • Your ears (and brain) have the ability to perceive sounds with frequencies as low as 10 Hz and as high as about 22,000 Hz. Sounds higher or lower than that range are inaudible. The range from roughly 10 to 22,000 Hz is the human ear's FREQUENCY RESPONSE.

Let's go back to our low frequency (10Hz) note:

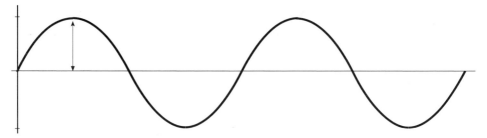

The amplitude of the wave (the distance from the "mean" or "zero" line to the top of the wave) defines a wave's *energy*. In general, this energy is perceived as **VOLUME**. (For technical reasons not worth going into, this energy is measured in volts.) The amplitude of the wave, then, corresponds to how LOUD the note is. Both the wave above and the wave below depict a 10Hz frequency. The only difference is the amplitudes.

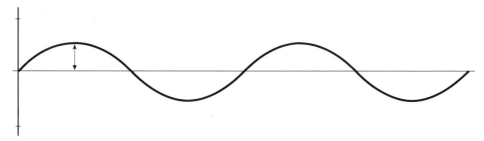

This second example is a "quieter" note of 10Hz. (the top example is pretty loud) but both are the same note: that is, they both have the same frequency.

In technical terms, a wave is continuously changing energy (amplitude) over time. If I asked you how loud this note was, you might have to ask me "when," because as you can see, the height of the line changes constantly.

To **DIGITIZE** an audio wave, you must be able to describe it with numbers—with coordinates, like any graph. (Remember that the x-axis or horizontal line is time, and the y-axis or vertical line is intensity or "volume.")

Let's choose a few places to check to see what the coordinates are:

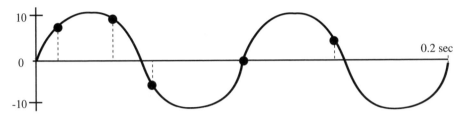

Now, we have selected 5 points in our wave, and by writing down the time and the "volume" at each of the times, we have created a small table:

TIME (x) (in seconds)	VOLUME (y) (in volts)
0.02	7
0.035	8
0.059	-5
0.1	0
0.145	4

If you replotted these 5 points on a new, clean graph, it would look like this:

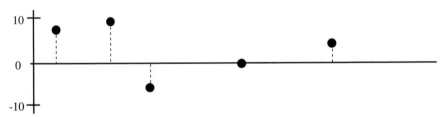

But other waves could be drawn through these points, like this one:

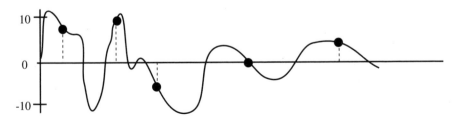

What we've done is, first, take a wave and convert it into numbers; this is called **ANALOG** to **DIGITAL CONVERSION** (often abbreviated A-D or A-to-D conversion). Then we took the numbers and re-generated an analog sound; this is called D-A conversion. As you can see, this second wave is NOT identical to the 10Hz wave we started with. What happened? As you may have figured, when you digitize a wave, you must check the amplitude (volume) of the wave VERY OFTEN…and for a variety of reasons, you want to do it at a constant rate:

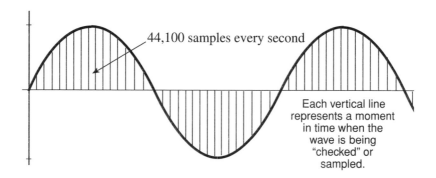

44,100 samples every second

Each vertical line represents a moment in time when the wave is being "checked" or sampled.

The more often you check, the more data in your table, and the more accurate your recreation of the wave will be. **How often you check** the amplitude is called your **SAMPLING RATE**.

You don't want to sample *too* often, however, because at some point you will be taking more samples than you need; realize that a computer doing all this sampling has to remember every point you sample, and more remembering means more storage needed. For sampling to re-create those extremely high frequency notes at the upper end of your hearing range, scientists (in particular, one guy—Joseph Nyquist) determined that you must sample the wave at least twice as fast as the "fastest" frequency (or highest note).

Since we can hear notes up to around 22,000Hz, the necessary rate of sampling for digital audio is usually at least 44,000Hz. This means that the computer (specifically, the A-D converter) is checking the analog music 44 *thousand* times each second to store the value of the wave at that moment. Compact discs (CDs) for example, sample at 44.1KHz.

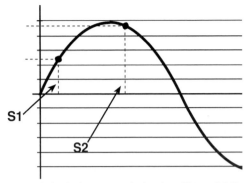

At both these sampled points (S1 and S2), the values on the y-axis fall in between two measured levels; the reconstructed audio wave will therefore have some error. By adding points to the y-axis, accuracy is increased.

As with any graph, you must determine just how precisely you want to measure the volume (power) at a given point in time; you do not have an infinite number of amplitudes on the vertical axis. If you start at "0" and the maximum power at the top of the y-axis is "10," how many volumes are in between? Ten? Fifty? One hundred?

The number of values on the y-axis determines how precise your data will be—the more values, the

more precise. You want to be accurate ENOUGH, but like sampling rate, there are costs in being too precise.

Since all computer counting is done using binary numbers, you have the following choices for how many values you want running up the y-axis: 2, 4, 8, 16, 32, 64, 128,

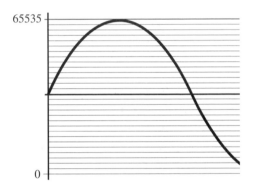

256, 512, 1024 . . . and so on. The same kinds of scientists who decided that 44.1KHz is sufficient sampling decided that 65,536 values was enough accuracy for re-creating very high quality audio. "65536" is 2 to the 16th power. In base two (called binary mathematics), this means that for every sample taken, a 16-bit number is remembered (some value between 0 and 65535). CDs have 16-bit audio. If you want to save space, you could sample less often (*e.g.*, 22KHz, 11KHz) or you could drop your accuracy from 16 bits to something like 8 or 4-bit audio. Reducing the sampling rate will result in the loss of some higher frequencies; reducing the number of bits will increase the amount of noise. Of course, you could sample more often, although the perceptual benefits are somewhat unclear—the DVD audio specification can sample up to 192KHz.

D I G I T A L   V I D E O

VIDEO can also be digitized, but doing so is much more complex than audio. But why? To capture video, a real-world scene must be sampled in both time and space.

Sampling in time is what a film camera does—then the projector shows the different still images in succession 24 times each second. Somewhere around 10 or 15 frames per second, a set of discrete images will begin to fuse together ("animate") and you will begin to perceive continuous motion.

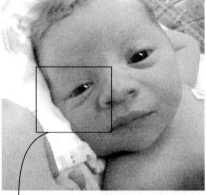

Sampling in space is the technology that captures each single image. For film, it is the microscopic grains of silver halide in the emulsion—each crystal responds uniquely to light. For video, it is the elements on the CCD chip(s) in the camera. A digitized image looks a lot like a TV image. The pictures must be broken down into tiny pieces, small enough that the viewer only sees the big picture and not the discrete pieces.

A sample of video—the smallest digital "chunk" of a video image—is called a **PIXEL** (short for "*pic*ture *el*ement"). If you look closely at your TV or computer monitor, you can see them. Technically speaking, a home television set has relatively big chunks—a fairly small sampling rate. A video signal consists of 525 lines per screen, with each line containing about 720 pixels. Thus, for each frame of video, you have about

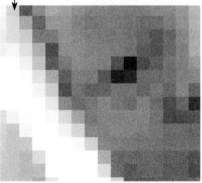

From a distance a video image looks contiguous, but a close-up illustration reveals pixels. The more pixels in a given area, the higher the resolution of the image.

378,000 pixels of information (720 x 525 = 378,000). Actually, only about 486 of the 525 lines are used for video picture, so a better count is closer to 349,920 pixels (720 x 486). By the way, home televisions only display about 400 of those 486 lines.

▶▶ Square vs non-square pixels: p. 175

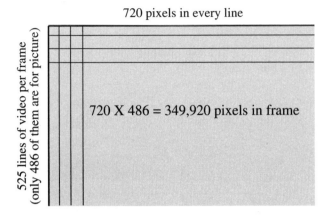

720 pixels in every line

525 lines of video per frame
(only 486 of them are for picture)

720 X 486 = 349,920 pixels in frame

But what is the information that we need to digitize? In our music analogy, it was amplitude (loudness) and frequency (pitch), that change in time. For video, there are also two kinds of information per pixel . . .

CHROMINANCE and LUMINANCE

CHROMINANCE is defined as the color part of the signal, relating to hue and saturation.

LUMINANCE is the brightness of the signal—the scale is from black, through grey, to white. A black and white television is only dealing with the luminance signal; it cannot pick up the chrominance.

There are different ways to encode chrominance and luminance information; the most common is through combinations of the colors *red, green* and *blue* (known as R, G and B). Both chrominance and luminance values can be computed from the relative quantities of a signal's R, G and B.

Just as for digital audio, a scale had to be created for the digital sampling of red, green and blue. And like the amplitudes at each sample for music digitizing, the scientists determined that 8-bit samples taken at each pixel of red, green, and blue would faithfully re-create a digital version of an analog image for broadcast on television.

FACTOID • Some computer screens and older capture cards utilize the older standard of 640 x 480 pixel dimensions for a video frame.

FACTOID • Although a broadcast-quality master is generally accepted in the United States as having 525 horizontal lines by 720 pixels per line, the resolution that is considered "film quality" is much, much higher: over 3,000 horizontal lines with more than 4,000 pixels per line (so-called "4K" resolution). When broadcast video is digitized, 8-bit color for each of the red, green, and blue signals is determined sufficient (256 shades per color); film quality is roughly equivalent to 12-bit color (4,096 shades each).

Although film projects at 24fps, rather than the 30fps for video, a digital FILM image would take up 100 times as much memory as a video image, for the same amount of time. This is one reason why, until the late 1990s, computers were not powerful enough to perform computer graphics that were usable in theatrical films. This also accounts for the amazing clarity of High Definition Video—a broadcast video signal format with over 1,000 horizontal lines, and 1000 pixels per line.

Let's pause for a moment to do some quick math.

For DIGITAL AUDIO on compact discs, we have already shown that the computer is sampling the music 44.1 thousand times a second, and at every sample, a 16-bit "chunk" is memorized. That means for every second of music:

44,100 x 16 bits   = 705,600 bits per second.

Remember that 16 bits is equal to 2 Bytes, so let's do this again:

44,100 x 2 Bytes   = 88,200 Bytes per second
                        = 86.1 KBytes/second—per channel

(Remember: stereo audio has TWO channels of sound).

To play a full minute of stereo digital audio on a compact disc . . .

86.1K x 2 x 60 seconds = 10,332 KBytes/minute
                       = 10.1 MBytes/minute

. . . **you need over 10 MB for every minute** . . . which means that your average compact disc is simply a computer disc holding an hours worth of digital audio, which equals about 650 megabytes. Because it cannot be erased, or used for anything else, we often forget that a compact disc is a piece of storage, not that much different from floppy disks or hard disks.

▶▶ HDTV: p. 280

But if you think digital audio takes up space, take a quick look at digital video. For each of our 349,920 pixels (in 720x486 video), we need 8 bits of information for each of the red, green and blue colors; that's 24 bits (or 3 bytes) PER PIXEL. (Actually, many video systems use 16 bits—8 bits for luminance and 8 bits for chrominance, but let's ignore that for now.)

349,920 pixels x 3 bytes = 1025 Kbytes PER VIDEO FRAME*.

Now, you need this data moving at 30 frames per second. . .

1025K/frame x 30 frames/second   = 30.0MB/second
                                               = 1.76GB/minute

…which is about 180 times as "large" as digital audio on a CD.

Because digital video data is so HUGE, there are only a few ways to make it fit in a cost-effective storage medium. Among the most common methods:

1) **Make the picture smaller**—if you must retain the full broadcast quality, you might want to shrink the image down—an image ¼ of a screen takes up ¼ of the data.

2) **Decrease the frame rate**—moving video at 30 frames per second takes up three times the memory as moving video at 10 frames per second. Although it is generally an unattractive alternative for editing, displaying fewer frames per second is a viable way to save space.

3) **Decrease the color quality**—rather than store the 8 bits for each R, G, and B, decrease the smoothness of the color palette. Every "bit" you shave off the R, G, and B signals decreases your storage further. Making the image black and white (removing all chrominance) saves even more. Color sampling like 4:2:2 and 4:1:1 are methods of accomplishing this savings.

---

* IMPORTANT: 921,600 bytes is not 921.6 KB because a kilobyte is actually 1024 bytes, not 1000. The computer business quietly modifies the rules of metric to meld with its own reality. It's a battle of metric vs binary.

A thousand somethings is supposed to be a kilo-something, **except** when it comes to computer bytes. Because computers count with binary numbers (not metric), a computer that counts to 1000 can also count to 2^{10}, or 1024.

Thus: a kilobyte (KB) is 1024 bytes, a megabyte (MB) is 1024 kilobytes, and a gigabyte (GB) is 1024 megabytes.

▶▶ 4:2:2 and color sampling: p. 165   ▶▶ Video on the web, compression: p. 290

4) Balance the number of bits allocated to each of the R, G and B colors to maximize efficiency of chrominance and luminance ranges. This is kind of technical, but there are "tricks" that can be done to get better color out of the same number of bits. For example, if you have 12 bits of color, instead of giving red 4 bits, green 4 bits and blue 4 bits, you might give red 4 bits, green 6 bits and blue 2 bits. For mathematical and perceptual reasons, this yields better ranges of color.

5) Use some form of VIDEO COMPRESSION. Video compression is a way to take the digitized information about an image and encode it so that it takes up less space. How good a particular technique of compression is often depends on the KINDS of images that are being compressed. Over the past few years, a number of compression techniques have been pioneered, in particular JPEG, M-JPEG, MPEG and Wavelet.

All the companies manufacturing nonlinear editing systems that utilize digitized source material must deal with all of these parameters in designing their system. The remaining factors—image size, resolution, color range, compression—must be dealt with in terms of the limitations of the system's memory storage, and computer processing power. Ultimately, the question must always be answered: "How good does an image have to be for offline editing?" or "Can this system be used for online editing?"

The earliest versions of the digital systems began with 15fps playback of small images, digitized at 4 bits. After only a few years of development, all frame rates extended to 30fps, sometimes with 2-field recording, and utilizing compression techniques that were much better. Early web streaming also utilized variations of all these methods to fit video through the narrow bandwidths of the Internet.

For the record, High Definition Video is even bigger than the 720x486 video we just examined. HD needs to be compressed to fit on DVDs or in channels for television broadcasting. How big is HD? If we're talking about 1080p/24 (*aka* "24P"), we get 2,073,600 pixels (1920x1080); with 8 bits of information for luminance and 4 each for Cr and Cb (*i.e.*, 4:2:2); which is 2 bytes per pixel.

2,073,600 pixels x 2 bytes = 4 MBytes PER VIDEO FRAME.

You need this data moving at 24 frames per second, so...

4MB x 24fps   = 96MB/second = 768 Megabits/second
               = 5.6GB/minute

Clearly, for this and other reasons, compression is critical.

C O M P R E S S I O N

Compression is simply the task of taking something BIG and squeezing it down into something SMALL. Today, when you say "compression" people often assume you are talking about digital video compression—however true that may be, there are many other things that need to be compressed: analog video, digital computer files, digital video...

Why would you want to compress something? To make it smaller, of course. You might want it smaller because it takes up too much physical space or because it is too big to move around easily. Since digital storage isn't cheap, and is generally priced per byte, if you can fit more files into the same space, you are saving money. There are many computer programs that compress and de-compress files: Stuffit Deluxe (from Aladdin), WinZip (from Niko Mak), to name two.

There's another major reason to compress something: when information must pass through a limited pathway. If computer data is like cars on a highway, the data bus size is the number of lanes in the highway. When a busy highway closes a few lanes to cars, you end up with a traffic jam. Computers often manipulate digital video but may have trouble moving all that data fast enough down the tiny bus "paths" that connect different parts of the computer. The data must be compressed to move smoothly and quickly.

Computers have limitations too. Hard disks, Jaz disks, floppy disks—each kind of storage device has a certain maximum speed that data can be "streamed" off of it. Hard disks are currently the fastest (up to around 100MB per second), Jaz disks are in between at around 8MB per second; floppy disks are terribly slow (at about 80K per second). So it's not just a matter of how much a storage device can hold, it is very important how quickly that device can move the data.

Once the data is moving off the storage device, it still may be limited by the pathways around the computer since they too have a limited number of "lanes." And the computer's chips can only handle so much work. For many other computer applications, the storage device, the pathways and the chips are all of sufficient capacity. But recall that digital video is **HUGE** and has special needs.

An uncompressed SDTV video picture requires 680KB per frame, which (at 30fps) translates into about 20MB per second.

▶▶ Video compression and storage: p. 174

PROBLEM: Let's say our Jaz disk can only "play" at about 8MB per second. Through some simple mathematics, it can be seen that you'd need to compress this 20MB/sec video picture down to 8MB/sec (written as 2.5:1) to get it to play from the disk. And that assumes the rest of the computer can handle it; many personal computers can't handle high quality video anyway. Often the processor and the pathways prove to be the bottlenecks more than the storage device.

There are many strategies by which an engineer can approach the compression problem. You want to shrink down the data as much as you can while losing as little information as possible.

Codecs

Remember the one about the two guys who'd been friends for so many years that they started referring to all their favorite jokes by number? ("Here's a good one—142!" "Ha, ha, ha!"*) Or maybe you're at a party, having a good time, but your date tugs on your sleeve and rolls her eyes in the opposite direction. She's bored to tears and wants to go home, obviously, though she didn't actually have to say it ...

These types of shorthand—which use minimal, mutually-understood signals to convey a greater amount of information—are **codecs**.

The word "codec" itself is a contraction, which stands for **COmpressor-DECompressor**. It refers to any of the specialized methods of "squeezing" digital data into a smaller file and then allowing it to be re-expanded again to (more or less) its original form once it's reached the destination. This is useful for, say, storing video on a DVD or sending it over the Web.

Codecs are algorithms (procedures, or sets of formulas) that can be implemented either in software programs or specialized hardware chips. A codec compresses data in one of two ways—*lossy* or *lossless*.

Lossless methods compress data by a certain degree, and then reconstruct it perfectly into its original form at the other end. (The *Star Trek* transporter, one would hope, uses a lossless form of compression.)

Lossy schemes, on the other hand, are generally able to "scrunch" digital data even more, but they do that by permanently throwing out certain

---

* A guy walks into the bar and overhears these two calling out numbers and sees them spiraling into paroxysms of laughter. He is so caught up in it that at a pause in the festivities he calls out "127"—whereby he is met with dead silence. One of the two original comedians swats the other on the back with a smirk and says "some guys just can't tell a joke..."

kinds of details, and settling for rough approximations instead. (The end results, though, might come out as appealing as reconstituted, powdered soup.)

Lossy compression can be accomplished either blatantly or with subtlety; just because a compression is lossy does not necessarily mean that the image will look terrible. It is possible to have loss so slight that most people can't tell the difference. This degree of flexibility has prompted the emergence of a few variations on the terminology:

> **Lossy:** the data has been altered and has lost some degree of quality . . . though the amount of degradation can vary widely, depending on the codec and its settings, and the nature of the original data.
>
> **"Perceptually lossless":** a lossy compression scheme where the degradation is supposedly subtle enough that you might not be able to tell the difference.
>
> **Lossless:** data compression that can be reconstructed exactly, without any loss in quality; *aka* "mathematically lossless."
>
> **Uncompressed:** digital data is stored/transferred as-is, without any codec whatsoever.

A Crude-But-Effective Demonstration:

Uncompressed: The quick brown fox jumps over the lazy editor.
Lossless: The quick brown fox jumps over the lazy editor.
Lossy: thquikbrnfxjmpsovrthlzyedtr

The original sentence is 3.2 inches long. By squeezing the typeface down, we have effectively shortened the sentence without losing anything; all the information is there—all the letters, spaces, the capital, the period. Now, however, it is only about 2.6 inches long. This is *lossless*, and the sentence has been compressed by 19%. For the *lossy* compression, we not only used a squeezed typeface, but we removed spaces and unnecessary letters and punctuation. Notice, however, that even with 59% of the original sentence thrown away, it is still fairly easy to figure out what it says.

[For an expanded discussion of compression styles, see *Compression Made Simple*, page 163.]

There are different types of codecs because there are different types of data. Each codec is tailored to a particular use, since it takes a different strategy of "ferreting out" in order to eliminate the redundancies that might exist within each form of information.

Many industries are interested in compression: whether to play video over telephone lines or the Internet, or movies from DVDs, or to broadcast HDTV;

type of data	LOSSY	LOSSLESS	UNCOMPRESSED
computer data		StuffIt PK-Zip Compact Pro	a copy of the original file
still pictures	JPEG JFIF	BMP GIF (depending on number of colors) TIFF (with LZW compression) RLE	PICT TIFF
moving pictures	Cinepak DV M-JPEG RealVideo Digital Betacam (with Bit Rate Reduction) Sorenson 4:2:0* 4:1:1*		ITU-R.601 NTSC/PAL D1 ITU-R.709 HDTV
audio	MP3 DTS RealAudio Liquid Audio SDDS Dolby AC-3	MLP	PCM AIFF WAV SD2

all require the video to be compressed and each has specific compression needs. Although there are many compression techniques, two are of particular interest to digital video (nonlinear) editing systems. Remember: JPEG and MPEG are not products—only standardized compression techniques.

JPEG (pronounced "JAY-peg") was created in the late '80s by the *Joint Photographic Experts Group*. They wanted to compress color images and their method is particularly good at doing that. The computer groups pixels into blocks of 64 (8 by 8), and then makes some generalizations about each area. Because it is only looking at a single frame, this is called an ***intraframe*** method. JPEG is commonly used to compress single images (and sometimes video) around 25:1 with marginal quality loss; even higher compression ratios (from 40:1 to perhaps 100:1) can be used,

---

* These formats are not "compression" per se, but are chroma sampling techniques that effectively reduce the amount of data in the color signal from the very beginning.

but with significant sacrifice of image quality. Most nonlinear systems use a variation of JPEG called *motion-JPEG*, or M-JPEG. This is just a series of JPEG still frames, strung together so that they can be readily played as "video" and easily edited.

JPEG is a compression scheme widely used in applications that are concerned with still images: desktop publishing, electronic photo processing, digital scanners, color laser printers, and the Web. It is also an inexpensive and simple method—the same JPEG chips that compress video can be used in reverse to decompress. This is called a *symmetrical* technique.

MPEG (pronounced "EM-peg") is a compression method specifically tailored for moving pictures (a kind of a JPEG spin-off, brought to you by the *Moving Pictures Expert Group* in the early 1990s). MPEG does not simply look around the single frame and compress it. It also looks at adjacent frames to see which pixels are changing and which are mostly the same. This is called *interframe* encoding (and, if you were wondering, uses "bi-directional prediction motion interpolation"). For example, if your face is on the screen, and you're just talking, probably the only part of the screen that is changing significantly is your mouth—the rest of you remains pretty much the same. By concentrating on the interframe changes, MPEG greatly reduces digital video data.

Another feature specific to MPEG is that it is also designed to handle sync audio. The JPEG techniques only deal with pictures.

MPEG is an *asymmetrical* technique. This means that compression and decompression are handled differently, by different sets of chips (or software). This makes MPEG well-suited to different kinds of applications: playback of pre-compressed video from DVDs, for example. MPEG is designed for moving images and sync sound, but has more trouble handling source material that is to be edited (an edit to a new shot introduces some pretty radical interframe changes). Although MPEG can compress better than M-JPEG, it is still relatively early in its use for the video in editing systems.

More About MPEG

There are a handful of variations on the MPEG standard: MPEG-1, MPEG-2, MPEG-4 (even the special audio-only variation of MPEG-1 called MP3). Each starts with the basics of MPEG, then tweaks the format to streamline it for one application or another.

MPEG-1 was designed to handle the storage and retrieval of moving pictures and audio on, primarily, CD-ROMs, at 1.2Mbps; it was standardized by the ISO in 1992.

While MPEG-1 is appropriate for many playback situations, it is not ideal. MPEG-2 was designed to solve the problems of MPEG-1, but most importantly, it was built as a scalable solution, with "low," "main," and "high" levels of resolution. It was approved by the ISO in 1994. MPEG-3 was invented to work for HDTV (1920x1080), but was eventually folded into the MPEG-2 specification; consequently, there is no MPEG-3 today. MPEG-4 is the newest variation on the method, originally targeted at low bit rate communication, and based on the foundation of Apple's QuickTime format. It is expected to be a standard for multimedia applications, but not for broadcasting. There are other variations in the MPEG family that are in various states of development, including MPEG-7 and MPEG-21.

DVDs and HDTV both utilize MPEG-2 compression. Because of the design of MPEG-2, differing codecs working at different resolutions create a matrix of compression levels. The profile chart on the next page shows the specifications for MPEG-2 uses.

The Almost-Technical Part of MPEG Compression

The *inter*frame nature of MPEG compression means that it not only looks within each frame to squeeze it down, but also at groups of sequential frames, to see how similar they are. Each consecutive handful of frames (called a "group of pictures," or GOP) acts like a little neighborhood of sorts, where the frames look out for one another. The leader of the group is the **I-frame** (for "intra" frame), whose role is to store a complete, self-contained JPEG-style snapshot of the corresponding frame of video. **P-frames** ("predictive" frames) are second in command; they only

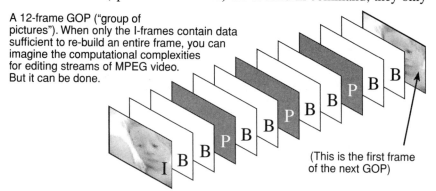

A 12-frame GOP ("group of pictures"). When only the I-frames contain data sufficient to re-build an entire frame, you can imagine the computational complexities for editing streams of MPEG video. But it can be done.

(This is the first frame of the next GOP)

remember what's changed in the image since that original I-frame was taken. **B-frames** (for "bi-directional") are go-betweens; they only deal with the smaller details that may transpire between I- and P-frames.

This sort of interdependency is very efficient, since only one frame out of the group needs to be memorized in its entirety. It's a lot like a digital "carpool" of sorts, where one person drives, and the rest just tag along for the ride. They all get to where they're going, but take up less space on the "highway" along the way.

When an edit is made to a stream of MPEG video, the computer must rebuild the GOP in its entirety to get an actual "frame" on which to make the edit (GOPs must always start on an I-frame). The length of a GOP is typically 15 frames long, but an edit may create a short GOP.

It takes only modest computing power to decompress an MPEG file, but encoding video into MPEG requires significantly more (remember, MPEG is "asymmetrical").

MPEG Decoder Complexity ("Profile")

Resolution ("Level")		Simple	Main	4:2:2	SNR	Spacial	High
	High		HDTV 1080p, 1080i 4:2:0 1920x1152 80Mbps	4:2:2 1920x1152 300Mbps			4:2:0 or 4:2:2 1920x1152 100Mbps
	High 1440		HDTV 720p 4:2:0 1440x1152 60Mbps		4:2:0 1440x1152 60Mbps		4:2:0 or 4:2:2 1440x1152 80Mbps
	Main	4:2:0 720x576 15Mbps	SDTV 4:2:0 720x576 15Mbps	4:2:2 720x576 50Mbps		4:2:2 720x576 15Mbps	4:2:0 or 4:2:2 720x576 20Mbps
	Low		4:2:0 352x288 4Mbps		4:2:0 352x288 4Mbps		

MPEG-2 Family of Profiles and Levels. All of these variations of MPEG-2 support different uses of the compression scheme. The "Main" level of complexity is the primary profile for ATSC transmissions (American broadcasting) for digital standard television (SDTV; pretty much equal to today's NTSC) and for the broadcast variations of HDTV. The 4:2:2 profile is used by television networks for the high-quality distribution of signals to their affiliates, prior to being broadcast.

DVDs are encoded with Main Profile at Main Level compression (known as MP@ML), as is SDTV.

COMPRESSION MADE SIMPLE
A LITTLE GAME

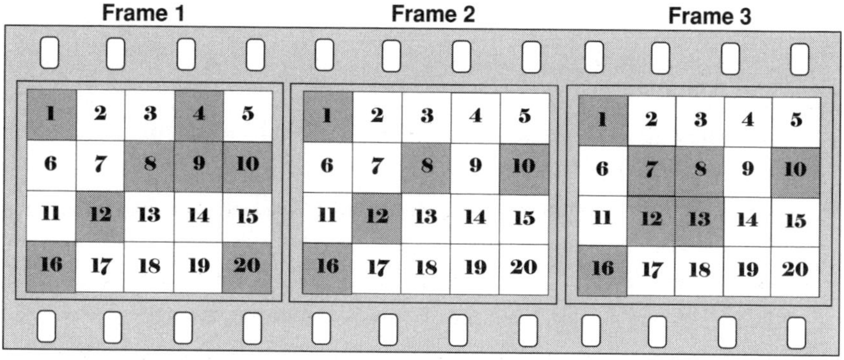

Frame 1 | Frame 2 | Frame 3

I want you to examine the 20 pixels in these three imaginary frames of digital film. Each has an identification number. Each is a color: either white or black. I have 3 frames of film similar to this one—a tiny clip from a movie—and I want you to communicate the patterns of colors in these frames over the phone to me in Santa Cruz, and you're being charged for the duration of your long distance call. Please describe the pattern to me as quickly and precisely as possible. Ready? What might you say?

No Compression

"Hi Mike. Frame 1: 1-black, 2-white, 3-white, 4-black, 5-white, 6-white, 7-white, 8-black, 9-black, 10-black, 11-white, 12-black, 13-white, 14-white, 15-white, 16-black, 17-white, 18-white, 19-white, 20-black.

"Frame 2: 1-black, 2-white, 3-white, 4-white, 5-white, 6-white, 7-white, 8-black, 9-white, 10-black, 11-white, 12-black, 13-white, 14-white, 15-white, 16-black, 17-white, 18-white, 19-white, 20-white.

"Frame 3: 1-black, 2-white, 3-white, 4-white, 5-white, 6-white, 7-black, 8-black, 9-white, 10-black, 11-white, 12-black, 13-black, 14-white, 15-white, 16-black, 17-white, 18-white, 19-white, 20-white."

[29.2 seconds.]

Lossless INTRAframe Compression—Run Length Encoding (RLE)

"Frame 1: 1-black, 2, 3-white, 4-black, 5-7-white, 8-10-black, 11-white, 12-black, 13-15-white, 16-black, 17-19-white, 20-black.

"Frame 2: 1-black, 2-7-white, 8-black, 9-white, 10-black, 11-white, 12-black, 13-15-white, 16-black, 17-20-white.

"Frame 3: 1-black, 2-6-white, 7-8-black, 9-white, 10-black, 11-white, 12,13-black, 14-15-white, 16-black, 17-20-white."

[18.5 seconds. In this case, about 3:2 compression.]

Comments: This an efficient way to compress a file and not lose any information. You'll notice that big areas that are all the same color will compress more because you just say something like "one through 500 are all white" and you've transmitted a lot of accurate data. When every pixel is different, the compressed image will not be much smaller than the original.

INTRAframe Compression—"kind of like JPEG"

"Look at vertical pairs; I'll give the top number and then average the pair:

"Frame 1: 1-grey, 2-white, 3-grey, 4-black, 5-grey, 11-grey, 12-grey, 13-white, 14-white, 15-grey.

"Frame 2: 1-grey, 2-white, 3-grey, 4-white, 5-grey, 11-grey, 12-grey, 13-white, 14-white, 15-white.

"Frame 3: 1-grey, 2-grey, 3-grey, 4-white, 5-grey, 11-grey, 12-grey, 13-grey, 14-white, 15-white."

[14.8 seconds. In this case, about 2:1 compression]

Comments: JPEG works by looking at blocks of pixels within the frame and averaging their values. This is like having fewer pixels, but when I go to draw my own frame from your JPEG data, I will be close but I'll have to guess to fill in missing info (because I was averaging, when my result is grey I can't tell which of the pair of blocks was white and which was black; I have no way of recreating the pair accurately). Consequently, this clearly shows a lossy method of compression—as is both JPEG and MPEG.

INTERframe Compression—"kind of like MPEG"

"Frame 1: 1-black, 2-white, 3-white, 4-black, 5-white, 6-white, 7-white, 8-black, 9-black, 10-black, 11-white, 12-black, 13-white, 14-white, 15-white, 16-black, 17-white, 18-white, 19-white, 20-black.

Frame 2: same except 4,9, and 20.

Frame 3: same as frame 2 except 7 and 12."

[10.9 seconds. In this case, about 3:1 compression.]

Comments: MPEG works great with moving pictures because, except for edits, each frame is very similar to the frame before it. MPEG encodes a keyframe every 12 or so frames of film, and then simply describes the differences between that key frame and the following frames. This is excellent for playing back movies, but is problematic in editing because wherever you make a cut, that frame may not be a key frame and thus the computer would have to pause and re-calculate what that precise frame would look like. This is the inherent challenge to MPEG editing systems. (Note: this particular interframe example is NOT lossy. Real schemes are obviously more complex and involve combinations of techniques.)

4:2:2 and COLOR SAMPLING

A familiar expression used in conjunction with digital video is 4:2:2, or 4:1:1 or even 4:4:4. These are all shorthand notations for different color sampling methods for digital video. The three digits are the relative sampling rates of the luminance (Y) and two chrominance components (red minus luminance, written R–Y is also called "C*r*"—and its sibling, blue minus luminance, B–Y, or "C*b*").

To understand what this is about, you must begin by visualizing the pixels in a digital image: 720 pixels per scan line for D1 (CCIR-601 or SDTV) video. In simpler discussions in this book, we explained that to sample a color picture, each pixel needed a luminance value and a pair of

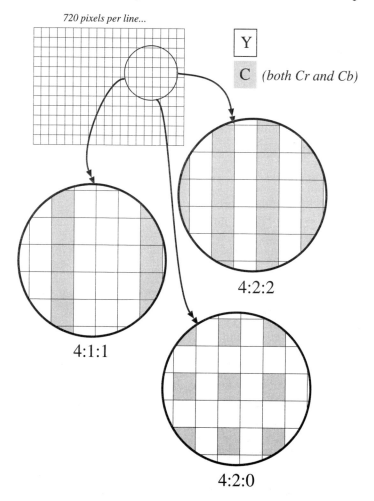

720 pixels per line...

Y

C *(both Cr and Cb)*

4:2:2

4:1:1

4:2:0

chrominance values: **Y, C*r*** and **C*b***. But in reality, it is not necessary to encode these three values for every pixel on the screen. Due to the nature of the human visual system and perception, we are considerably more sensitive to luminance than chrominance. Consequently, to faithfully encode the picture, we do not need Y, C*r*, and C*b* for *every* pixel. Every pixel does need Y, that is true. But it was determined that even for D1-quality video, you only need half the number of chrominance data points. If every pixel gets a Y value, then only every other pixel needs a chroma value (both C*r* and C*b*). If you look at the ratio of Y to the C*r* and C*b*, it will be 4 Ys for every two C*r*+C*b*—and you write this 4:2:2.

High quality professional video with full color, is 4:2:2—D-1, D-5, Digital Betacam, Digital-S, just to name some popular formats.

(In the world of high-definition video, with more pixels per frame, the sampling has to happen *faster* than with NTSC; hence, the HD equivalent to 4:2:2 is called "22:11:11".)

But the brain is very forgiving. It is possible to save space by sampling the chroma even less than this and not give up too much in the way of color resolution. If, for every four Y samples, you take *one* C*r*+C*b*, you've cut the color detail in half but you still produce an image that is almost identical to the original. This sampling is **4:1:1**. It is used in the consumer and prosumer DV formats (DVCAM, DVCPRO, and a few other component video formats). 4:1:1 seems like it would produce lousy images, but the truth is, it doesn't.

There are variations on these color sampling methods. 4:2:0 does *not* mean that for every four Ys, there are 2 C*r*'s and *zero* C*b*'s. This bastardized terminology is trying to communicate a sampling like 4:1:1, except that (for all practical purposes) the color samples skip every other scan line. It's not superior color, but it's reasonable, and as long as space saving is important and the signal is not going through multiple generations (losing fidelity with each), it is very useful. DVD and MPEG-2 Main Profile both sample with 4:2:0 color.

When all the digits are fours (*e.g.,* 4:4:4) it simply means that Y, C*r* and C*b* are represented in *every* pixel, which isn't required for distribution, but is often advantageous in-house for video requiring heavy processing—as used in film or high-end video effects.

When there is a fourth unit in this system, (*e.g.,* 4:2:2:2 or 4:2:2:4), the final digit refers to the presence of an alpha channel, a transparency bit that goes along with the video. This is a useful format for creating special effects and executing some kinds of graphics. While the extra bit isn't required for delivery systems, it is a production format for high-end film and video.

R A I D   T E C H N O L O G Y

Because of the data transfer rate limitations of hard disks for the highest quality of video (or for simultaneous playback of multiple video tracks), other strategies have been developed to increase the functionality of the medium. Called RAIDs—for *Redundant Array of Inexpensive Disks*—a number of disks are used together in parallel to simulate a much faster single disk. Note that a RAID can be designed from any disks that are considered "inexpensive" and today most RAID drives are comprised of non-removable hard disk sets. Also, RAIDs are not necessarily bundled products in neat packs; any bunch of disks can be turned into an array by use of appropriate software. A bunch of disks that are not being used together as a RAID are called a **JBOD**—"just a bunch of disks."

RAIDs are a prerequisite when multiple editors want to access source material simultaneously in a networked environment. Since each channel of uncompressed video needs 20MBps, and each editor needs two streams for real-time effects (dissolves, wipes, *etc.*) RAIDs must be configured to provide 40+MBps for each workstation connected to the network.

There are different RAID "levels" which describe various technical attributes of the array. RAID level 0 (written "RAID 0") is the most common for video playback and editing, and is designed for high-bandwidth.

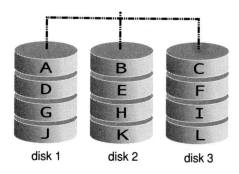

RAID 0: Digital data (but for our purposes, video frames A through L) are not recorded onto a single disk, but spread out over a set of disks. This improves transfer rates. A failure of a single disk, however, will devistate the entire data set.

disk 1   disk 2   disk 3

RAID 0 works like this: digitized material is "striped" across 2 or more disks (4 or 5 are common): little consecutive pieces of data are recorded onto different disks. For example: if a hard disk has an access time of 11ms and a data transfer rate of about 5MB per second, a RAID of 5 hard disks might produce an effective access time of 2.25ms and a data-transfer rate as high as 25MB per second. Hard disk arrays require extremely high bandwidth pathways to move the data—the usual SCSI interface is insufficient for many products. UltraSCSI, UltraWide SCSI, or IEEE 1394 interfaces that provide 40MBps+ data transfer rates are required for drives with this bandwidth.

Another use of RAIDs is in eliminating the risks associated with loosing valuable data. Called "fault tolerance," RAIDs can be configured to mirror everything written on one drive onto a second "backup" drive. RAID 1 is a fault-tolerant array which is not used to increase bandwidth, but to increase protection. RAID 1 is used in many financial computer set-ups.

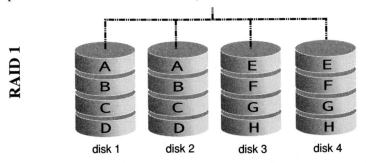

RAID 3 is an array that is also sometimes used in video production, in particular with live streaming of video. Like RAID 0, RAID 3 stripes data across a number of disk drives, but it also has an extra drive that manages error correction codes, to give higher efficiency and lower risks associated with a single disk failure. This is one degree of fault tolerance in a RAID.

A significant problem of arrays with no fault tolerance (like RAID 0) is that failure in any one of the disks in the array will cripple the *entire* array, resulting in the need to re-capture the entire N gigabytes of video (where N is large). A number of high-end products choose to provide *proprietary* arrays, specifically designed both to create appropriate video bandwidth and proper professional protection of data.

Computers handling real-time animation, film effects, HDTV/high-resolution graphics, and TV commercial playback, all often need to supplant regular drives with RAID technology.

QUICK OUTLINE OF RAID SCHEMES

TYPE	Basic Description	For...
RAID 0	Striped Disk Array without Fault Tolerance	video
RAID 1	Mirroring and Duplexing	financial
RAID 2	Hamming Code ECC	geeks
RAID 3	Parallel Transfer with Parity	video
RAID 4	Independent Data Disks with Shared Parity Blocks	geeks
RAID 5	Independent Data Disks with Distributed Parity Blocks	web
RAID 6	Independent Data Disks with 2 Independent Distributed Parity Blocks	geeks
RAID 7	Optimized Asynchrony	geeks
RAID 10	Very High Reliability/Performance (like RAIDs 1+0)	web
RAID 53	A variation on RAID 3	geeks

DIGITAL STORAGE CHARACTERISTICS

A ll digital storage media have certain characteristics that define their usability and performance. These fundamental traits are:

- STORAGE CAPACITY
- DATA-TRANSFER RATE (maximum and sustained)
- ACCESS TIME

Storage capacity simply refers to how much data can be stored on the media. This quantity is expressed in terms of bytes: megabytes (MB), gigabytes (GB) or terabytes (TB).

It is important to note that many digital storage media—in particular magnetic disk media—must be prepared before they can hold data (it is not entirely unlike "blacking" a videotape prior to use). This preparation is called *formatting*; formatting not only prepares the disk, it actually records some information onto the blank disk. The actual amount of format information depends on how big the disk is. For example, 3.7GB hard disks, once formatted, actually can only hold about 3.0GB; 37GB hard disks format down to around 35GB.

Data transfer rate is how quickly data can be moved onto or off of the medium. It is actually the result of a number of processes, noteworthy among them are the *internal data transfer rate* (how fast the read/write head can get data on and off a disk) and the *sustained data transfer rate* (the speed of material through the drive controller). Data transfer rate is important because for video to play, you need to move a great deal of data off the digital storage device very quickly. If you can't move it quickly, you need to make the digital data size of each frame smaller so that you can pump them out at 30fps. There are, of course, various ways to accomplish this (decrease the frame rate, make each frame image smaller, *etc.*), but each choice will compromise image quality to some degree.

Sometimes devices are advertised as having very high data transfer rates, but these rates might be the *maximum* speeds; for smooth video playback you need a sustained (or average) rate that is pretty high. *This sustained rate is the important item for a storage device.* It is also important to note that transfer rate is slightly different for reading (pouring out) and for writing (recording in).

Related to data transfer rates are the rates at which the data flows *around* the computer, through cables and between components. Here you have protocols like FireWire, USB, Ethernet, SCSI (for cables and connectors); ISA, EISA, NuBus, PCI, AGP (for motherboards); and IDE, EIDE, ATAPI (for internal storage devices). Any of these "bottlenecks" can further limit the speed at which the data flows through computer hardware.

Access time is how long a certain device takes to locate a chosen piece of information. Access time is actually composed of *seek time*—the time it takes the disk to cue from one track to another (and with the worst-case scenario a seek from the inside to outside radius of the disk)—and *rotational latency*, basically how fast the disk spins. On linear tapes (*e.g.,* DAT, DV, Exabyte), access time is a critical factor, as it determines how "random access" a medium can be.

Computers also need to access material. When you ask for a shot or a file, the time it takes for the computer to find it, load it into memory and present it to you is generally considered part of "access time." This time, however, is miniscule when compared to the shuttling of videotape or even disk access. For reference, consider that a complete video frame takes 1/30 of a second, or 33 milliseconds. Hard disks can access a "frame" in less than 10ms.

The size of an uncompressed (ITU-601) video frame is about 680KB. If you need 30 of these every second, then your video-computer must move data at a rate of 20MB per second. Suppose that the specification of your hard disk tells you that it only moves data ("sustained transfer rate") at about 10MB per second, regardless of how much data it HOLDS (its capacity). This is pretty slow, and is one example of why compression techniques are so important to digital video and computers—not simply to make the economics better for video storage, but because some devices simply are not able to manipulate data that is so large and must be moved so quickly.

This close-up of a hard disk and it's read/write head captures the superfast speed of the head sweeping across the disk surface.

B   A   N   D   W   I   D   T   H

Image Quality, Compression, and Transfer Rates

No talk of computers goes on very long without the word "bandwidth" coming up. This is true whether the discussion is about video post-production, digital video, or the Internet.

As people who describe this stuff are fond of saying: *If digital data is water in a big bucket, and there is a hole (and pipe) at the bottom that allows you to fill a glass when you want some, bandwidth is the diameter of the pipe. If you have a lot of bandwidth, it doesn't take long to fill a glass. If you have a tiny little straw for a pipe, water is going to dribble out, taking a long time to fill your glass. Even if the bucket is huge. Even if your glass is tall. It's all about the pipe.* Get it?

Every time you see a cable connecting two computer components (a camera and computer; a hard disk drive and the base tower; two computers on the Internet; *etc.*), think **bandwidth**: *how big is the pipe connecting these things?*

Bandwidth, like velocity, is measured in terms of an amount of something that moves over a unit of time; traditionally, this is as a number of **bits per second**. If the object being discussed is a mass storage disk or drive (as opposed to a cable or network), bandwidth is described as **bytes per second** (8 bits to a byte, you recall). This bandwidth is also sometimes referred to in the context of a drive's "sustained data transfer rate," but it amounts to the same thing.

Be careful to note whether you are discussing **MBps** (bytes) or **Mbps** (bits). Many publications and spec sheets confuse the two:

Ultra SCSI supports data rates of 20MBps:
20M Bytes/second x 8 bits/Byte = 160Mbps

Baud rate, the rate at which a modem transmits data, is a number related to bandwidth, but it is not exactly the same thing. There are a number of factors that determine your modem's connection rate, including the line quality, the modem of the computer you're "talking" to, and so on. Not all the bits being modemed are *your* data; for every 8 bit "word," some of those bits are header information, so you're really only getting to use some of the 56Kbps a 56K modem offers.

Disk rotation. One disk characteristic that impacts transfer rates (and effective bandwidth) is the speed at which the disk spins: faster spinning translates into more bits per second flowing off the disk. This can be most evident in discussions of CD-ROM drives, which began increasing rotational velocity (2X, 8X, 16X) as years progressed, effectively increasing the bandwidth of the media, and increasing both the speed at which material could be copied off a disk and the potential for multimedia applications on CD-ROMs to deliver streaming video. Mass storage devices often refer to their ability to deliver increasing bandwidth by the irritating factor of RPM, the *revolutions per minute* the disk provides. While it is an interesting factoid, the only relevant benchmark is the sustained data transfer rate or effective bandwidth (either, in **MBps**).

Below is a little table that gives crude apples-to-apples comparisons of bandwidths of various devices, cables, tape formats and compressions.

Tape and Media Formats:

MP3 (stereo audio) 128Kbps
DVD-Video and SDTV 1.5–9.8Mbps
Television (6MHz) broadcast channel 19Mbps
HDTV compressed for broadcast 19Mbps
MiniDV (or DVCAM) 25Mbps
Television (6MHz) cable channel 38Mbps
DVCPRO and Digital-S 50Mbps
DV100 and Digital Betacam 100Mbps
Uncompressed video (ITU-R BT.601) 160Mbps
Uncompressed HDTV (24P) 778Mbps

Cable and Modem Transmission:

Typical Palm *wireless* Internet connection 8–19.2Kbps
A standard 56K modem .... ~56Kbps
ISDN 128–144Kbps
Serial (and "AppleTalk") 230Kbps
SDSL modem 384Kbps
ADSL modem (typical download/upload) 1.54Mbps/128Kbps
ADSL modem (*max* download/upload) 8Mbps/1Mbps
DS-1 ("T-1") 1.544Mbps
DS-2 ("T-2") 6.312Mbps
USB 1.5–12Mbps
SCSI-1 and SCS1-2 .... 4Mbps
Ethernet (10Base-T) 10Mbps
Fast SCSI 10Mbps
Apple AirPort wireless LAN (IEEE 802.11 HR DSSS) 11Mbps
Fast Wide SCSI and Ultra SCSI 20Mbps
DOCSIS Cable Modem (*max* download/upload) ..40Mbps/10Mbps

SCSI-3 and Ultra2 SCSI...40Mbps
Fast Ethernet (100Base-T)...100Mbps
IEEE 1394 ("FireWire").................................... 100–400Mbps
Gigabit Ethernet 1.24Gbps
FibreChannel 266Mbps–4Gbps
Optical Fiber (OC-48 through OC-255)............2.488–13.21Gbps

Mass Storage Transfer Rates (MBps x8):

CD-ROM (1x) 1.3Mbps
DVD-ROM (1x) 10Mbps
Zip disk...9–19Mbps
Jaz disk 64Mbps
Hard disks..80–800Mbps

If you want to maintain a given image quality on, for example, MiniDV, with 25Mbps, then your cables and storage devices will be required to manage at least 25Mbps to provide real-time playback of this material. Clearly, you cannot shoot on MiniDV, transfer to your new multi-gigabyte drive and expect to play the video, when the drive is connected to your PC by a SCSI-1 cable. *It just isn't going to flow.* Similarly, you can clearly see why compression is so critical to play any video from the Internet when a typical modem runs at under 56Kbps.

To achieve real-time effects, a system must be able to play two or more "streams" of video simultaneously: multiply the image bandwidth by 2 (or more) to see what demands this functionality places on a computer.

For cable, network, and modem rates, realize that what is *technically* feasible and what a typical user will experience in the real world are often megabits apart. Cable modems, for instance, might have a maximum downstream bandwidth of 40Mbps, but actual users may only be able to achieve 750Kbps due to the nature of the network or other limitations. Keep maximum, theoretical, or ideal world values in check with actual user's experiential data.

MULTIPLES of 1000	1,000 (thousand) ....10^3	=	kilo-	(K)
	1,000,000 (million)..10^6	=	mega-	(M)
	1,000,000,000 (billion).....................................10^9	=	giga-	(G)
	1,000,000,000,000 (trillion)................................10^{12}	=	tera-	(T)
	1,000,000,000,000,000 (quadrillion)....................10^{15}	=	peta-	(P)
	1,000,000,000,000,000,000 (quintillion)..............10^{18}	=	exa-	(E)
	1,000,000,000,000,000,000,000 (sextillion).........10^{21}	=	zetta-	(Z)

VIDEO COMPRESSION, BANDWIDTH and STORAGE

Compression ratio relative to uncompressed HDTV	SDTV	Format	KB/frame (approx.)	MB/sec (@30fps)	Mbps	Storage/GB (min:sec)
100x		70mm IMAX (15-perf)	400,000			0:00.1
9x		35mm film (4-perf)	38,000			0:01
1:1		**Hi-def 1080i uncompressed**	4,050	120	950	0:08
2:1		Hi-def 720p uncompressed	1,800	53	420	0:19
		PAL	800	20*	160*	0:51
6:1	1:1	**NTSC** (480i) UNCOMPRESSED "D1" or CCIR / ITU-601	680	20	160	0:51
	2:1	Digital Betacam	375	12	95	2:00
	3:1	M-JPEG: "AVR 77" BetaSP				
	4:1					
	5:1	DV, MiniDV	125	3	25	4:30
	6:1					
50:1	8:1	MPEG-HDTV (1080i, 720p)	80	2.4	19	7:00
	12:1	S-VHS or Hi-8	50			11:20
	16:1	MPEG-SDTV/DVD (480i)	40	1.2	9.8	15:00
	20:1		30			19:00
	30:1	M-JPEG: "AVR 4" VHS	20			28:00
	120:1	DSL / T-1	6	0.2	1.5	80:00
	500:1	STREAMING VIDEO 300Kbps stream	1	0.04	0.3	450:00
	2500:1	56K modem	0.2	<0.01	0.05	2000:00

NOTES: This table represents compression on the vertical scale, **not image quality**. Picture quality varies depending on the compression scheme used, and not simply based on the resulting amount of data per frame; for example, MPEG looks better than Motion-JPEG at the same bit rate.

Formats listed in gray are approximate digital equivalents for analog media.

* PAL has more scan lines (and hence, more data) per frame than NTSC does. But with fewer of those frames per second (25 instead of 30), it amounts to the same overall data rate as NTSC.

P I X E L S

"SQUARE" VS "NON-SQUARE"

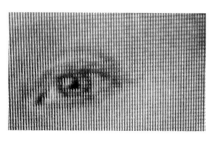

As you know, a pixel is the fundamental picture element of digital images. Analog TV monitors technically don't have pixels, but red, green and blue dots which, in groups of three, approximate a picture element. But for the moment, let's discuss both analog and digital displays as having pixels. This is, for all practical purposes, the case.

If you design a bitmapped graphic (for example, in a desktop application like Photoshop), and then import it into your video editing application, the image will generally end up looking "squished." The problem represents one of the fundamental differences between computer video and broadcast video—each uses pixels that are different shapes.* Why? It's because the two types of video come from different pedigrees—computer display resolutions (640x480, 800x600, *etc.*) use nice, round numbers that are even multiples of 4x3. NTSC and PAL video gear, on the other hand, is based on the different "D1" (or "ITU-R BT.601") standard for professional video, which uses awkward numbers like 720x486 (for NTSC) and 720x576 (for PAL) that aren't really multiples of anything.**

Some savvier graphics applications (After Effects, for example) understand these "non-square" pixels and will handle the conversion work for you, but many don't. In that case, you'll need to compensate somehow, either when exporting from the graphics application, or when importing into the nonlinear editing system. Since the particulars will vary depending on the system in question, you'll want to check the documentation for your software, or ask a knowledgeable user. Knowing that there is such a difference in the first place, though, means you're already halfway to mastery of pixel nightmares; it also will give you an appreciation for the

---

* OK, technically, pixels aren't really shapes at all, but infinitesimally small points spaced evenly across the screen. But since the practical upshot is the same, we're going to overlook this little detail.

** 720 pixels across was chosen as a middle-of-the-road compromise, in order to create a single standard that could be used to handle both NTSC and PAL using similar digital hardware. The result? A number that's pretty non-intuitive in both cases.

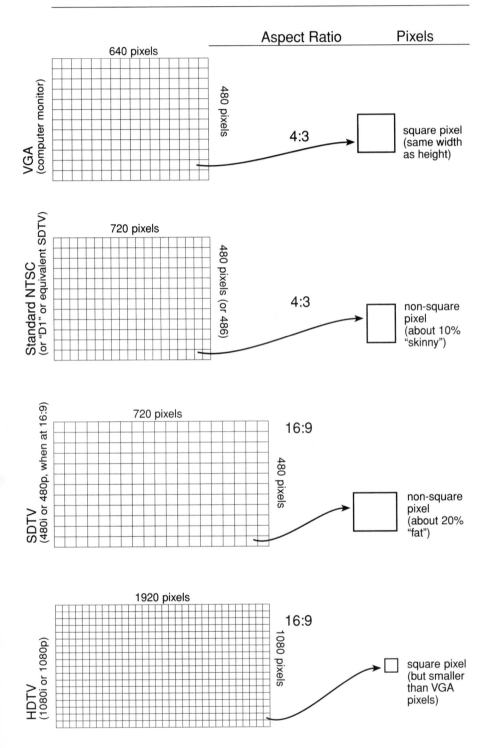

Aspect Ratio   Pixels

VGA (computer monitor)
640 pixels
480 pixels
4:3
square pixel (same width as height)

Standard NTSC (or "D1" or equivalent SDTV)
720 pixels
480 pixels (or 486)
4:3
non-square pixel (about 10% "skinny")

SDTV (480i or 480p, when at 16:9)
720 pixels
480 pixels
16:9
non-square pixel (about 20% "fat")

HDTV (1080i or 1080p)
1920 pixels
1080 pixels
16:9
square pixel (but smaller than VGA pixels)

In graphics application

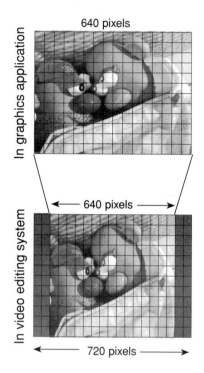

640 pixels

In video editing system

←— 640 pixels —→

←——— 720 pixels ———→

work being done on your behalf by those developing standards for digital television.

The organizations and individuals who created DTV were keenly interested in a truce between the computer and the broadcast video worlds (the much-used rallying cry is "convergence!"), so they've simplified things once again by using nice, square pixels for HDTV.

If you take a closer look at HDTV, notice how much smaller those pixels are, as compared to standard TV pixels. That's precisely the idea, of course: smaller pixels mean finer detail.

The other issue with regard to pixels is that size measurements cannot be made based on absolute dimensions (inches, points, *etc.*) on a display because the actual dimensions are in pixel units. 24-point type in a word processing program may be some other size on a VGA display; it depends mostly on the size of the display and its resolution. Another time this is particularly noticeable is in frame sizes for video when played on different computer monitors. A 640x480 video window (a professional resolution for broadcasting or graphics) may fill a TV monitor, but on a typical 1024x768 pixel computer monitor, it only

takes up about 60% of the window; it is even smaller on a high-resolution monitor. At right are two examples of 4:3 displays; notice the significant discrepancy in the relative "size" of a 640x480 window. Also

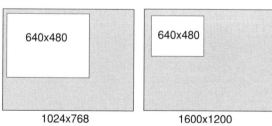

640x480

1024x768

640x480

1600x1200

realize that streaming web video that plays on computer displays is often small: 320x240 would be a quarter of these imbedded windows. The impact of these variations on framing, titles, and even editing itself is important.

A S P E C T   R A T I O   B A S I C S

The shape of a frame, whether film or video—analog or digital—is traditionally described by the ratio of its width to its height. This ratio is called the **aspect ratio**. Television (whether NTSC or PAL) is 4:3, which means it is four units wide by three high. Therefore, if an image is 10 inches long, the height would be 7.5 inches; if it is 640 pixels long, it would be 480 pixels high. Another way of writing this ratio is 1.33:1.

Traditional film as projected on a big screen typically has an aspect ratio of about 16:9 (actually 1.85:1). This is far more "horizontal" than television. Some say film was originally presented in this wide format because it was more like the proscenium of a stage, as you would see it from the audience during an opera, for instance. Regardless, the issue of aspect ratio becomes important when one wants to play a movie on TV, but it's the wrong shape. It's also an issue when a director shoots DV, for example, and hopes to transfer it to film. Or when producing a TV show to be seen by audiences in both NTSC and HDTV. Aspect ratio is a hot topic when the subject of a digital, universal master format is discussed.

Here are the two principal aspect ratios for moving pictures. The darker, more "vertical" rectangle is 4:3; the grey horizontal rectangle is 16:9.

If you put a 16:9 image on a 4:3 display, it will only fit if you shrink it down about 25%

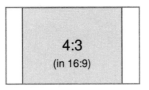

and leave blanks along the top and bottom of the screen. Similarly, a 4:3 image will only fit on a 16:9 display if you shrink it down and leave blank spaces along the sides. Thus, there are compromises that must be made in order to convert material from one format to another.

Methods

Shrinking the image down and leaving bars at the top and bottom of the screen is called *letterbox* format; it's the most common way to transfer film to tape, without having to modify the image *per se*.

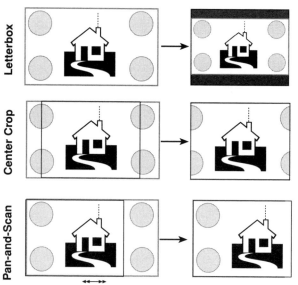

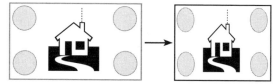

But if you decide to avoid the bars and simply crop the image, and ignore any aesthetic considerations, you might perform a *center crop*—which can be remarkably unsatisfying because it chops off characters and action not in the middle of the screen.

In practice, the most common technique, performed in telecine and nonlinear online sessions by Hollywood studio clients, is called *pan-and-scan*. In this case, the operator moves the 4:3 frame around on the 16:9 image, keeping the most important material in-frame at any given moment.

Finally, you could convert shapes by using an *anamorphic* compression (meaning that the image is squished along one axis but not the other). But people tend to dislike having their images squished. Research has shown that a squish of 5% is undetectable, and 7% is detectable, though not objectionable. But these manipulations are a far cry from the 25% squish or stretch needed to convert these aspect ratios fully.

For the most part, going the other way (changing a 4:3 image to 16:9) involves variations on these conversion methods—called *pillarbox*, *middle cut*, and *tilt-and-scan*.

With frame conversions in both directions, newer technology makes it possible to mix methods. For example, a *little* letterbox plus a *little* center crop, plus *a tiny bit of* anamorphic squeezing can all be used together as a compromise, better approximating the original source. These **blends**, as they are called, reduce the necessity of having to make aesthetic judgments for each individual scene. All methods will continue to be needed as long as 4:3 and 16:9 coexist, in their uneasy truce, in the name of convergence.

LIGHT and COLOR

BASIC PRINCIPLES

It is probably foolish to attempt to explain the basics of light in an introductory book such as this; the topic is enormously complicated and profoundly technical. Still, everything we do in production and post-production is about light: understanding optics and light makes it possible to shoot, light scenes, create moods; light hits CCDs and gets encoded using only red, green and blue sensors. These coded light instructions get compressed, in analog or digital, into luminance and chrominance and ultimately, we watch video on monitors or as projected film, and our brains perceive color and motion. We even debate how much of the spectrum to allocate to HDTV broadcasting. It's all in a discussion of *light*.

Furthermore, we make our job more challenging by trying to discuss basic issues of color in a monochromatic book. But, as simply and as straightforwardly as possible, we will outline a few principles that relate to light that should be useful and provide a good foundation for deeper inquiry.

ER

Electromagnetic radiation (ER) is a form of energy that is radiated from many sources in the universe, but for us on earth, it comes primarily from the sun. In the vacuum of space, this radiation moves at a constant velocity (written as *c*). Scientists have two separate ways of describing this ER, the *quantum* theory and the *wave* theory. Suffice it to say here that, as a wave, this radiation may have a frequency from very slow to very fast, creating waves that are very long or very short. *Frequency* is measured in cycles per second of the wave (unit: Hz); the associated *wavelength* of the ER can be a kilometer or longer; it can be a meter or a micron or a tiny fraction of a micron. The sun radiates a broad spectrum of wavelengths that bathe the earth with ER.

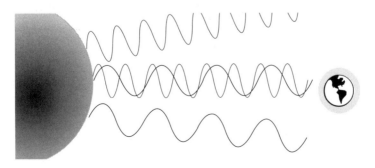

While all kinds of ER reach the surface of the earth, the wavelengths between 400 and 700nm (*nanometer* = a billionth of a meter) reach us in high quantities. Most lifeforms on earth developed ways to sense these wavelengths, and some living things (*e.g.*, plants) harness some of them for energy.

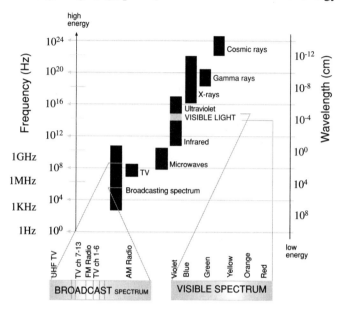

Human evolution led to the development of ER sensors in our eyes: the **rods** and **cones** in the retina.

Seeing

Humans perceive wavelengths between 400–700nm as "visible light." When the wavelengths of ER from 400–700nm mix together in their naturally occurring proportions, we see colorless light we call "white." A wedge of glass (a *prism*) can break up this white light and reveal the individual colors we associate with the differing wavelengths. On the short end (~400nm) there is blue-appearing light; at the long end (~700nm) there is red. When we combine the wavelengths of light appropriately, we can re-create white light. This is an **additive** property of colors.

Back in the retina: It is the cones that enable us to see colors (rods only "see" black and white and at low light levels). Cones come in three varieties: those with a peak of sensitivity to ER in the 400–450nm (blue) range, those that are most sensitive to the 480–525nm (green) range, and those tuned to the 575–650nm (red) range, give or take. When patterns of these rods and cones fire and send signals to the brain, it is the brain that produces a perception of light and color. There is nothing about the ER itself that is colored. *Color is a subjective sensation*.

Colors are such familiar "properties" of the objects around us that it comes as a bit of a shock to be told that colors are subjective, figments of the mind. There is no color *in* a "colored light." No color *on* a "colored object." There is no "color vision" in the sense there is a "size vision." Size is a real physical property of objects and their relationship to each other. We have no vision *of* color, but we have vision *with* color.

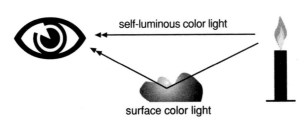

self-luminous color light

surface color light

Properties of self-luminous color light are the same whether the light goes through a filter before reaching your eye or not; surface colors are complicated by the properties of the object and the light's interaction with them and with surrounding objects.

A color, then, is a particular kind of sensory experience—a response to a stimulus. Color is complicated to discuss because there are, for all intents and purposes, two modes of appearance of color stimuli and each has somewhat different properties. One mode is a **self-luminous** color (like a light); the other is a **surface** color (like paint).

Both self-luminous and surface colors can be described using three *dimensions*, and for each a sensory scale can be established with a given color standing higher or lower on this scale. Two of these dimensions apply to all colors: *hue* and *saturation*. Hue is the *coloredness* of something, like "red" or "blue." Saturation is the intensity of the hue; it is the *concentration* of color. *Brightness* is the third dimension, but only applies to self-luminous colors, which can be anywhere from bright to dim. Surface colors have no inherent brightness; the third color dimension for them is called *lightness,* which indicates a color is lighter or darker than its surroundings. Lightness is a relative term.

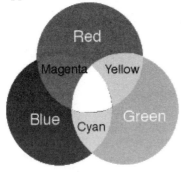

Red

Magenta   Yellow

Blue   Cyan   Green

Perceptually, we can describe colors we see as combinations of hue, saturation and brightness, or of hue, saturation and lightness. But physiologically (and technically), we can create colored lights by mixing ratios of the three *additive* primary colors: **red**, **green** and **blue**.

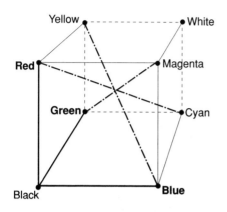

Color space is the gamut of colors that the eye can perceive, based on the relative firing of the retinal cones. If we graph the human perception of colors with axes for each of the red, green and blue cones, we see a rendering of color space that can illustrate a few properties of colors. While traditionally color space is drawn as a cylinder (of hue, saturation and lightness), it can also be represented as cubic, as it is at left.

Wavelengths of light are absorbed by some things and reflected by others, although most objects do a little of both. Look at the cube illustration. If some object is hit with "white" light, and absorbs the red wavelengths, what will be reflected for a human to see? white − red = *cyan*. If something absorbs red and blue, *green* will be reflected, and therefore *seen* [white − red − blue = green]. FYI: Chlorophyll in plants absorbs ER in the 400–500nm range (blue), and a little more around 650nm (red). The rest—the energy in the 500–650nm range (green) is reflected. This is why most leaves look green.

Colors have "opposites," that is, the color on one corner of the cube is considered the opposite (or *negative*) of the color in the corner farthest away from it. Thus, red is opposite cyan, yellow is opposite blue, white is opposite black. This can be written mathematically as R = − C and Y= − B (or B = − Y). A light color plus its opposite will always make a non-colored (*e.g.,* "white") light. Opposite colors, which when mixed together become colorless, are called "complementary."

The way colors mix together to yield other colors depends on whether one is dealing with colors of light (self-luminous) or colors of paint (surface colors). *For our purposes here, we are mainly concerned with self-luminous colors which emanate from video, computer screens and light bulbs.*

Generating Light

In addition to ER from the sun, visible light can be radiated from some objects. The two ways an object may emit light are called *luminescence* and *incandescence*.

Luminescence occurs when an object produces visible light from reacting with electricity. When certain gasses have electric currents running through them, the reaction of the electric current with the electrons in the gas's electron "shells" causes energy to be released as light. This is the functional basis for neon lights, cathode ray tubes and fluorescent home fixtures. (By the way, *fluorescence* is a kind of luminescence.)

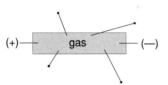

In luminescence, energy is released as light when an electric current "excites" the electrons in a gas.

Incandescence is created by heating (*technically*: molecular agitation) an object. When solid (or some liquid) objects are heated, they may produce light (until they melt or burn up). When you put a black iron poker in the fire, as it heats up it starts to glow: first reddish, eventually getting "white hot." This is incandescence. The visible energy output in this light is expressed as a *temperature*—in degrees Kelvin (273°+ degrees Centigrade).

Theoretically, if you heated an imaginary block of perfect carbon until it got "white hot," the temperature would reach about 5,500°K. This is very hot. Recognize that nothing on earth can really get this hot, as most things would vaporize before they reached this temperature. The surface of the sun (the "photosphere") is around 6,000°K. But we are talking theoretically. And in theory, the block of carbon will glow *reddish*-white as it gets up to around 3,000°K. At around 6,000°K the light would look about as *white*-white as a person would like. If you could heat it more, to 8,000°K and 10,000°K, it would look *blue*-white. The hotter the solid, the more short wavelength (bluer) light is emitted.

This theoretical heating has lead to a practical way to describe the color or shade of a white light—that is, we denote its *color temperature* (the temperature of the theoretical carbon block). It does *not* mean that a light bulb is burning at 3,000°K. It is important to know that the color temperature scale can be applied to *any* light, not just incandescent lights.

Incandescence is about how objects glow when heated. Although incandescent light bulbs glow because the electricity is turned on, the glowing is caused by the heating of the filament by the resistance in the wire.

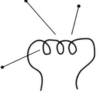

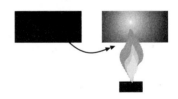

A color temperature of 5,500°K will look white, but it is still slightly yellowish; this is fine for black and white photography because 5,500°K makes for lighting that produces warm natural flesh tones. Tungsten light bulbs radiate light with a color temperature of 2,700°K, giving them a decidedly yellowish tint (tungsten illuminated objects cannot yield good color rendition if recorded on "daylight" color film). TV standards organizations settled on 6,500°K as "white" for NTSC signals. (6,500°K is the color of natural light on a somewhat overcast day.) Color TV sets, however, have pre-set white settings of the bluer 9,500°–11,000°K, which you might notice when you see the bluish room illumination of darkened homes watching TV in the evening.

But remember: color temperature has nothing to do with *actual* "temperature" (in radiated heat energy) from a light source. Thus, a 2,700°K tungsten bulb may be yellowish, but you can change its color temperature by inserting a colored filter (*gel*) between the bulb and the object you want to illuminate. A blue gel on a tungsten bulb will *raise* the color temperature of the light even though it does nothing to the actual temperature of the filament.

Though the human brain will adjust automatically to see white as "white"over a broad range of the color temperatures, film and video do not, as they capture the red, green, and blue proportions of the light in the scene. It is important to force a video camera to identify a given color temperature of white as "white" so that it can skew the other quantities of RGB to make things "look" right. This critical action is called **white balancing**.

While we believe that humans likely "see" colors pretty much the same, we can never really know for sure if two people perceive the same color in the same way since color is a private subjective experience. Moreover, even a fixed color can change in appearance. A red card is a red card is a red card (to paraphrase Shakespeare), but not if it's located next to a black card one time, a green card the next, or illuminated by a light bulb the third time. Colors look different depending on the colors around them. So "getting a color right" may be a somewhat futile exercise. This is an important caveat in our discussion of color "accuracy," particularly when debating the relative experiences of various electronic image compression schemes. One can devise objective tests to measure important aspects of a color signal, color bars, voltages, *etc.*, but the truth is that the perception of the faithfulness of color reproductions (in video or film or on computers), is pretty much subjective. Period. If it looks good, it *is* good. There is no other important test.

ER + Sound Waves = Radio

Sound waves, compressions of the molecules in air that we perceive as sound, are *not* like light waves (ER). Sound moves far slower than light; sound cannot travel in a vacuum. But when we want to transmit sound waves farther than they might naturally travel we only have a few choices.

We could simply make the sound waves with higher power (louder) so they will travel farther before their energy dissipates. This is the nature of loudspeakers, amplifiers, and, well, yelling. Or, we could encode the pressure waves into another medium for transport, a medium that travels better. This is the nature of broadcasting.

Sound can be encoded ("modulated") *onto* ER waves (the "carrier wave") either by fixing the ER amplitude and modulating the frequency, or fixing the ER frequency and modulating the amplitude. This second choice, *amplitude modulation* or AM, uses an ER carrier wave at a fixed frequency (let's say 1500KHz) but has a small dynamic range of sound, among other limitations; the alternative, *frequency modulation* or FM, uses a small *range* of ER carrier wave frequencies (200KHz wide—although commonly named for the frequency at the midpoint of this range, like 94.5MHz). FM not only has better dynamic range of sound, but can be encoded to include pairs of stereo channels.

The broadcast of video, like the broadcast of audio, encodes the signals onto ER carrier waves, spanning a small range of the ER spectrum (6MHz wide), and then modulates sound using FM techniques and picture using AM techniques. Both together make up a broadcast video signal. What is important here is that while sound is described with amplitudes and frequencies that can look a great deal like the descriptions of ER, it is only because both have wave properties; even though ER can be used to transmit sound, ER and sound waves are not the same thing.

Thomas Edison, the inventor of the lightbulb (1879), heated up thin wires by passing a current through them. He was looking at what materials incandesced better than others. The story goes that he tried 6,000 different materials before he settled on the carbonized cotton thread filament that generated the light in his early bulbs.

T Y P E

A Crash Course in Letters and Titles

Hopefully, you begin this section already quite familiar with **fonts**. But let's review a couple of key concepts that will help you with many of your video titling needs. First know, just for the record, that fonts are measured in *points*. A point is 1/64th of an inch. Now, here are two sentences; the first is written in a *serif* font (Times), the second in a *sans serif* font (Helvetica).

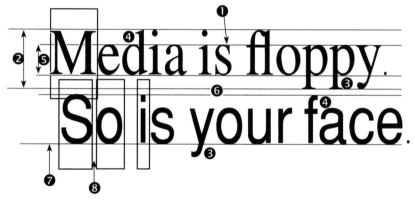

Serifs (1) are little tips that finish the ends of the main strokes of some letters. Serifs have been shown to increase legibility in printing. *Sans serif* fonts (literally "without serifs") are good for short lines and headlines.

Font size (2) is an unfortunate term, which is more relative than absolute. It is *roughly* the distance from the bottom of a **descender** (3) to the top of an **ascender** (4). Relative font size can be determined by **x-height** (5) which is height of a lowercase "x." Still, two fonts that are the same font size often do not look the same size. This book is set in 11-point type.

Leading (6) is the space between lines of type. It can be measured simply as the number of points between lines (measured from the bottom of the descenders on one line to the top of the ascenders of the next line). 10–20% of the font size is considered nice leading. Leading here is 13.2 pt.

Baseline (7): The line the text is "typed" on.

Letter space (8): Space between letters, which can be tight (with letters touching each other) or very loose, as in the headline "TYPE" above. The adjustment of letter space is through **kerning**, which can be done automatically or manually.

Font Physiology

Readability, important in the print world, is crucial in video. Unlike type in books, on-screen text is only seen for a brief moment—the viewer is under a time pressure to read it. Then, too, video presents the text in low resolution (as compared to print), and frequently with a host of competing color interference and backgrounds. This is why font size, kerning, sharpness, contrast, leading, and so on, are important to successful use of on-screen type.

On-Screen Type Guidelines

1. Make sure the text is present long enough for slow reading. It is not that readers are necessarily *slow*, but they may not have been expecting to read—a title or caption may pop up while they were looking at another part of the screen. Viewers need time first to pick up that there *is* text, then have time to read it.

2. Keep it short. Be concise. It is not easy to read on screen, either online or on TV, and people tend to dislike it. It's a caption, not a novel.

3. Font size. Just because you can easily read 12-point type in a book does not mean it will play on video. *Points* are virtually useless measurements in video; the only thing that matters is pixels. A graphics monitor may display 100–150 pixels per inch (there are 64 points to an inch). Experiment.

4. Contrast. Any typographer can tell you that it is harder to read white on black than black on white. True in print but *not* in video. *Keep text light on dark backgrounds.* Use a soft *drop shadow* on light letters to maintain contrast, particularly over changing video backgrounds. Try to avoid harsh color contrasts that may be difficult to *see*: red on green, for instance, will make you go insane.

5. No jaggies. Ensure that any font you use is *anti-aliased* with the background, and is generated from a font style that has screen/outline fonts, like PostScript. Bitmapped fonts with jaggy edges are unprofessional and hard to read.

6. No serifs. If the font is large enough, the little decorations will likely show okay; but in general, and particularly if the text will end up going through a video-to-film conversion, avoid serifs. Similarly, some font styles have letters with strokes that are thick in some places and thin in others. Thin strokes, particularly in small fonts, can be hard to read on screen.

Bitmapped and Outline Fonts

With all the talk of compressed and uncompressed, digital and analog, there is yet another way two graphical files can differ; it has to do with the nature of the data. It comes down to this: bitmapped (raster) or outline (vector).

Bitmapped images are composed of data that describes the pixels, with color values (RGB) for each of the 640x480 (or whatever) number of pixels in the frame. **JPEG, TIFF, GIF, PICT, TARGA,** and **Photoshop** are all bitmapped file formats that are concerned with the pixels *in* the image. Consequently, when you see an image on a display of some kind or want to print it on your printer, you will have concerns about image resolution, lines per inch, all that stuff. Like on the Internet.

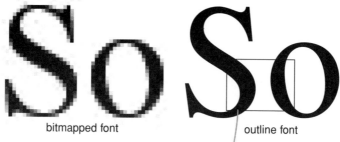

bitmapped font   outline font

On the other hand, drawings and illustrations can be made with mathematical tools that draw lines (*vectors*), make shapes, and fill in the shapes with colors or gradations. This is *object-oriented*, and the basis for Adobe's PostScript language. Objects (or drawings or fonts) in this format are described with formulas and can be printed at any resolution. This is why they are also called *resolution independent*. Instead of drawing a series of pixels between points A and B, and remembering each pixel's color, a vector remembers only where point A is and where point B is, relative to each other and to the frame. It also knows to "draw a line between them" when asked. No matter what resolution the display, or how enlarged the graphic is, the line will be generated crisply in the best resolution possible. **EPS** and **Illustrator** files contain PostScript vector file formats.

A close-up of an outline font can reveal the spline-based curves that make up the shapes

When importing images into editing or compositing workstations, recognize the quality-advantage of object files, and the resolution issues of bitmapped files. This is also true in webpage design, where there is the trade-off in download time and legibility between the native fonts on the browser's PC, or bitmapped illustrations of text (*e.g.*, JPEG, GIF) that make a site look a certain way.

Chapter 4

Editing
Primitives

"Editing is not so much a putting-together
as it is the discovery of a path."

—Walter Murch (1996)

Editor: Apocalypse Now, The English Patient

WHAT IS AN EDITING PRIMITIVE?

The term *editing primitive* derived originally from computer programmers designing editing systems. Before a program (software) can be written, the system's designer needs to identify what the system will do and how it will do it. These are the "primitives." There are only so many things an editing system needs to do. Editing is pretty simple: cutting and pasting. There are variables, and there are a number of computer functions and database manipulations, but most editing systems deal with similar concepts in pretty much the same way.

Although it seems that every editing system is unique, they all share certain commonalities that once understood need only be "translated" in order to switch from system to system. Once you know one editing system, it becomes considerably easier to learn others.

The entire digital system metaphor—two windows, a timeline below, a bin for source/sequence material, and tools for marking in/out points, shuttling, and performing a variety of pretty standard edits—is common to almost all systems. It is important for editors to recognize these primitives.

In product literature it is not uncommon to find many of the functions listed below presented as if they were unique to that particular product (like "nonlinear," "frame accurate," *etc.*). In reality, most systems are:

- ☑ Flexible—can be used on projects from home videos to television shows.
- ☑ Fast—at least faster than cutting on either film or tape.
- ☑ Simple—with simple "intuitive" interfaces for the computer-savvy.
- ☑ Cost-effective (if used correctly).
- ☑ Designed with 2–8 separate channels of audio input/output.
- ☑ Able to trim or extend edits numerically.
- ☑ Able to undo and redo most edits/functions.
- ☑ Able to handle drop- and non-drop timecodes.
- ☑ Able to do film- ("ripple") and video-style ("non-ripple") edits.
- ☑ Able to copy an edited sequence (like creating dupe-workprint).
- ☑ Able to control a videotape deck for digitizing and batch-digitizing.
- ☑ Able to output the master cut to videotape.

In many cases, the key difference is simply the interface: how does it *look*? how does it *feel* to use? how do *you* feel when you use it? can you work the way you prefer to work? There are pretty interfaces that suck; and good interface design is more than having an "exciting new look."

" E D I T I N G   T I M E "

If you have never edited "the old way" you will have no concept of time *savings* using nonlinear equipment. You can only save time relative to what you expected to spend using alternative methods. It is perhaps time to stop discussing nonlinear time savings. One might ask *how fast can you edit using nonlinear equipment?* as if speed was a goal. Unfortunately, if not a goal, speed is still an evaluation metric for equipment. Expensive hardware can produce more "real time" effects, which might take many times longer to execute using a personal computer. High resolutions may cause actions to take longer than low resolutions. But editing itself, the art form of telling a story with pictures and sound, the craft of cutting lots of material into less material, can be done practically at the speed of the ideas themselves. But those ideas do take time, and rushing them will likely decrease the quality of the output.

There are four noteworthy time nozzles to get a project completed:

(1) Logging, capturing/digitizing

If material requires a film finish, whether for HDTV telecine or some contractual obligation, the timecodes on the source tape must be related to film numbers. Telecine logs facilitate the process, but regardless, getting a tape log in order, and organizing the files on the editing system, all take time and thinking. For video finishes, timecode tapes are all that you need, and little logging is actually required beyond the work in the capture.

For most systems, capturing is a real-time venture. Whether a 30-minute DV tape or 90 hours of telecine film transfers to DigiBeta, all material must get fed into the editing system. It is true, however, that video is simply digital data and input shouldn't *need* to be like a video dub, and could be more like a file transfer. Expect more and more equipment to provide for faster-than-real-time editing input as the decade progresses.

(2) Organizing

All captured files are best managed by organizing into proper bins with proper names, and a little time spent in this organization will be highly leveraged in more efficient cutting (and avoid the all-too-common loss of a shot "you knew was there somewhere...").

(3) Editing

To edit you must watch source material, *all of it*, and think about how to put the pieces together. The time it takes to edit varies widely, depending on the complexity of the project (how many edit decisions are there?, how much footage must be pored through?) the editor's experience, and a few key aspects of the process (how many times do you want or need to go back over the work to refine it?). As they say, editing is all about *re-editing*. Once a first cut is done, re-editing (and refining) can continue until someone says *stop*. Different production genres tend to allot themselves different amounts of time before locking picture.

A feature film and a TV movie (called an *M.O.W.* for "movie of the week") may both have 200,000 feet of film, but the feature could take twice as long to lock as the MOW. Even with roughly the same number of edits. A 5-minute music video might take as long to cut as a 1-hour episodic television show.

(4) Output

Aside from any intermediate "approval" versions, when the project is done (and "locked"), the editor or assistant will perform an assembly out of the system in real time. It's important to factor in some initial set-up time (patching, checking levels, bars and tone). The process may also involve making a number variations of the edited master from the system (a "texted" master, a "textless" master; "mixed" audio, "split" audio). Lastly, there is a little time needed to *spot check* each tape.

Editing Time Approximations

An experienced editor, particularly one working with a clear script, can do an initial pass through long-form material at a rate of around one reel per day (or roughly a minute of a edited narrative per hour of editing). That is, as long as everything is already in the computer, well-logged, and the cutting is in scene order (*i.e.,* from scene 1 to scene 100). This is enough time to produce a reasonable first cut. No effects. No compositing. Simply racing through to put a narrative film together. Cutting along with dailies limits how much you *can* cut each day. Low-budget film productions might schedule the picture edit for 5–10 weeks; television movies for 4–8 weeks.

A 30-second network commercial, with its 900 frames, might be allocated two days to edit, or a week. In commercials every frame will be scrutinized; the shooting ratio itself is extraordinarily high—four or more hours of material for 30 seconds of commercial is not unusual.

Music videos run about 5 minutes in length, and require little (if any) sound editing. The pacing can be slow and narrative (like a movie) or fast and furious (like some commercials). Anywhere from three to seven days is reasonable.

So, how long does something *really* take? It takes until you are done. It has little to do with the equipment.

E D I T I N G   F L O W C H A R T S

THE POST-PRODUCTION PATH

The familiar method of working on film is known by most people in the film industry. The fact that this knowledge has been so well established for so long is what makes changing it so challenging.

However, electronic tools can be inserted into the post-production process in a number of different ways, and some of them barely change the flow at all. It is also possible to *entirely* revamp traditional means of posting film, and utilize all digital equipment. The degree of change is up to each production.

Following this page is a series of flowcharts. The first (#1) shows steps necessary to prepare a film-shoot for an digital editing session; the second (#2) depicts finishing a project after edit decisions are complete and picture is locked. The process is highly different when a film shoot is to end with a video delivery, as shown in the next flow chart (#3). And different again (#4) if the delivery is both SDTV and HDTV formats. The charts are generalized to present options for editing on any number of digital systems. Remember, these are only some of many variations that can be used; television, theatrical feature films, low-budget independents and network commercials will all modify this process based on the desired end-product, budgetary issues and creative preferences. The key understanding is that a production or facility will augment and invent the process based on what equipment is available and the specific needs of the project. For example, the HDTV finish could have begun with an HDTV telecine and then an HDTV online—much like the SDTV finish depicted (#3); however costs of these equipment are high enough that SDTV telecine, film cutting, and other seemingly unnecessary steps significantly reduce expenses. Economics as well as technology drive these pathways.

A fifth flow chart is included here (#5) that shows a very simple pathway for a DV or MiniDV *online* on a (low cost) nonlinear system. The system can perform all the effects and editing at DV resolution, output to a DV *master*, which can be delivered for compression for online streaming, for upconversion to a SDTV broadcast format (Digital Betacam, if needed), or even put into a video-to-film transfer—for film festivals, for example. The DV master could also be projected itself, in any number of video projection formats. This flow is also applicable at higher-than-DV resolutions, but for *low-infrastructure* online editing.

1. Originate on Film, Finish on Nonlinear Editing System

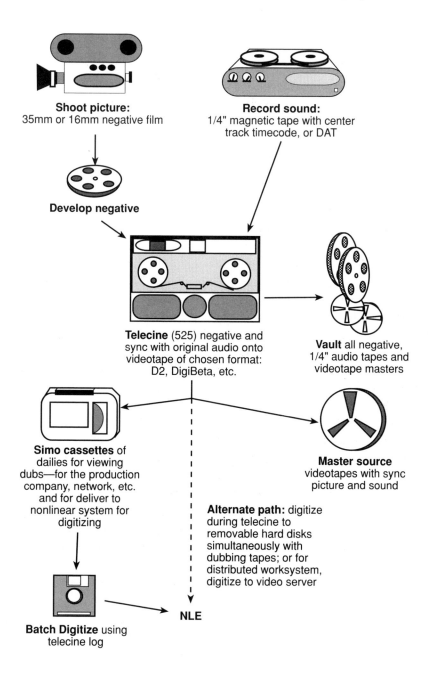

Shoot picture:
35mm or 16mm negative film

Record sound:
1/4" magnetic tape with center
track timecode, or DAT

Develop negative

Telecine (525) negative and
sync with original audio onto
videotape of chosen format:
D2, DigiBeta, etc.

Vault all negative,
1/4" audio tapes and
videotape masters

Simo cassettes of
dailies for viewing
dubs—for the production
company, network, etc.
and for deliver to
nonlinear system for
digitizing

Master source
videotapes with sync
picture and sound

Alternate path: digitize
during telecine to
removable hard disks
simultaneously with
dubbing tapes; or for
distributed worksystem,
digitize to video server

NLE

Batch Digitize using
telecine log

2. From Nonlinear Editing System to Film Finish

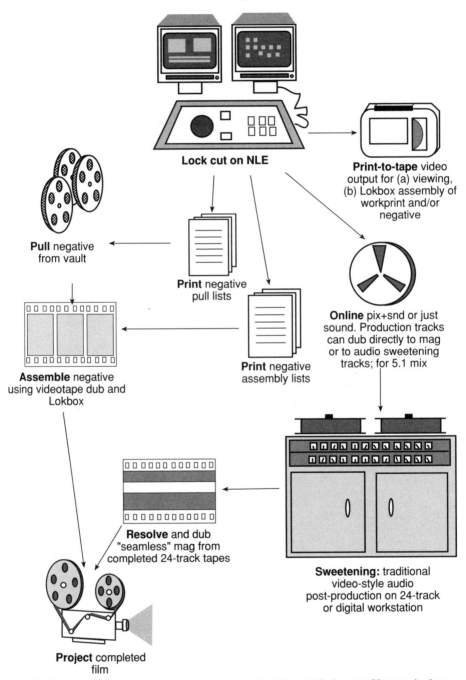

Lock cut on NLE

Print-to-tape video output for (a) viewing, (b) Lokbox assembly of workprint and/or negative

Pull negative from vault

Print negative pull lists

Online pix+snd or just sound. Production tracks can dub directly to mag or to audio sweetening tracks; for 5.1 mix

Assemble negative using videotape dub and Lokbox

Print negative assembly lists

Resolve and dub "seamless" mag from completed 24-track tapes

Sweetening: traditional video-style audio post-production on 24-track or digital workstation

Project completed film

NB: This path assumes a cost savings by post-production without workprint costs. Many productions will choose to print workprint, sync traditionally with code numbers, and telecine workprint. In that case, this path ends with assembled workprint, which can then be used to cut negative.

3. Offline Nonlinear Editing System to Video (SDTV) Finish

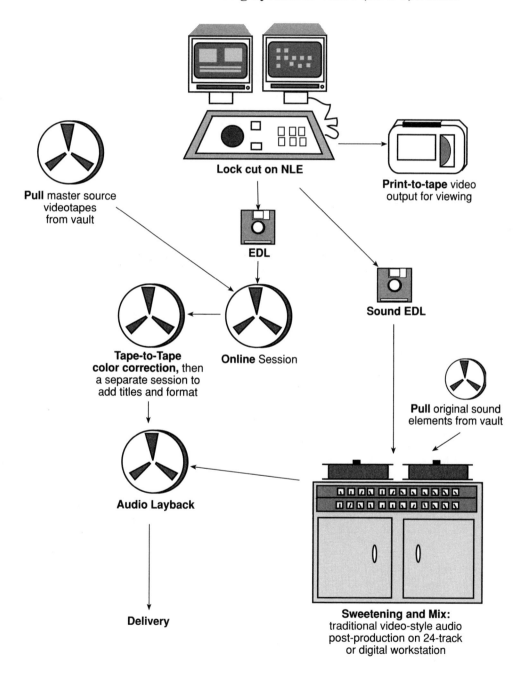

Lock cut on NLE

Pull master source
videotapes
from vault

Print-to-tape video
output for viewing

EDL

Sound EDL

**Tape-to-Tape
color correction,** then
a separate session to
add titles and format

Online Session

Pull original sound
elements from vault

Audio Layback

Delivery

Sweetening and Mix:
traditional video-style audio
post-production on 24-track
or digital workstation

NB: The icon for online and master tapes is shown as a 1" tape reel, which is archaic; however, a D2 tape reel or other high-quality format is usually a big cassette and uninteresting for an icon.

4. Offline Nonlinear Editing System to Video (HDTV) Finish

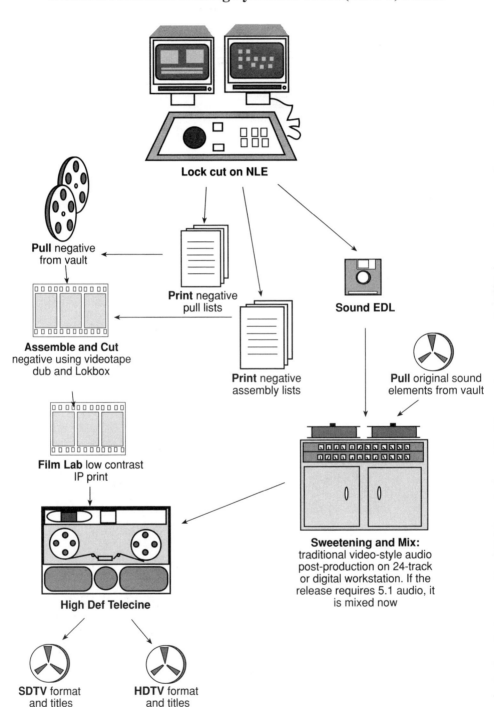

Lock cut on NLE

Pull negative
from vault

Print negative
pull lists

Sound EDL

Assemble and Cut
negative using videotape
dub and Lokbox

Print negative
assembly lists

Pull original sound
elements from vault

Film Lab low contrast
IP print

Sweetening and Mix:
traditional video-style audio
post-production on 24-track
or digital workstation. If the
release requires 5.1 audio, it
is mixed now

High Def Telecine

SDTV format
and titles

HDTV format
and titles

5. Online Nonlinear Editing System to Video (to Film) Finish

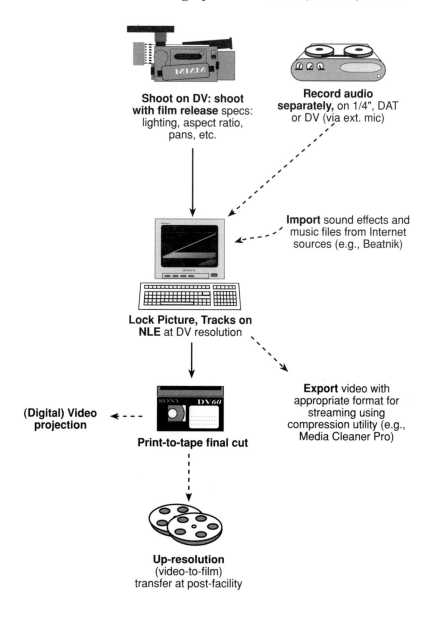

Shoot on DV: shoot with film release specs: lighting, aspect ratio, pans, etc.

Record audio separately, on 1/4", DAT or DV (via ext. mic)

Import sound effects and music files from Internet sources (e.g., Beatnik)

Lock Picture, Tracks on NLE at DV resolution

(Digital) Video projection

Print-to-tape final cut

Export video with appropriate format for streaming using compression utility (e.g., Media Cleaner Pro)

Up-resolution (video-to-film) transfer at post-facility

▶▶ Video on the web: p. 287   ▶▶ Video-to-film transfers: p. 309

C   O   M   P   U   T   E   R   S

When you set up a computer to edit, there are a handful of key elements, crucial to functionality, that require a modicum of computer understanding to sort out. Nothing pushes a computer to its functional limits like moving video and graphics in real time. And while no computer manufacturer will ever say it out loud, problems with editing systems are minimized when the computer is *dedicated* for the task. *Yes,* it is realistic to run editing software on a PC along with word processing, spreadsheets and graphics, and *yes,* hard disk and memory economics make it completely feasible; but for professionals (and even prosumers) the constant risks of file and driver conflicts, viruses, and ever-changing software and operating systems increase the potential for catastrophic failures. Dedicated computers tend to be more reliable. Computers that are specifically designed for a given function (and do nothing else) are called *closed architecture* systems; on the other hand, off-the-shelf computers (*i.e.,* Macs and PCs)—are called *open architecture.*

A computer is characterized by a monitor (to watch), a keyboard (to type on), a pointing device (like a mouse or trackball), and some kind of box containing a hard disk and (probably) fans. Also in this box are computer boards holding many chips, some variation of RAM and ROM, and a CPU. The physical tangible part of a computer, in particular the circuitry, is called *hardware.* The intangible instructions for how a computer works and does tasks is called *software.* The manufactured box of hardware and software is called a *platform.* But you knew all this, didn't you?

CPUs

Virtually all computers have at least one CPU, or *central processing unit.* This chip is the brains of the computer—it gives a system its language, and this in turn allows software to have a given style and look.

All CPUs have a "clock speed" that relates directly to how fast the chip can perform its calculations. Clock speeds in personal computers range from roughly 500MHz to over 1GHz ("Hz" here means clock pulses per second). Editing itself is not all that computationally complicated. It doesn't take a supercomputer to ripple an edit list or even to stream high-resolution video. The faster the clock speed, however, the faster that graphics—which require an extremely large number of calculations—can be visualized (or rendered). The ability of a computer to stream video pertains to its internal pathways (buses) and not CPU speed. Realtime effects require not only multiple streams of video but extremely fast and powerful CPUs.

MEMORY

A computer's active memory is called RAM, or *random access memory*. This set of memory chips is a workspace where information is quickly accessed and used. Editing generally takes place in RAM. It is often called "volatile storage" because it is temporary and will be erased by a loss of power—say, if you unplug your computer. RAM needs to be large enough to provide sufficient space to store all information necessary for editing, as well as for the application itself. Exactly how large is purely a function of the size of editing software and how much data. A system that can handle a large sequence or production needs a great deal of RAM. This is particularly true for the systems that can manipulate the individual pixels of the images, and not just the sequence of frames. Linear systems move very little data around for editing—they are only machine controllers, shuttling and cueing video decks. While personal computers might have typical RAM configurations of 32MB to 96MB, digital video systems require more. Does 256MB of RAM make a system "better" or "faster"? More RAM can make an application more "responsive." And, up to a point, rendering can be accelerated by more RAM, particularly if effects involve numerous layers. It is widely accepted that there are diminishing returns with RAM beyond the suggested allocation for each product. If your work requires multiple applications open at the same time (for example, Media 100 and After Effects and Photoshop), then their RAM requirements need to be added together. Always remember that the basic system software on any computer gets first dibs on RAM and must be figured in all calculations.

STORAGE

Computer users make a distinction between forms of digital storage: "memory" refers to RAM; "storage" means longer-term, semi-permanent storage (and not chip-based) for information. Hard disks and Jaz disks are two examples of computer storage media. Hard disks are protected from the environment and will maintain data after the power is turned off; RAM won't. Information tends to flow from the hard disk to RAM when you "load up" a project or open a file. When information is "saved," it is copied from RAM back to a place on the hard disk.

Hard disks need to hold much more than RAM does. They are not chips the way CPUs and RAM are, but rather magnetic disks that can store high densities of data. Typical internal hard disk sizes range from about 6GBs to more than 40GB; external drives can be purchased in larger

▶▶ Open and closed architecture systems: p. 322

sizes. The largest sizes are most practical for "large" data storage—like digital video and graphics. Most data manipulation and word processing requires considerably less storage. As with RAM, every few years sees a dramatic increase in the density of data that can be stored on magnetic media. Although hard disks are often internal to computers, external drives and removable media are staples in editing applications.

There are a number of ways storage can be connected to your computer. Because moving video is bandwidth intensive, it is important for the implications of your storage choices to be considered. SCSI connections are plenty fast enough for moving video. Faster UltraSCSI connections aren't required for single streams of video or compressed video, but may be essential for uncompressed (ITU-601) and multiple channels of simultaneously streaming material. FireWire (IEEE 1394) connections are designed for video and can move up to 400Mbps with ease. Compare the bandwidth of the video you require with the capabilities of your drives and your connections.

RAM is faster than the fastest hard disks and information is moved in and out extremely quickly. Unfortunately, the cost of RAM (per byte) is high (relative to disks) and, pending technological advances, the chips are still volatile.

ROM, or *read-only memory*, is like a tiny permanent hard disk printed on a computer chip. Like a hard disk, ROM can provide information and instructions to a computer system, but unlike a hard disk, its contents cannot be modified. ROM is fundamental to all computers and thus all editing systems, but it has little relevance to the investigation or comparison of systems. More flexible forms of ROM, in particular PROM (Programmable ROM) or E-PROM (*Erasable-Programmable ROM*) chips are often part of the ever-changing hardware in editing systems.

THIRD-PARTY CARDS

The more that open architecture platforms are standard for post-production software, the more that third-party specialized hardware cards will be required to adjust the basic platform for the task at hand. Many of the high-bandwidth, high-acceleration needs of graphics compositing and high-definition video manipulation cannot be handled by the CPU and internal architecture of the basic PC; it simply doesn't have the power to manage these enormous demands. Consequently, companies like Matrox and Pinnacle and ICE provide cards that slide into base systems and aug-

---

◀◀ Digital media characteristics and bandwidth: p. 169

ment functionality by shifting the data away from the CPU and bus pathways of the host, and onto a dedicated swatch of hardware designed specifically to handle the job. There are also cards that add machine control to a typical computer, cards that capture and/or digitize video (*e.g.*, from Canopus, Digital Origin, DPS, Matrox, Pinnacle), provide high-end audio functionality (*e.g.*, Creative Labs, Yamaha), and so on.

The primary issue of third-party hardware, aside from concerns of compatibility, is simply one of *responsibility*. Open architecture, software editing tools, and third-party hardware together can result in fabulous and affordable functionality, but also can mean a great deal of finger pointing when a problem arises, ("Nope, that crash can't be from our software, it's probably that card . . .")

HEAT

When electricity flows through computer chips, they heat up. If they get too hot they can stop functioning. Consequently, all computerized equipment has a complex high-tech method for self-cooling: *fans*. Without them, systems would overheat and stop working. All computers have fans placed near sensitive hardware locations. Videotape equipment that contains both computer and mechanical components need extensive cooling. Videotape facilities and edit bays are often carefully cooled for this reason. Never forget to keep computers and editing equipment at their proper working temperatures. Also remember that the insides of computers are designed with *airflow* as well as temperature in mind; changing the orientation or blocking vents can cause problems.

Laptops, of course, do not have fans, and the heat generated from super CPUs can be extraordinary. While laptop computers use various heat *sinks* to capture and re-distribute heat, the volatility of these mobile computers can be riskier than environmentally controlled configurations for high-end production.

SAVE and AUTO-SAVE

Since RAM is a temporary workspace, and is so vulnerable to power loss, computers have means for automatically saving the contents of RAM onto the hard disk. When you select "save" on a computer, this is what is happening. The material is still in RAM, but a duplicate is copied onto the hard disk. Most systems also have an "auto-save" function that will periodically save the data in your RAM for you, in case of accident. This way, the only material that can be lost in a power outage is the newest material edited, after the last "save." You would want a computer to save often, but since it does take some time (often only seconds, but potentially much longer in long-form editorial, for example), you probably wouldn't want an auto-save after every edit. Typical systems allow you to select how often you want work saved: every 15 minutes or every 10 edits, or some such factor. It only takes losing your data once for that lesson to be hard learned. *Note*: the data being saved is the editing data and not the media files themselves. The concept is that media files are both enormous and easily recapturable from source tapes; not so for edit and log data.

BACKUPS

RAM is only temporary memory; hard disks are "semi-permanent," and are relatively safe for the storage of information. But accidents do happen. Some of these accidents will be computer-related: strange power surges, "glitches" on the magnetic media, mechanical problems with the hard disk's "read" head . . . and some of these accidents will be human error: "Whoops! You didn't want to delete Act II?"

For these and other reasons, it is always a good idea to have a copy of your work kept separate from the computer, just in case. This copy is called a *backup*. The most common backup strategy is to use Zip or floppy disks (for small things like edit information) and digital tape (for large volumes like digital source).

Every day, at the end of the day, an editor may want to copy the entire project (the edit decision *data*, not the digital video/audio source material itself) onto a removable disk for safekeeping. Doing this certainly limits an accident's severity. Beyond this the degree of safety you need is subjective. On some projects, the *Hess Backup Methodology* works well: a separate floppy disk is kept for each day of the week (and one for "lunchtime"): thus on Wednesday you back up on the "Wednesday" disk, and don't re-use it until the following Wednesday. This, along with a single disk for lunchtime backups, gives you pretty complete security—with a half-day maximum from any crash to the most recent backup. Many users do not bother with this degree of coverage: they consider it satisfactory to back up on a single disk twice a day—at lunch, and at the close of the edit session.

It is actually pretty rare for a serious malfunction on a professional editing system to lose an editor's work; but on those odd occasions, the backup is priceless.

FILE SYSTEM ORGANIZATION

The organization of data in editing systems is based on *hierarchical file systems*. This is the computer's familiar layout where files go in folders, which go in other folders, which go in other folders, *ad infinitum*. How a computer is logically organized doesn't necessarily correspond to the computer's physical *hardware*—in which computers are comprised of one or more hard disks, any of which might have one or more partitions.*

Digital editing systems, with their enormous demands on storage, often utilize a number of hard disks, each holding many gigabytes. Whether editing systems and their associated storage are shared or used by a single editor, organizing projects, media files and other data onto those drives and partitions is an important aspect to project management. The implications of these choices can affect system performance, ease in changing between projects, backing up and archiving, and simple data searching.

When you store files on your computer, where do you put . . .

1) the (huge) digital media files relative to the (tiny) project files?
2) the digitized audio relative to the digitized picture?

1. When media is digitized (or captured), large media files are created on larger hard disks. These files generally are raw media, and do not contain interpretive, descriptive, log (or *meta-*) data. Projects, however they are created and named, generate small local files (icons, "clips," *etc.*), that relate metadata to media data. Organizational style is a function of (a) the conventions of the genre (reels? acts?); (b) the editor's personal preference; and (c) the system itself. In general, projects get organized like this:

The NLE computer has a "small" system hard disk that contains the operating system, utilities, applications, and so on. This drive (say, less than 6GB) will also have a folder for each project-production. Also connected to the computer are media drives—hard disks (either internal or external; local or networked)—dedicated to digital media. These drives

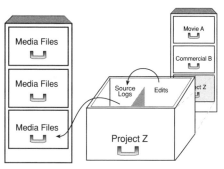

---

* There was a time when a computer could not interact with a hard disk that was too big—consequently, large disk drives needed to be partitioned into smaller sections, each of which the computer "recognized" as if it were a separate drive (although you knew it was one physical device). There are other reasons to partition a drive (multiple users; multiple projects), but today it is often by choice.

(probably as big as you can find 'em; say, 40GB) are the targets for captured files. When these files are created during capture/digitizing onto the drives (and probably given arcane techno-names), they are automatically linked to simultaneously-created aliases (master clips, picons, *etc.*) that are placed in the appropriate locations in the project folder. These get normal names—usually scene/take numbers or English descriptions.

2. Storing of digital audio. When hard disks have relatively low bandwidth (relative to the image size), playback can stutter or fail altogether when audio and video are both trying to stream from the same drive. This was a problem in earlier nonlinear days, but today most drives provide sufficient bandwidth. By far the greatest rationale for separating audio and video is economic, not technical. High performance hard disks and RAIDs are often required to play uncompressed video. Using these drives for data as "simple" as audio might be simply a waste of high-cost real estate.

When you do your "edit," you're not changing the actual media files, of course; those remain unchanged. Your edited *sequence* is merely a set of pointers, a hyper-EDL of sorts . . . that point to the original aliases . . . that point to the media files that contain the actual video and audio. Those media files can be moved, archived—even deleted, if need be, to free up space; and then *batch re-digitized* (or captured and relinked) from the original sources later on, based on reel number and timecode.

Thus, the project logs and editing metadata are what contain your actual "work."

Relational Databases and Metadata

One of the advantages computers have in the editing world is their fantastic ability to organize and manipulate data. Editing is a data-intensive exercise. For videotape, data include computer-read timecodes, record points, list tracing, and so on. For film editing, there are ample key numbers and code numbers that must be logged and tracked. Code numbers are linked to both their original key numbers and the scene numbers of the dailies. Reels are built and re-built, and scenes are moved around. Bins of trims hang in every corner of the rooms; although the editing itself is not number intensive, locating shots that have been cut and moved and hung up and changed . . . without losing them . . . is an ordeal fit for any computer.

A database is a special file where discrete pieces of information are linked up and made accessible. A Rolodex is a kind of database. Information of all types can be stored on "cards" that contain any number of smaller pieces of information, called *fields*. The fields in your average Rolodex might be *name, address* and *phone number*. The fields in your average film log might be *scene, key number* and *description*. Traditionally, film assistants build different books of related information, but until the 1990s they had to do this entirely by hand. Now computers make this organization easier.

Most nonlinear systems have a *log*, which is a database of information ("metadata") about all the source material. For a given strip of film, a database might include any or all of the following:

▸▸ Scene/take numbers
▸▸ Shot description and/or key words for search fields
▸▸ Start film key numbers and/or code numbers
▸▸ Associated camera roll, lab roll, and sound roll numbers
▸▸ Nagra timecode number
▸▸ Source videotape (after telecine) timecode number, *AND* 3:2 pulldown sequence information
▸▸ Date shot
▸▸ Film type (35mm, 16mm, 3-perf, 4-perf, 70mm, Vistavision, *etc.*)
▸▸ Film frame rate (24fps, 25fps, 30fps, *etc.*)

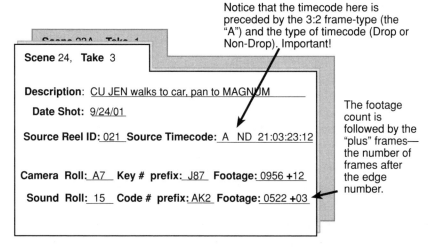

Notice that the timecode here is preceded by the 3:2 frame-type (the "A") and the type of timecode (Drop or Non-Drop). Important!

Scene 24, Take 3

Description: CU JEN walks to car, pan to MAGNUM

Date Shot: 9/24/01

Source Reel ID: 021 Source Timecode: A ND 21:03:23:12

Camera Roll: A7 Key # prefix: J87 Footage: 0956 +12

Sound Roll: 15 Code # prefix: AK2 Footage: 0522 +03

The footage count is followed by the "plus" frames—the number of frames after the edge number.

Although an editor might only need a few of these pieces of information, the database contains *all* pertinent information. Some fields, like shot durations, do not need to be input as they are calculated from other database information.

One feature of most "film-style" systems is that database information is generally "buried" in the system, so that the editor does not have much contact with numbers and data that are unnecessary or even a hindrance in the creative process. Systems do not always have a way to display film numbers and other filmic data. Some have virtually no displayed numbers at all, short of the scene/take labels.

A *relational database* is a specific subclass of index you might need. It is often manifested in an editing system's source log, and relates both timecodes (from the telecine transfer) and film edge numbers. By interlocking these pieces of information, along with 3:2 type, a computer can translate timecodes into film numbers, or vice versa.

The equivalence between videotape and film is "locked up" at *index frames.* These frames are designated in telecine usually by a hole-punch made at a key-numbered film frame. By clearly marking this index frame, you can locate it in video, and log it into the database. The numbers from this point, continuing forward in the source, will be tracked. KeyKode and KeyKode readers in telecine facilitate this process.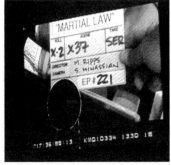

Index frames are usually the first frame of a scene, or sometimes on the clap board's sync point. As long as this frame can be easily identified, and falls near the head of a shot before which no material will be edited, any might do.

Since most systems edit and count at 30fps, they must contain a mathematical formula, or *algorithm*, for converting the videotape timecodes into related film numbers. This is done using the database. Systems that digitize, edit and count at 24fps do not need an algorithm to convert an edited sequence to edge numbers, but it will still need one to generate (30fps) timecodes. Regardless, all systems that output film information and tape information have some kind of relational database and internal algorithm.

Systems without a relational database often use 3rd party software products that do this same thing. There are a number of programs available to convert timecodes to film numbers, and they are equal to built-in software when given the correct data.

The expansion of the Internet has augmented demands on computer databases. Most websites are built upon industrial strength relational databases—called *enterprise level* systems. As video media becomes more prevalent on the Internet, editing system databases may find themselves being built on network-friendly enterprise level hosts.

▶▶ Video on the web: p. 287

NOMENCLATURE

REELS, SCENES AND TAKES

Since one of the key fundamentals to good editing is good management of source material, how you name and label material is the cornerstone to good management. If you call every shot "temp" you're never going to be able to locate that one great shot from last Tuesday in the pile.

Reel Numbers

Assuming you are shooting on videotape, source tapes need reel numbers. There are no rules here, only guidelines, and they depend on the kind of project you are working on. Reel names should have two elements, the name (or code number) for the project, and the sequential number of the specific tape reel. Thus, a TV show called "Dragons Live in Caves" episode 101 might begin with a prefix "Dragon101" or "D101" or perhaps "DLIC101." To make tapes manageable, tape length should be either 30 minutes or 60; it is tempting to purchase 2-hour tapes due to economics, but the added labor of shuttling around a 2-hour tape to locate a shot (for digitizing or capture) is probably not worth it.

So, a five hour shoot will have five 1-hour tapes: D101-001, D101-002 and so on. By using a 3-digit reel number, you also maintain some consistency with more traditional online editing systems that may require a number in this format. Every video tape you ever shoot should not only be labeled with a REEL NUMBER, but should have a paper log sheet that outlines (either specifically or generally) what is on the tape and where, with timecode references, dates shot, and other useful data. No matter how ubiquitous computers become, a paper logsheet for a tape reel is a valuable item.

Particularly useful is to coordinate reel numbers with the hours of the tape's timecode. Thus, reel 002 begins at hour two (02:00:00:00) and reel 012 begins at hour twelve (12:00:00:00).* This is especially practical when numerous tapes are in use, and you want to ensure the reel you inserted that was labeled 12 is actually reel 12 (the source timecode is the key). Unfortunately, this clever method does not work if (a) the project requires more than 24 source tapes and/or (b) for some reason you cannot preset the timecode being recorded. Consumer camcorders in particular often fail to provide this decent functionality.

---

*Don't forget that actual tape start times back off these points by a few minutes, to allow for bars and tone, etc. A start time of 02:00:00:00 generally begins at 01:58:00:00

Finally, if the project is destined to go through online in a traditional bay, many online systems have strict reel naming limitations. Older editing systems once required only 3-digit reel names, 000 up to 255, with a possible appended "B"—although this paltry offering has been improved upon, many traditionalists persist in using this naming scheme. Today, it is not uncommon for a reel name to be limited to 5 or 6 alphanumeric characters and no punctuation. Check with your chosen online system for options.

Scenes

While computers are not always well designed for organizing filmic material, what follows is the reasonably official hierarchical numbering scheme used in film/video production to name scenes.

Scene 1
Scene 2
Scene 3
Scene 4

Now, you insert a scene between 2 and 3:

Scene 1
Scene 2
Scene A3 ◄
Scene 3
Scene 4

Every scene of a movie is called, obviously, a "scene." When you move the camera around to cover that scene from a different angle, the camera in a new position—whether for all or part of the scene, that is known as a "set-up." Each set-up gets its own number, related, of course, to the scene in which it is a part. While the establishing (or "master") shot is usually labeled without a letter suffix, don't be concerned if a production does not contain a scene labeled "Scene 1." It only means that the first set-up was probably "A," thus Scene 1A. Regardless of the *order* shot, scenes are properly ordered like this:

1   1A   1B   1C   2   2A   2B   2C   A3A   A3B   A3C   3A   3B

When there are more than 26 set-ups, and you are looking down the barrel of scene 18Z, the next set-up will be 18AA, then 18AB and so on. Also know that *2nd Unit* and *B-roll* material is either not labeled with a scene number or it may be summarily labeled with an X, like X100.

REVIEW: A letter* after *a scene number is the set-up code, a letter* before *a scene number is an inserted scene.

Takes

Every time you shoot the action again from the same set-up, it is a new *take*. There might be a couple or there might be dozens (maybe hundreds) of takes, but only a few selected ones make it to post-production ("printed takes" or "circled takes"). Takes get named in a pretty obvious and ordinary way, and here (as in life) are written after the scene number, separated by a dash ("1A-1"), a slash "(1A/1"), a dot ("1A.1") or other small character:

```
1-1
1A-2
1B-1
1B-2
1AA-5
2B-1
2Z-3
2BC-1
A3A-1
A3A-2
```

Descriptions

Key word and description fields are important for searching out shots that have odd content. But in many productions it is common to "describe" shots with the basic camera info, possibly culled from the shooting script. These comments are traditionally the work of the assistant, who watches the dailies, looks at the script, and adds comments to shots for the editor. It is imperative in the editor-assistant relationship, that the assistant has a good understanding of what is going to be called what. For instance, should words be in CAPs and others in lowercase? It might be decided, for example, that you want shots in lowercase, but character names in caps, as in "ws room" and "cu MIKE, or vice versa. Many editors are picky that spacing and spelling be consistent..

```
1       WS Room Master
1A      CU Jonah
1B      CU Mike
1C      CU Jen
1D      OS Mike to Jen
1E      OS Jen to Mike
1F      PU – CU Jen
2A      EXT HOUSE DAY – ESTABLISHING SHOT
2B      EXT HOUSE NIGHT
A3A     Long Shot Driving, l-to-r pan
```

In the computer, all the takes would line up neatly, making it easy to see a text-based list in a graphical way. A real log might look like this:

▶▶ Abbreviations: p. 399

```
1C/1    cu JEN
1C/2    cu JEN*
1C/5    cu JEN
1D/2    os MICHAEL to JEN
1D/3    os MICHAEL to JEN
1D/4    os MICHAEL to JEN*
1E/1    os JEN to MIKE*
1E/5    os JEN to MIKE
2A/4    EXT HOUSE DAY – es
2A/9    EXT HOUSE DAY – es *
```

It is easy to see how many takes there are of each set-up simply from the visual shape of the list. Notice how the assistant added asterisks (*) to the "good" takes that were probably identified in the dailies session (by the director or editor) as preferred.

Whether to use the description field or not is an editor's prerogative. Many editors choose to leave these fields blank. In this case, they might use the script instead for this information, or use written timecode notes. Either way, skipping text descriptions speeds the process along from logging to editing.

In the case of documentaries (or other unscripted material), certain naming consistencies should be adopted so that you don't call one take of the forest "trees," the next one "woods," and a third one "jungle." Scene numbers are not just a quaint counting scheme; they are useful shorthand for identifying material that goes together, regardless of when or where it was shot. The script will provide information about what a shot covers, and you will probably have written notes about which takes are better than others (that's one of the things you note when watching dailies).

Slates are quite handy for reasons having nothing to do with syncing pix and sound; they provide a visual label for an upcoming set-up, so you can find a shot while fast-forwarding through material. Even the most low-budget, amateur production in the world will benefit greatly in post from the use of head slates (FYI: a slate used at the end of scene rather than the head is called a "tail slate" and is shot upside down).

Computers

Some computers sort alphanumerics in a particularly uniformed way: 1 is before 2, 2 is before 3, but 40 is before 5. This is because the computer is only looking at the *first character* and sorting on that, and then after the first pass, it looks at the second character, and so on. You notice this when 1, 10 and 100 are before 2. *Smarter* sorting computers will get the "values" correct in the sort, but don't know how to deal with alphas BEFORE the first number (A3A)—computers will likely put A3A before scene 1, and alongside A24A, but not near scene 3 or scene 24, as the case may be. Sometimes these alphas will all be lumped at the end. (NB: single alphas AFTER a number tend to get sorted correctly, but computers, again, will have no concept of sorting 18X before 18AA).

It comes down to whether the computer sees this alphanumeric as "text" or a "number." Technically, these are "text" and need to be sorted as such. But even then, there is no guarantee that it will be sorted in the "film-style" way required. The labeling method used for a century on films in the United States is not well suited to computer sorts.

As a filmmaker using a computer, you have limited choices: don't use the prefix A or change your scheme. A common work-around is simply to place all the shots relating to a single scene in a dedicated folder. This way, whether the shot is sorted correctly or not, you have *manually* organized material that edits together into a common location.

Thumbnails

For small projects, or for projects without a large number of circled takes per scene, *thumbnail* views (tiny picture icons) of source material can help identify what is what; it easily helps an editor locate the shot of the trees and the shot of the sky or the close up of Jen. But in more ambitious projects, 27 takes of "CU Mike" will have the *identical thumbnails*,

and managing a screen full of little images can sometimes be of less use than it first appeared. Combining strategic use of folders and bins with thumbnail views can help, but ultimately, the balance between work to set-up, and time to start editing needs to be managed.

THE PRIMITIVES OF EDITING

Let's look at a piece of 35mm film, 4 perforations per frame. The beginning of the film (the *head*) is at the top, the end of the film (the *tail*) is at the bottom.

Rather than draw the film up and down, we can lay it down on its side, like this, with the beginning on the right and end on the left:

Actually, it really doesn't matter where you place the HEAD or TAIL of a piece of film except that film editors and video editors see things differently. Film editors are familiar with *heads* of a reel being on the right—on a flatbed, the *supply* roll of film is mounted on the left plates, threaded through the prism heads, and the *take up* reels are on the right:

Technically speaking, videotape runs exactly the same way: a cassette tape has the take-up reel on the right (thus the "head" on the right). Still, when you're an English-speaking video editor, watching moving images on a monitor, it feels like *just the opposite is happening*—that images move from left to right and thus the head of a sequence is on the left. Like reading English. Even knowing what is *technically* going on doesn't change these feelings.

Editing systems tend to locate the head of a shot *on the left*—as if you were reading the English language, and not like working on a flatbed. Since most systems display film with heads left, **this book does the same**, although it is important to recognize that this is neither universal nor necessary.

The head/tail issue surfaces at two points of the electronic nonlinear interface: 1) editing transitions, and 2) working with timelines.

HEAD ← TAIL

Of course, just like linear electronic systems, nonlinear editing systems do not actually CUT or TAPE anything. Linear editing is a process of recording selected pieces of source material from one videotape to another. If you want to make one shot longer, you need to go back to the source tape and record the desired shot longer.

Film editing, on the other hand, involves actually cutting out pieces of film from the source dailies. If you have a 100-foot shot, and you use 4 feet from the middle of the shot, you'll have two pieces of film left over—these are called the "head trim" and the "tail trim."

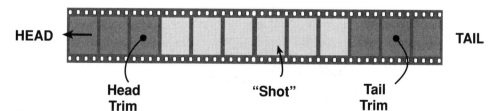

HEAD ← TAIL

Head Trim   "Shot"   Tail Trim

Every time you cut a shot out from a strip of film, the remaining trims must be carefully put away. They are usually hung together in a *trim bin* and labeled with *trim tabs*.

Digital nonlinear editing is not like film editing nor is it like videotape editing. When you make an edit, the editing system remembers that edit location and connects it to another edit location in the computer's "mind." Electronic "pointers" direct the computer to "move" from the end of one

◀◀ The film story: p. 24

shot to the beginning of another. Let's examine these three methods more closely:

FILM. Three shots are cut from dailies and taped together. To extend a shot, you rip off the splice tape, go back to the head or tail trim to find more film, cut it in, and re-tape. All the following shots push down.

VIDEO. Three shots are recorded from source tapes onto a separate videotape master; no splices are visible. To extend a shot, you re-record the shot you want to change, making it longer, and then tell the computer to re-record all the following shots after it.

NONLINEAR. Three shots are connected with *pointers* (computer links in software) and are not actually recorded in physical form. To extend a shot, you go to the "splice" you want to adjust, tell the system to allow you to trim it, and add the new material. The computer re-connects the pointer from the new end to the next shot. Nothing else has changed:

Since the material you see edited together is only a *virtual* (simulation) of actually connected shots, and since there is no "master" tape *per se* (only source media are used to simulate a master), THE FULL LENGTH OF EACH SHOT IS ALWAYS AVAILABLE AT EVERY SPLICE POINT without your ever needing to go back to the actual dailies (or head/tail trims).

One way to visualize this is the *folded paper method*. Think of a piece of film as a piece of paper:

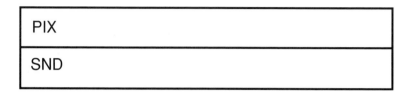

At one end you cut the head, and further down you cut the tail:

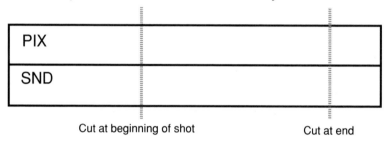

Cut at beginning of shot   Cut at end

Rather than hang up the head trim and tail trim in a film bin, you simply fold them back, out of the way:

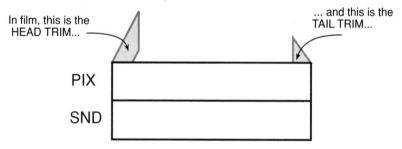

Before splicing two shots might look like this:

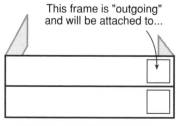

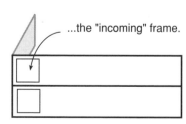

FIRST SHOT (the "A side")   SECOND SHOT (the "B side")

Thinking about Re-Editing

Once you have organized a series of clips into some kind of sequence, the re-edit begins—this is perhaps the core of what editing is. No matter how elaborate the re-edit, the basic components are two:

1. Put new material in. (*Inserting, Replacing*)
2. Take old material out. (*Lifting, Deleting*)

From these two simple actions, all the more complicated editing manipulations can be generated. The most fundamental manipulation of editing is a *trim*—change the transition between two shots.

There are a few ways to think about editing: **shot-based** (*this flows into this flows into this...*) and **transition-based** (*this is the frame that ends this shot, which connects to* this *frame that begins this next shot...*). Shot-based re-editing involves the re-arranging and re-ordering of existing shots. Thinking in a shot-based way means marking an in- and out-point on source material and splicing it into a sequence, a direct descendent of video-style editing in which you must choose these points before committing to an edit. Although a "commitment" in nonlinear editing is less severe than in tape or film editing (you can always change the out-point if you end up disliking it, easily), it is a style that can be a little more frustrating to the seasoned film editor. Many times you do not want to decide on the end of a shot until you have the previous edit working to your satisfaction. Transition-based editing acknowledges the nature of shots, but is focused on the attachment points—the edits.

Trims

All nonlinear systems allow you to modify transitions between any two shots already edited into the master. It simply involves going to the desired transition point and "trimming" it. Although names vary from system to system, both shortening and extending shots are done in the same way. This can be confusing, since the word "trim" tends to sound like it is about shortening. The idea that shortening and lengthening can be done the exact same way—and that both might be called "trimming"— seems counterintuitive to this name. *Get over it.*

Nonlinear editing systems have at minimum two *kinds* of trims. One kind allows the editor to adjust each side of the cut — the outgoing and incoming sides — separately. This is a "film-style" trim and will virtually always change the duration of the program. The second modifies both shots *simultaneously*—for every frame that is added to one side, the exact

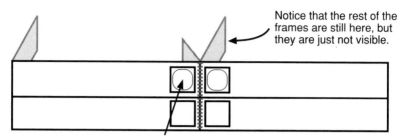

Notice that the rest of the frames are still here, but they are just not visible.

This is the outgoing frame, showing on the left side of a transition display. With a cursor we can highlight it if we only want to add or subtract frames from this side of the transition.

same number of frames is lost from the other side. This is a "video-style" trim, and will never change the duration of the program.

Trimming usually takes place in the window for the edited sequence, splitting the sequence to present the two frames at the splice. The above illustration shows two frames on either side of a transition—the last outgoing frame is shown below on the left, and the first incoming frame is on the right. While ideally you'd want to see these two frames with as much detail as possible, digital system tend to make them smaller, nested in a "transition" window.

Once in a transition editing mode, you choose which *side* of the transition you want to adjust, usually using the cursor to select one (or toggling a button). All systems give you a way to select which side you want to control. In the above illustration we are indicating the A-side of the transition—the tail of the outgoing shot. Switching between the outgoing and incoming sides of a transition with ease is key in a well-designed system.

 A **film-style trim**—known by many names (non-sync trim, "yellow" trim, ripple trim, *etc.*)—will change the absolute length of your program, either longer or shorter. Either way, the film-style trim will default to *rippling*—a video term for shoving down or pulling up all following edits, as if you were actually working on film.

 The **video-style trim**—also called many other names (synchronizer trim, slide-cut, rolling trim)—modifies both outgoing and incoming shots *simultaneously*. For every frame that is added to one side, the exact same number of frames is lost from the other side. On film, it is as if both shots are interlocked in a synchronizer. This way, there will be no durational change in the length of the show; in video terminology, no following record points are moved, nothing is rippled. (Rather

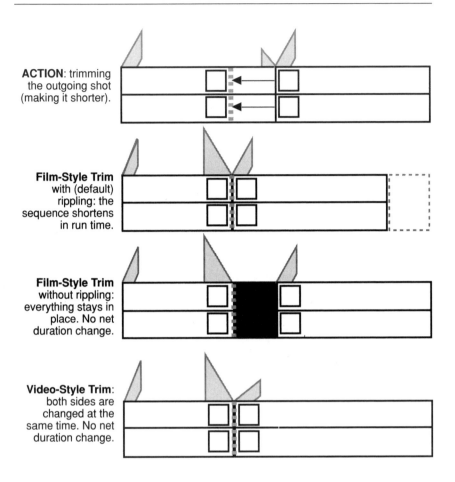

ACTION: trimming the outgoing shot (making it shorter).

Film-Style Trim with (default) rippling: the sequence shortens in run time.

Film-Style Trim without rippling: everything stays in place. No net duration change.

Video-Style Trim: both sides are changed at the same time. No net duration change.

than selecting the outgoing or incoming shot, you needn't designate either side of the transition since both will be modified—or you might just designate the splice point itself.)

Most systems use a transitional-type editing standard in their RE-EDIT-ING modes, but fewer of them are transition-based in *first cut* situations.

Split Edits

When you trim just the picture without changing sound, or vice versa, you will end up with transitions in picture and sound at different locations. In film these are called *pre-laps*, or *overlaps*, or *L-cuts* (named for the shape they form). In videotape, they are *split edits*—where sync picture and sound are split apart at a transition, but continue and will return to sync later in the shot. Here is a split edit, with picture leading sound:

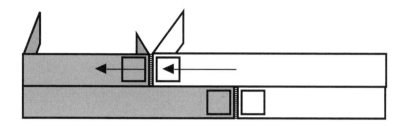

Although there are many ways to create these overlaps, the easiest for an editor to perform is a video-style (*synchronizer*) trim of the picture alone, without modifying any sound. As you lengthen or shorten one side of the transition, the other side will be adjusted the same amount in the same direction. This way, all the following edits in your sequence will never be pushed out of sync. A film-style trim of picture-only (or sound-only) will usually create a situation where the picture runs a different length than the sound.

Because of the way shots are created and modified, students of non-linear are often directed to create split edits by first making a straight cut (in both picture and sound), and then modifying the picture. Although this is not a rule, cutting for sound first and the overlapping picture later is the most common and perhaps the easiest method of making overlaps.

Inserting

If *trimming* is defined as going to an existing transition (*splice*) and modifying either the outgoing or incoming shot (or both), then INSERTING can be defined as going to source footage, finding a new shot, and placing it anywhere into the already-cut master.

Inserting does not necessarily have to come directly from source material; previously edited sequences (whether just a few shots or entire acts) are often inserted into a cut master.

Inserts, like trims, can be designated as "film-style" or "video-style." A *film-style insert* places in a new shot and shoves all following material down in time, like in film. The master is automatically rippled. A video-style insert rolls over the existing master, obscuring old shots with the new one. If 3 feet of film (2 seconds) are inserted, then 3 feet of existing master is taken out. Durationally, you have not changed the timing of the program.

Some systems see the video-style insert as somewhat dangerous (a "red" insert) because you are always going to lose existing master material, and if you insert too much, you might "record over" something you wanted. Thus, it can be risky. Ironically, other systems see it the opposite way: video-style inserts as the most safe (a "green" insert) since there is absolutely no way to throw a sequence out of sync by performing an insert of this kind.

From two different nonlinear systems (above and below), here are icons representing a film-style insert (left) and video-style insert (right).

On systems where losing sync between picture and sound tracks is risky, easy to do, or hard to fix, film-style trims and inserts can be *dangerous*. Video-style editing, then, is seen as safe.

Film-style inserts can also be characterized as "4-point" edits. For these, an in- and out-point must be determined on the new shot as well as

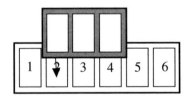

ACTION: A new shot (in grey) is being inserted into the master

RESULT A:
A **film-style** insert will break open the master and drop in the new shot— pushing the rest of the sequence down in time.

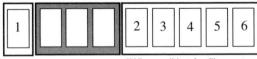

*When editing for film, note that this insert "steps on" the negative cut margin.

RESULT B:
A **video-style** insert places each frame of the new shot **over** a frame of the old shot. The total length is unchanged.

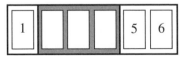

▶▶ Negative cut margins: p. 261-262

in the master. In general, the computer will calculate the fourth point if the editor designates any three.

By the same logic, video-style inserts are sometimes described as "3-point" edits. Here, an editor either chooses the exact dimensions of the new shot (an in- and out-point) plus where the shot should go in the master—or vice versa (the exact location is specified, and a beginning or end point is chosen in the new shot). Either way, 3-point edits can "roll in" or be "back-timed" over existing material in the master.

Clip-Based Variations: Slip and Slide

Transition-based editing has certain powerful features, but clip-based editing can often be intuitive. *Slip* and *slide* (regardless of how they are named in a particular system) combine a series of tedious steps (trims) into slick clip-like manipulations.

Slipping looks at a shot in the cut sequence and says, in effect, "the *place* where I put this is fine, but the *shot itself* is a little off." Slipping synchronously trims the head *and tail* of the source clip at the same time, so every frame added to the head is taken off the tail (or the other way around). This keeps the clip a fixed duration, and will not change anything else in the edited sequence.

Sliding allows a shot in the cut sequence to "float over" the sequence, *moving* to a new record-in position, *without affecting* the shot itself. Visualizing the shot floating over the sequence is conceptually clean, although in truth, the computer is performing trims on the record in/out points simultaneously to execute the move. An ideal use of sliding is in re-positioning a picture cut-away over another shot, when the cut-away itself is fine but the timing is off.

The power of these kinds of functions is that they keep edit lists in sync, but allow modifying transitions of one "side" of a cut and not the other.

Lifts

In editing, pulling a frame or sequence of frames out of the edited master is more than a delete. Very often you want to move a sequence, to "hang it up" for use later. In most cases, this is known as a "LIFT." Different systems are going to manage *lifts* in different ways. When the lift is simply marking some frames and deleting them, one would need to mark an IN and OUT in the edited master, and DELETE. This might be done by selecting all the shots between the in/out points; it might be done by selecting a lift and/or delete button.

As with any deleted shot (just as with inserted shots), the editor needs to decide if the following shots will be rippled (slid down to fill the hole), or left in their positions relative to time.

To hang the *lift* up for later use, a system should make it easy to place it among other items of source material. In most cases, however, computers will see any attempt to move a series of shots (a cut sequence) back into the source bins as an opportunity to move a series of *individual* shots, and effectively lose the nature of the edited chunk. In cases where this cannot be done in a single step, the lift would be inserted into its own sequence, and that sequence, in turn, can be managed as a single entity.

The ease and number of steps this common function commands sets the stage for how much an editing system is designed for "editing" and how much it is simply a computer moving data around. Lifting is a powerful tool, accessed sometimes as a single kind of functionality, and other times as a set of different tools.

Ripple

Although differentiating between film-style and video-style for trims, inserts, and lifts would seem to be sufficient, there are a few cases of the film-style edit that require an additional decision. "Rippling" is actually a specific video term for adjusting subsequent EDL events when shots are added or subtracted—and thus changing the record times.

For nonlinear editing, although there are no event numbers *per se*, rippling is simply the unlocking of shots to time, past a given splice point. By definition, video-style edits do not involve rippling. But film-style edits that shorten master durations might ripple or not ripple (the alternative to rippling is slugging with leader to maintain sync or time).

Undo/Cancel

Most computers, and therefore editing systems, are able to undo the last function performed. All systems have some kind of UNDO or CANCEL button. If you make an edit you don't like—whether by accident or on purpose—you can press this button and return your edited master to its previous condition. Different systems have their own unique methodologies for going back to prior states.

A few systems allow for canceling a number of events by repeatedly selecting the cancel function or browsing a "history" of prior actions. This can be extremely powerful, allowing an editor to work backward through a troubled edit list, edit by edit, until the source of the problem is undone.

On the other hand, it can be confusing. In a few systems there is no way to know for sure exactly what each UNDO is canceling. It can undo work without your having a clear understanding of what is being changed. Also, trying to undo a *single* edit might involve removing work that you want to keep. Still, a few levels of undo is a very useful feature.

Numeric Trims

On editing systems that allow transitions to be precisely adjusted using timecode or just numeric values, lengthening or shortening shots can be a little confusing. Here is a shot from our EDL:

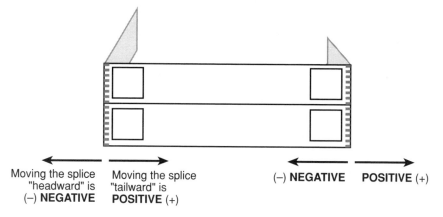

Moving the splice "headward" is (–) **NEGATIVE**   Moving the splice "tailward" is **POSITIVE** (+)   (–) **NEGATIVE**   **POSITIVE** (+)

If we wanted to make the shot 1 frame LONGER, we could specify the tail of the shot, and "trim the out-point 1 frame" which means we would add one (+1) frame at the "out" timecode. Similarly, to shorten this shot 1 frame we would trim 1 frame from the out-point ("–1" at the tail). This is logical if you think of the "1" as the value, and the –/+ as the direction.

At the head, to make the shot one frame longer, you would want to start the timecode of the "in" one frame earlier, in other words trim minus one frame (–1 frame). In the same way, to shorten the shot, you would want to trim "+1" frame. What this means (look at the above diagram) is that to move an edit point to the LEFT—at either the head or tail—you use a **minus** sign. To move an edit point to the RIGHT you use a **plus** sign.

Mistakes are often made because people **incorrectly** interpret the syntax to mean "you lengthen a shot by ADDING (+) frames and shorten a shot by SUBTRACTING (–) frames." *This is not true.*

This is a fundamental concept in linear videotape editing, but it is often confusing to film editors or computer users editing for the first time. Nonlinear systems generally allow for visual and physical modification of shots *in addition* to numeric trims, and never rely on the typing of numbers as the only way to modify edits.

Video Layers

Horizontal editing involves tools and functions that help in storytelling, in particular tools for keeping picture in sync with sound. Nonlinear systems also include functions for *vertical editing*, the manipulation of stackable layers of images that can interact in a number of ways. Most functions needed for video layer maintenance are comparable to those for video and audio—establishing synchronized relationships, rippling, and so on.

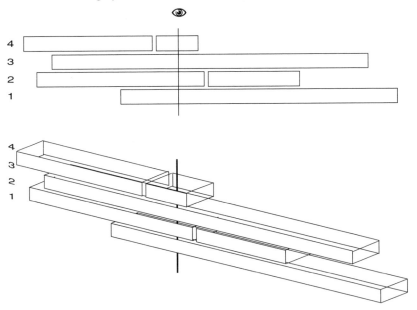

Video layers are metaphors for stacks of opaque or translucent strips, with your eye on the top. Most systems stack from bottom to top, with your eye above the stack, looking down. Thus, an opaque image on top will obscure anything beneath. A few systems invert the stack, placing your eye on the bottom looking up.

To have maximum flexibility to manipulate video objects, each object is ideally maintained on a different layer. This makes them easy to fly

▶▶ Horizontal and vertical nonlinearity: p. 377

in, rotate, and augment transparency, blur, color and motion path. Some graphics synthesis tools create images in layers (like Adobe Photoshop), which can then be maintained as independent layers when the elements are imported into multi-layer editing systems (or *compositing systems*).

Keyframes

The concept of keyframes for editing is a descendent of traditional (drawn) animation, where key moments in the action and story are fully rendered by artists, and then the "in-between" frames are filled in to complete the smooth action. While understanding how keyframes are selected and managed is more related to performing effects, animation and storyboarding than to editing, they have become ingrained in the editing tools and today are a fundamental aspect of system functionality.

On a timeline (whether the EDL timeline, or a special one dedicated to effects management), frames of the material can be selected as **keyframes**. At these points, attributes of an effect are described. For instance, at keyframe A, the editor may want a title to start and to appear small; by keyframe B, two seconds later, the editor may want the title to be full screen. With these two keyframes chosen, and with the attributes defined at each point, the computer can generate the in-between frames, taking the title from small to large over the defined period.

Technically, this *is* animation. And all kinds of effects can be animated in this way: transparencies, sizes, shapes, titles, ripples and other video effects. A traditional dissolve is a specific case of keyframe animation, where at the beginning of the dissolve one shot begins to increase in transparency while another shot lying beneath is gradually exposed.

Attributes being animated, and their values, at a given point in the timeline.

Keyframes in a mini-timeline.

Splines and Bézier Curves

A *spline* is a mathematical method for drawing and describing a curve, popularized by Pierre Bézier in the 1970s. Splines (and their best-known type, called *Bézier curves*) provide the underlying structure of outline fonts (like PostScript) and illustration software (like Macromedia Freehand and Adobe Illustrator) and for 2D/3D animation software. They're also found in editing systems. Drawing curves is part of the control of animation paths as well as the "rubber band" tools that control audio and other timeline attributes.

Without getting all mathematical, the most direct path between two points (or keyframes) is a straight line. Here's a rubber band line controlling a change in audio:

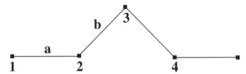

The audio (beginning at keyframe 1) goes along at a given level (a), hits keyframe 2, and immediately the level rises (b) until keyframe 3, and then abruptly drops down back to its original level at keyframe 4. This is fine for many applications, but in fine detail, the abrupt changes are jarring. Animators call these harsh line changes "dog legs." For gradual changes in the line—and therefore in the corresponding audio or motion—you need to smooth out the kinks, and turn the corners into curves. Splines allow this.

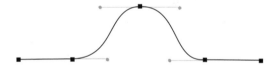

In this second case (above), Bézier *handles* have been pulled out of each keyframe creating small line segments. The resulting curve is *tangent* to the

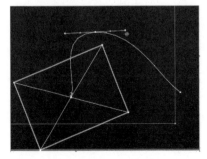

line segment at the keyframe; the angle of the segment drives the direction the curve takes; the length of the segment directs how far the curve moves in that given direction. Good spline control provides for smooth (and professional-looking) effects and fine-tuned audio.

(At left) A path of a motion effect in an editing system with handles extracted from a keyframe.

3 - A N D   4 - P O I N T   E D I T S

3 and 4-point edits are the core of any editing system. They were a staple of linear videotape editing and have evolved in the nonlinear realm. The aforementioned *points* are of the "mark in, mark out" type. A shot cut into a sequence can always be described by *four points*—source in, source out, record in, and record out. In other words, every shot (source) begins on (for example) frame 1 and ends on frame 31 (two points), which fully describes that shot—and then you need to denote where it goes: it starts in the edited master 1 minute in (record in) and runs for 1 second, the duration of the shot (record out).

On a timeline, we can illustrate it this way:

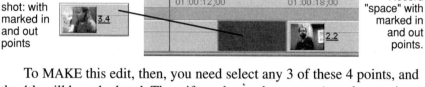

Source shot: with marked in and out points

Record "space" with marked in and out points.

To MAKE this edit, then, you need select any 3 of these 4 points, and the 4th will be calculated. Thus, if you know the *source in* and *out* points, and you know the *record in*—a computer can determine the *record out*. It will simply start the shot at the *record in* point, and roll it in until it reaches the end of the shot: presto! That will be the *record out*.

If you know the source in and out, and only know the record *out* point—you know where you want the shot to END, the computer can 'backtime" the shot into the edited sequence.

Of course it also works the other way: if you have a little "space" in your EDL (a record in and out point), all you need to do is find a point in the source and drop it in—forward or backward

What happens, for instance, if you have selected *both* a source in and out (your shot) and have a little space in your EDL where you want it to go (a record in and out)? What happens if they define different durations? There are three possibilities. First, of course, is that a system will summarily announce that the edit cannot be done. *Too many points.* Second, the computer can change the "shape" of the space in the EDL—rippling the sequence. It attaches the source-in frame to the record-in frame, attaches the source-out frame to the record-out location, then smooths everything. In this case, the record-in and -out are indicating *actual* frames of material, and not simply a timecode on a tape to which frames are recorded.

The third variation is a "fit-to-fill" function, which will speed up or slow down the source shot to fit into the allocated *hole* in the edited sequence. This 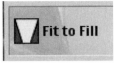 is a fancy computer function that can be practical, but is easily abused. Also, not every "fit" can be re-created for film optical.

Three-point editing is the basis for most nonlinear editing. A standard default for editing systems is to allow the editor to *mark in* and *mark out* in the source, and if the computer does not see a specific mark in on the record side, it defaults the record in point to either (a) where the timeline is parked (*i.e.,* the frame you are looking at on your master display), or (b) the last frame of your sequence.

While 3-point editing is pretty equal among systems, how a computer deals with four points can vary. For example, if a 1-second shot is chosen on the source side (2 points) and the mark in and out on the record side are 5 seconds apart, the computer could stretch the 1-second shot into the 5-second space; or it could replace the 5-second space with the short shot, rippling the sequence (with a film-style insert).

S Y N C H R O N I Z A T I O N

M ost systems work hard to keep picture and sound in sync—if they arrived that way. Many editors view going out of sync as a nightmare that is best avoided. Truth is, moving shots in and out of sync with comfort is critical to editing. Digital systems make it remarkably easy to see if a sync shot is slid out by a particular edit. The out-of-sync shot is usually highlighted in some way, and the number of frames out of sync is displayed on the timeline and/or in the shot itself.

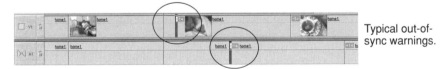

Typical out-of-sync warnings.

Important: if you move out of sync on purpose and then begin to resync, start at the FIRST point where sync was lost, as all shots following this point are equally out of sync. Otherwise, further efforts to resync will continue to reshift following shots. All systems allow shots with no built-in sync relationship or shots with adjusted relationships to be "re-established" as a single synchronous element.

E F F E C T S

An Introduction and Overview

Effects, or "F/X" are manipulations of images used in editing. They may be simple, like fades and dissolves, or they may be complex, like morphing. An effect may combine shots or manipulate a single shot.

All nonlinear systems, by definition, handle one video stream in real time. This is sufficient for traditional (*horizontal*) editing, but not for performing effects in real time. To perform a video effect, a computer has to manipulate *two or more video streams* at the same time. Two streams allow for real-time basic effects: wipes and dissolves and simple superimpositions. For complex effects, or effects that require more streams than a system can process, the computer must render effects to preview them (*see* Chapter 8: *Vertical Nonlinearity*). In other words, it must calculate each frame of the effect and generate them out of real time. The effect is literally built by the computer frame-by-frame and stored on a hard drive. While it may be transparent to the user, the system "cuts in" this new rendered effect in the place of the original shots. (This process is analogous to the ordering of film optical effects and then cutting them into the workprint or negative). Since the effect is playing back from a new rendered source, if you change any part of the effect, that change (and sometimes the whole effect) will need to be re-rendered. This will take some time and may discourage editors from performing many effects, at least until the sequence is more completed.

Basic Effects

Basic effects are **dissolves**, **fades**, and **wipes**—all different ways to replace one picture gradually with another. In the case of a *fade*, one picture is replaced with a "black" picture (or the other way around, in the case of a "fade up"). A *dissolve* is when an outgoing picture fades out at the same time as an incoming picture fades in. One way to visualize what is happening is to have one picture track overlapping with another picture track, and then you slowly (or quickly) switch from one track to the other. The classic type of dissolve is a "center point dissolve" which means that when the outgoing image is halfway gone, the incoming image is halfway in. A *wipe* is when one picture replaces the next via a moving geometric shape—a circle, a box, a line, whatever. Whichever type of simple effect is desired, the same principle applies: you need to have independent control of two separate channels of video and then some way of combining them to get output.

Complex Effects

More complex effects are **superimpositions** and **keys**. A *superimposition* (or "super") is when you see multiple sources at the same time. A *key* is where a shaped hole (called a "matte") is cut into a background and is filled with another picture. The shaped hole, however, does not always need to be a solid geometric shape. Any way you can choose some portion of an image to separate it from the rest is a viable way to cut out a key matte. A common way is to have the computer locate a single color ("chrominance") or brightness level ("luminance") in your picture, and have that cut out. If you want to cut a person out of the background in a picture, the easiest way is to make the background easy to identify and select: a evenly lit solid color. It would also help if that color is nowhere else in the picture. So, to do a "chroma key" you probably need to set up a brightly lit green (or blue) screen behind the person, and then cut out the green (or blue). The classic example of this is the TV weatherman standing before a map. He is actually standing before a green screen and watching himself on a little side monitor to see where he is pointing. If he accidently wears a tie with a little green in it, you can see the map in his tie. Again, you need at least two video channels to do a key: one for the background (in this case, the map) and one for the foreground (the weatherdude). But you also need the shaped hole (remember, the *matte*) to put in the middle. In computers, this is called an **alpha channel**.

Another complex effect is called a "DVE," short for *digital video effect*. The term *DVE* originally referred to the dedicated hardware (pioneered by Ampex) that could perform these 2D and, later, 3D effects. Nowadays, these and other manipulations are done with software (or software/hardware combinations) on open-architecture systems. In any case, a frame or series of frames are independently manipulated with respect to the background; the entire frame may be resized (bigger or smaller), positioned (left, right, up, down), rotated (spinning), flipped, and so on. Objects on separate layers can be animated with keyframes and Bézier curves, with motion attributes selected for each keyframe. These "DVE moves" are also sometimes called *motion effects*, but shouldn't be confused with variable speed motion effects (described forthwith).

Titles and Character Generation

Systems all allow users to type text onto the screen, adjust font and size (and perhaps other characteristics), and add effects (like drop shadows).

◄◄ Typography on video: p. 187

While the user may not realize it, these titles are actually keys cut out of the background—and consequently can generally be filled with moving video.

Variable Speed Motion Effects

Motion effects are the result of changing the speed of the film or video as part of an edited sequence. These effects may include speed-ups, slow-mo, and freeze-frames. There are various reasons that editors may want to create motion effects: perhaps you have a shot of me walking to the door, but I seem to get there too quickly. You want to do a slow-mo, but you don't want it to look that way; you simply want the 3-second shot to take 4 seconds. If you tell the system to take a 3-second shot and automatically fit it into a 4-second hole, it is called **fit-to-fill**. Motion effects are generally easy to create electronically in offline, trickier to re-create in a linear online bay (depending on how fast or slow the VTRs can physically go), and difficult to re-create as film opticals for theatrical projects.

Paint

Computers are readily able to paint the pixels you see on the screen. You might be drawing an entire work of art from scratch, starting with a blank screen and filling it with your painting. If you are an animator, you would have to paint every single frame—a daunting task; remember, film is 24 frames per second. Alternately, you could use your paint tools to alter an *existing* image, perhaps a frame of film or video. In this case, you are only changing colors and modifying the frames. This method is used to produce many effects in film and television, among them *rotoscoping (*frame by frame painted additions traced over live action); this is how the phasers in *Star Trek* were added. Painting software is often referred to as "2D paint" because you are only modifying the flat image of the screen. When software allows you not only to paint frames but to control the frames and the 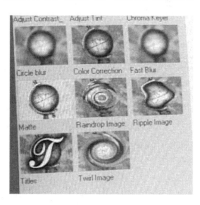 objects within them over time, the paint system is said to also do animation. Dedicated tools have been in use for decades that facilitate high resolution (uncompressed) digital film or video painting—the most notable being Quantel's original Paintbox (introduced around 1984).

CGI – 3D Effects

Rather than paint existing live action objects (or modify them), sometimes it is cheaper or necessary to draw a *new object* in a kind of architectural design program. CGI, or *computer generated images* (or *imagery*), are complex 3D objects initially built simply as a skeleton or points and lines (called a "wireframe"). Once the wireframe is completed, more points and curved lines are added to increase detail; then a skin is applied to the "bones," commonly through a method called *texture mapping* (when a 2D-painted swatch is literally wrapped around the wireframe). The finished realistic 3D CGI object can be keyed into a number of different environments and its components animated. This way, rather than draw each frame of an animation (the old 2D way), the object need only be moved around the frame over time, and each frame is like a snapshot. Many television commercials use CGI to animate Coke bottles, beer cans, bears, dinosaurs, cars, and even people ("synthespians").

Plug-Ins

A *plug-in* is a small software module written in a standardized format, which can be used inside a more complex software product to accomplish some effect. Adobe Photoshop pioneered the use of plug-ins—they packaged their image manipulation product with a few special effects (blur, sharpen, *etc.*), and then they allowed third-party vendors to create their own (like "lens flare" and "mosaic") and plug them into Photoshop. Editing software adopted this idea and allowed for these Photoshop plug-ins, as well as many others, to fit into their special effects functionality. When a product works with plug-ins, the user has a great deal of flexibility to expand the type and range of effects.

Color Correction

Traditionally, color correction is the video engineer's domain, and an online-related task. But TBCs, "scopes" and other tools for color correction have been implemented in software and placed in editing systems that are used for both offline and online. As an effect, color correction allows interesting mood changes in chroma and luma, adding sepia or duotone looks, and adjusting for poor shooting. Be careful: it is all too easy (and way too common) to apply the power of NLEs in such a way that later "finishing" work is more difficult (and thus costly) to fix. If the color correction work is to be replicated in online, communication beforehand between offline and online editors is crucial.

EDITING ERGONOMICS

Once upon a time, perhaps before you began editing, there were Moviolas and flatbeds and VTRs. These mechanical devices had knobs and levers and buttons that actually, physically, *did* something. Push a button and it engaged a motor or turned on a light. Turn a knob and you varried the speed of the motor. Couldn't have been simpler.

Next thing you know, along comes "virtual" editing, where you still push a button (seemingly) but nothing really happens except a bunch of numbers ricochet around inside a computer. Turns out those pretty on-screen "buttons" you push and "knobs" you turn aren't really physically *connected* to anything.

Yes, keyboards are pretty well designed . . . for typing. In about a minute of even slow typing, you use about 90% of the keys and all your fingers get a little workout. But in editing, there are only about five keys you need to press about 50% of the time, and probably only another ten that you use the next 30% of the time. A typewriter keyboard (and mouse), while extraordinarily cost-effective and ubiquitous, can nevertheless be a reasonably lousy way to move video around and edit it.

Of course, the *computer* doesn't care where the instructions come from to tell it to bounce the numbers around in the right way—but *people* do. In the old nonlinear systems there were often inventive physical controllers for selecting and manipulating things. Partly, this was due to reluctance on the part of professional editors to give up an old way of working. Call it inertia.

Indeed, many early nonlinear equipment manufacturers were acutely aware of the desire for a "physical" user interface (*e.g.,* EditDroid, Lightworks), while others found that putting a few keys on the keyboard to do similar functions was a reasonable compromise (particularly with the high cost of designing and manufacturing a physical console). Still others (such as Avid) attempted to satisfy their users with the former, but eventually fell back on the latter. As a consequence, most editors have never had the experience of using an ergonomically designed controller.

Play Control and "Rock n' Roll"

Needless to say, the thing you do more than anything else when editing is *move your video around*: forward, backward, fast, slow. You need to watch it move, watch the details, compare it to other shots, try them on.

▶▶ System interface historical overview: p. 324

Detail from the original EditDroid console (1984), above, features a KEM-style motion controller, with jog and marking functions close at hand. The revised Lightworks console (2000), below, has a smooth Steenbeck-style shuttle (with a pair of levers for left- or right-handed editors) and easy-to-manage key-editing buttons radiating ergonomically outward.

It goes without saying, then, that an editing system must allow for really facile shuttling through your source material and your edited sequences. Yet, oddly enough, this is one of the most overlooked basics of desktop systems. Editing is not simply the attachment of frame X to frame Y. Ideally, you need to be able to "experience" that cut point as it moves back and forth at various speeds. You need to watch how pacing flows through a series of shots. In short, you need great shuttle control.

The highest sophistication was embodied in control devices with feedback, like real-life KEM-style knobs (as on the EditDroid) and Steenbeck-style shuttles (as on the Lightworks), which provided for smooth acceleration and slick control of playback motion. As editing systems proliferated on PCs, though, the familiarity of editors with this degree of control began to wane, and has become a largely-forgotten remnant of the tactile heritage of film.

In lieu of a console, a growing standard is the "JKL"-style keyboard control, popularized in the Avid, and introduced into more systems every year. In this scheme, the keyboard's **J**, **K**, and **L** function as **Reverse**, **Stop** and **Play** buttons, respectively. By pressing the keys in combination, an editor can "rock and roll" back and forth over a cut, watching how the shots segue, in a method approximating what is time-honored in the editing tradition.

Some systems allow this kind of play control even in transition editing, so that a transition can be refined and "sculpted" very quickly. Many others don't.

For desktop video, however, there are alternatives to the basic keyboard. A small number of third-party vendors offer various "control surfaces" specialized for the tasks of post-production. Some are built solidly with well-tooled mechanisms and buttons with good tactile feedback; others are cheaply assembled and are little more than cosmetic additions to the junior edit bay.

Besides controlling playback, there are other good uses for external controllers: performing DVE manipulations, for

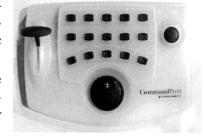

example. From an ergonomic perspective, a well-designed *joystick* is ideal for this moving-things-around-in-three-dimensions, quickly and precisely.

Responsiveness

These types of physical controllers are designed to respond to user input fluidly and **in real time**. While manufacturers casually tend to use the term "real time" only to refer to a system's freedom from having to render a particular effect before seeing it, they are only talking about what happens *once the effect has already been constructed*.

To create that sophisticated effect in the first place, though, you have to perform numerous painstakingly detailed adjustments. The graphical user interface (GUI) and the mouse are notorious for feeling "sluggish" for these

kinds of manipulations—this isn't typing, after all. And, while setting up a complicated effect, that GUI might have a delay (several seconds or more) in responding to you after *each* one of your "clicks," sometimes causing immense frustration by missing subsequent commands.

A good part of this occasional sluggishness is actually in the hardware. A DVE "board," for example, may simply not be as fluid and responsive as a similarly feature-laden stand-alone DVE unit

For color correction, sets of trackballs are efficient for controlling values in multivariant color space.

(that you might see in an online bay) that's really designed for more hands-on use. So when a manufacturer does provide a physical controller (or allows for one to be added on), this is an encouraging sign that may mean the hardware is genuinely designed to keep up quickly with your *continuous* moves, often not the case with mouse-driven desktop systems.

JL Cooper's
"Media Command Station"
is an example of one third-party manufacturer's approach to augmenting open architecture editing systems with a multi-faceted tactile user interface.

Functionality is only part of the editing system story. Ergonomic issues are visceral, not esoteric, and need to be considered in designing and choosing all post-production tools.

M O N I T O R S   a n d   W I N D O W S

The placement and use of video monitors was at one time a somewhat complex thing—and seemed to vary with virtually every video and nonlinear system. Systems generally had a two-monitor set-up: source material on one and master (cut sequences) on the other. In many cases these two video monitors were not sufficient, and a third monitor was required exclusively for editing data.

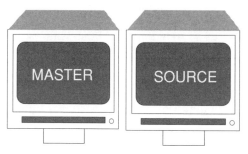

Most commonly, source material was shown on the right and the edited material (master) on the left.

Confusion with the layout of video monitors came from a number of situations. In the first place, film editors who worked on flatbed tables were used to source rolls being loaded up on the left side of the table and the take up reels on the right. This led to problems in transition editing because in a film editor's mind, the head of a shot was on the right and the tail on the left. Video editors, when working with transition editing, and not used to having a strip of film before them, tended to visualize the film running left to right, the way one reads English, and wanted monitors to represent this. Thus, film editors and video editors moving into nonlinear editing would request different monitor functionality and configuration.

Viewing all source and master material in a full-screen mode seemed ideal from an aesthetic viewpoint, but it could be logically difficult. Digital systems solve many of the confusing monitor layouts simply by working with smaller images. Rather than needing banks of monitors for views of multiple shots (whether in a true "multicamera mode" or simply viewing different takes), digital systems used "windows" and "picons" imbedded in the display to show smaller views of shots.

Digital systems are configured minimally with one, and preferably two graphics monitors, neither of which are specifically "source" or "master." Images are presented in windows that can take on many sizes and can be located in either monitor. Typically, there is a central editing monitor upon which everything is or at least can be accessed: timelines, source clips, transitions, *etc.* But managing all these overlapping windows and many source

clips can be daunting, and a clear advantage is reaped when two monitors are used in tandem. In the traditional nonlinear two-monitor set-up, the monitor on the left provides a location for spreading out small thumbnail images of each source take. The right monitor is the editing and system environment, and will have imbedded in it locations for "source" and "master." But an editor may configure any set-up to his or her liking.

When involved in transition editing, the master window itself is often subdivided to show outgoing and incoming frames. It is somewhat ironic that in the adjusting of a transition, perhaps a key moment in the editing process, the images themselves are relegated to a size only a fraction of the original image. Somewhat better are systems that at least maintain the full size of the source/master windows and use them for the outgoing and incoming frames.

Consumer editing systems often default to source material displaying on the left, with cut sequences on the right. Users can configure these windows in any way that facilitates learning or using the systems. Whether on the left or right, full screen or thumbnails, configurable or rigid, editors easily adapt to the locations of the elements of the editing system.

Most software will default with source on the left, edited sequences on the right, but as with many other features, editors can often set preferences to their liking.

When digital systems use "multisync" monitors, it means that the monitor can present either the computer data, or actual video (like a TV

▶▶ System interface historical overview: p. 324

set). It is generally advisable for any editing configuration of computer monitors to be augmented with an NTSC display, which will present the streaming video in a WYSIWYG way. (Computer displays tend to have difficulty displaying the smooth streaming of video even if there are no bandwidth problems, due to frame rates and sync signals.) Clearly, if the intended output of the system will only ever be as streaming web video, there is no need ever to evaluate video in NTSC.

A typical professional nonlinear edit bay, with two computer monitors and one NTSC monitor. Also notice the keyboard and mouse controller in the front, and the audio console on the right.

T I M E L I N E S

All editing systems use a metaphor for video and audio tracks called a *timeline*. Although timelines in various forms have been used for eons, their application in nonlinear editing, specifically in making computerized film-style editing more accessible, was first introduced on the EditDroid in 1984:

Since then, most systems have adopted similar or their own versions of the timeline. Here is a typical example:

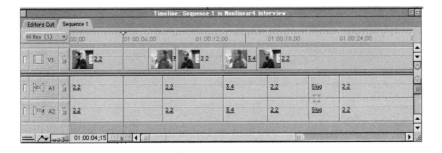

Horizontal timelines are not the only way time can be represented; however, virtually all systems have self-standardized them as running from side to side on the editing screen. If the timeline were designed to depict a flatbed film editing table, time zero ("heads," or the beginning of a sequence) would start on the far *right*-hand side (like the example from the EditDroid, above). Unfortunately for film editors, virtually all systems provide for time zero to be on the *left*-most side ("heads" on the left). This would be consistent with the transitional edit method of having outgoing A-side frames on the left, and incoming B-side frames on the right.

A timeline is actually a series of "tracks" running together as if in a synchronizer. Picture, usually on the top, may be labeled "video," "V1," "picture," or some such tag. Audio channels lie underneath, labeled "A" or "SND." Systems often offer many more (sometimes unlimited) "virtual"

▶▶ System interface historical overview: p. 324

tracks of audio, but may be limited to playing only *some* simultaneously, in real time. Also, if the system has special effects tracks (for keys, titles, *etc.*) there may be one, dozens, or hundreds of additional video tracks that will run above the main video layer.

In audio, another feature that offers a powerful editing tool is the display of the actual audio waveforms rather than simply a graphic bar representing sound. These audio waves, made possible by the system's use of digitized audio, allow for accurate visual guides to individual parts of sounds. One of the earliest uses of this display was in the SoundDroid prototype, from 1985.

Sync picture and a pair of stereo audio channels (A1 and A2) showing audio waveforms. Particularly when zoomed in close, waveforms can be an invaluable tool.

The existence of a timeline is a significant advantage over text displays. The advantages are obvious: editing electronically has removed the last vestiges of physicality in editing—there is nothing to hold onto, no edits actually taking place, nothing being cut. A timeline provides a virtual-but-viewable object to which editors can relate their work. As well as appearing as a visual guide, the timeline makes computerized editing a more geometric exercise—like film—in that you are moving shapes and changing them until they fit in appropriate ways. It usually makes the computers easier to use.

But there are still subtle—less tangible—drawbacks. A timeline, like the icon, is still just a computer phenomenon. Having a timeline generally means that an editor is dependent on it for the execution of edits. An active timeline means that the computer is not in the background, but central to the editing process. You must "point and click" on the display to perform

◀◀ Digital audio waves: p. 145

certain editing functions. It should be recognized that it is possible to edit without manipulating a timeline, although systems are no longer designed with this goal in mind. Today all systems have timelines and are considered simpler to use than systems without them.

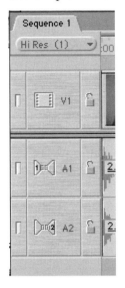

Aside from where heads and tails are found on a timeline, there are a number of features commonly associated with this kind of graphical display

• **Track enabling.** Usually along the border of a track are controls that identify where edits will go, what layers are visible, whether or not a track is "locked" (unable to be changed) and so on. The paradigm is adopted from the maintenance and enabling of layers in graphical software like Macromedia Freehand, Adobe Illustrator and Photoshop, among others. Being able to identify, isolate, and control tracks is essential to editing.

At left is a example of video ("V1") and a pair of audio tracks ("A1" and "A2") and their enabling controls.

At the bottom of this timeline display are a pair of scaling tools (right), the run time indicator of the now-line, and (left) a track width controller, for increasing the vertical height of the tracks.

• **Zooming in and out.** Although the horizontal and vertical displays of the timeline are bounded by the screen, the timeline in its entirety may be enormous. Consequently, editors often want need to scale the display in various ways. The smallest functional unit of picture editing is *the frame*; this is the closest users would ever need to expand. By zooming in, very short edits can be easily identified and manipulated; this couldn't be done from farther out. With point-and-click editing, zooming in (and out) is essential to locate relatively short shots. The "farthest" unit an editor might want to see would be the sequence in its entirety—30 seconds, 10 minutes, 2 hours . . .

Timelines have a scale that is generally user-configurable—where the span can be compressed or expanded depending on the view that is appropriate for the work being done. In general, wide views are best for finding your bearings, sensing pacing, and moving around. For actual editing, ranges of 3 to 15 seconds are about right.

The top timeline has a scale that covers only 12 seconds; shots and transitions are clear and readily manageable. The bottom timeline range covers more than 5 minutes, and the edits, although mostly straight-cuts, blur into a reasonably non-useable form. Still, it provides a reference as the entire sequence is played.

• **Scale units**: Although most electronic editing systems are based on video timecodes, those that track edge numbers allow the scale to be converted to feet and frames. Record times (run times) are usually displayed along the axis of the timeline, enabling quick identification of relative lengths of shots and approximate durations in a cut sequence. Record start time is generally a preference to be set, but traditional record times start at 1 hour (01:00:00:00), and not zero.

• **Identification**: A graphic display is just that—graphic. In addition, some method for identifying the source information for a single shot is usually provided somewhere on the display. Text (scene/take information, source in/out points, descriptions, video event numbers) can be located in affiliated displays, or if room is available, actually inscribed in the timeline itself. In either case, this information is important to have in order to use the timeline effectively. Scene naming nomenclature should be considered in conjunction with the system's method for displaying names in the timeline.

• **Color and shading**: Sometimes color or shading is employed to help distinguish shots originating from different source dailies. This is an ideal way to identify the relationships between picture and sound, especially in complicated editing where there are stolen lines, split-edits, and frequent differences between source picture and track.

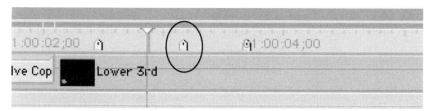

• **Cue points and marks**. An editor can usually place special marks in the timeline for quick cueing. The number of available marks may be limited, but at least some are essential for syncing material up or locating chosen locations in long shots and sequences.

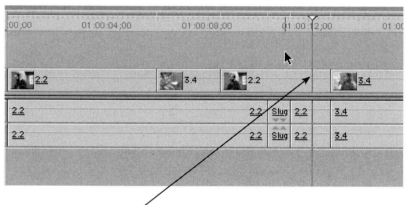

Final Cut Pro courtesy of Apple Computer

• **"Now" line or "play" line**. There is always some indicator on the timeline as to where the virtual master is currently positioned. The timeline will either move past a fixed "head" location (like celluloid on a film table), or the *now-line* will run back and forth over the timeline. The moving now-line is the most common paradigm.

A secondary but important feature of the now-line is its ability to track the rolling master wherever it plays. Some systems only update the timeline when the master stops playing; better systems actually scroll the timeline or now-line as the master plays. This feature is beneficial to editors, but does involve more computer work (screen refreshing). Also, moving the now-line rather than the timeline forces the timeline to "jump" when the now-line reaches the edge of the screen—and can in some cases be confusing.

• **Durational accuracy**. Timelines represent shots with lengths that are proportional to the shot's duration. One clever augmentation of timeline mechanics is to put the timeline through a "lens" of sorts, which compresses the shots farther away from the now-line, and expands the shots toward the center. This distortion is quite efficient, as the only time you ever need to have durational accuracy is at the point where you are working, and *access*-only to the material farther away. This implementation also saves the editor from the jarring and incessant shifting of the timeline during searches and playing.

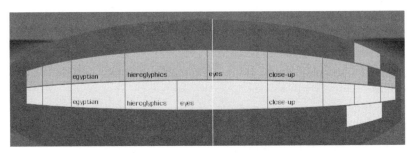

Judgment courtesy of Eidos Technologies

• **Variations**. Although most timelines run horizontally, some systems run them vertically. The traditional audio "cue sheet" is a vertical timeline, as are many digital audio workstation displays that mimic these.

Also, many timelines can be converted to unique presentations of the edited sequence, among them "head views" and "head-tail" views that show small images of the actual frames at the splices, adding to the graphic nature of the computerized editing system. Also possible, but generally more eye-candy than required editing functionality, are "filmstrip" views, displaying every frame of a sequence. Compositing systems and graphically heavy post-production utilizes this frame-by-frame work more than editorial.

In any format, timelines can provide for simple moving of shots, re-ordering existing sequences, and easy deletion of unwanted shots. Clear and easy-to-follow timelines are a real asset to a nonlinear editing system. Before investing time (or money) in one system, its functionality regarding timelines should be investigated.

Beware of timelines that use or force you to alternate ("checkerboard") material between separate "A" and "B" tracks in order to do transition effects. What if you later decide to re-arrange the order of the shots? While some systems will manage the rippling of dual tracks intelligently, others require you to do a lot of cumbersome **re**-checkerboarding.

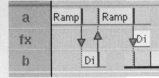

FUNDAMENTAL EDITS

There are virtually infinite ways to edit material. But there are a small number of basic editing "situations" that come up so frequently that anyone who tells stories in video and audio should become familiar with them. The ease (or difficulty) of these functions is often glossed over when products are tested and demonstrated, in favor of the flashier abilities associated with effects. In truth, most editing time is spent doing a small number of techniques that, while not technically complicated, are mental work in the art and skill of editing, and really should be easy to do. Desktop editing systems tend to lump these important functions in with other, somewhat more obscure tools; find out how to do these functions easily and you will be able to edit competently on any system.

Let's illustrate with a short edited sequence. We are concerned more with the geometry of the shots and not so much the aesthetics at this point.

Three shots are edited together. The first, a close up of Mike (called 2.2—scene 2, take 2) is edited with a straight pix+snd (our shorthand for "picture and sound") cut to shot 3.4, a close up of Jen. Then, when she stops talking, it cuts back to Mike and shot 2.2, again with a straight cut. Once this foundation is laid, and the flow is working, some basic refinement takes place:

1) Cut-away. A pix-only insert (here, of Jen). This lets us see reactions. Sometimes it is more important to see someone listening than someone talking. An excellent practical use of a cut-away is to bridge a *jump*

▶▶ Practical editing methods: p. 353

cut. In this example, if Mike is talking too slowly, a chunk of pix+snd can be cut right out of the middle of the shot—creating a jump cut. Normally, this would be too jarring to leave, but a cut-away that covers this jump can be exceptionally useful.

2) Slug with leader. A sound-only insert of "leader" or "black." This is a time-tested fast way to lose a noise, a line, a flub, or some other bit of the audio track. No "time" is lost, just a piece of sound. A more sophisticated version of slugging with leader is slugging with **fill,** when a snippet of ambience from the scenes being edited is used to slug. It makes for a smoother track.

3) Overlap. Executed by a pix-only slide cut (every frame added to one side of the edit is removed from the other, keeping everything in sync). In this case, the picture edit precedes the sound edit. When picture and sound edits happen 10–40 frames (or more) apart, conversations tend to flow more naturally, and feel less "ping-pongy."

Honorable mention: **Lift, with ripple.** A cut in picture and sound together that tosses out a chunk of material, sometimes a whole shot and sometimes only a few frames, and then pulls all the following shots up to fill the space. Lift, with ripple, is a little nip and tuck at an edit point (and sometimes in the middle of a shot to make a jump-cut for later fixing) and it's a key tool for the first re-edit pass, to set the pacing of the story. It is most effective and streamlined when it is done before picture and sound edit points begin to diverge from each other, due to various inserts and overlaps.

Inserts and overlaps, picture only and sound only: these permutations are sufficient to do 85% of all narrative editing. You could edit Academy Award-winning films your whole life, and never need to do a traveling matte or a motion effect. *These are your daily tools.*

◀◀ Primitives of editing: p. 216

DIGITIZING and CAPTURING

With the generation of digital editing systems comes the art and science of digitizing compressed (and uncompressed) video—a necessary evil on the evolutionary path to all-digital disk-based shooting, but one that is likely to be a component of post-production for a number of years.

Digitizing is the process whereby the source video is converted from NTSC (or PAL) video into digital data and stored on some type of magnetic storage media, usually hard disks. *Capturing*, like digitizing, is the moment when source material is brought into the editing system from videotape, usually in real time. Unlike digitizing, capturing is the process used when the source material is already in the digital domain—having been shot on digital videotape. Capturing is more "data transfer" than anything else, although the methods of video capture *feel* more like the digitizing process than they *feel* like file copying.

Digitizing and capturing are the processes that precede all other editing activities; they may be separate from or part of the "logging" of the source material. At a purely technical level, digitizing begins with feeding the video source into a digitizing card that is connected to the computer; the card samples the video, assigns numeric values to each pixel, and records them onto a disk.

Video can be digitized or captured via an *"assembly" method*, or may be grabbed via a *"dub" method*. The *assembly method* involves creating a type of *off*-offline edit with a videotape deck and an editing system (or a remote computer). The editor or assistant rolls through a videotape marking in- and out-points of the desired shots on the tape, creating a kind of videotape EDL for raw source footage and associated data. This special decision list is then used by the editing system to perform a **batch digitize**, an automated process for digitizing each shot as designated on this EDL. It is a sort of auto-assembly, complete with an edit session, a source tape, and a master (in this case, a digital random access master). Batch digitizing allows material to be logged remotely, without tying up the editing system with the somewhat laborious logging session—you know: selecting the heads and tails of each source clip. It is done with increasing frequency as the cost of the digital editing system increases in cost.

If tying up an editing system is not too costly, an editor may use the system as a tape controller and mark ins and outs, then record (*i.e.*, digitize) each desired shot. Regardless of whether digitization is done immediately or later, this method requires pre-selection of desired source material. If source material originated on film, and was then telecined to videotape, often the "EDL" for assembling comes as a *telecine log*. This telecine EDL provides the data for the auto-digitization of the source clips. When there is a large amount of source material relative to selected shots, the assembly method is most efficient.

The *"dub" method* involves the real-time transfer of the entire source videotape (or large portions of the tape) onto the digital editing storage media. While this method allows "immediate" availability of all source material, it is generally considered more wasteful of valuable digital disk space with extraneous material. Still, it removes the necessity of the "off-offline" edit (the laborious pre-selection of shots). The dub method is rel-

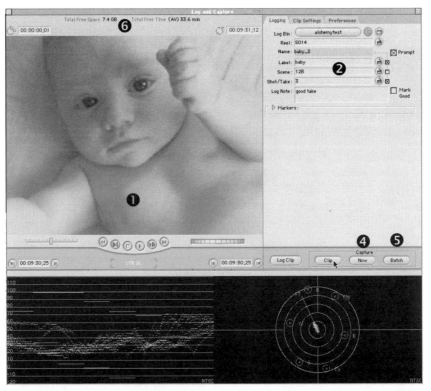

A log/capture window will generally have tools for (1) watching the video with motion control beneath, (2) entering log label information, (3) software waveform monitors and vectorscopes, (4) options for capturing on the fly or (5) via an EDL in a batch digitization or capture. The window also has (6) various fields showing mark points, timecodes, resolution/space on drives. In all capturing, along with proper head-roll and roll-out space, good timecode on the source tape is a critical factor.

atively inefficient except in systems that make it exceptionally easy to discard extraneous pieces of source material at will. With this functionality, source material is removed from the digital storage device as soon as it is determined not-useful rather than prior to capturing. As the cost for disk storage decreases, on-the-fly capturing via this method certainly increases in expediency. To dub in video, it is important that source material have continuous unbroken timecode, or recapturing later (or online) will be virtually impossible.

The dub method can be critical when source timecode is unavailable. Though some systems may abort capture when discontinuities in timecode are discovered, sometimes these failings are only "noticed" if the tape is cueing, and breaks during a capture are missed. The result is that poorly timecoded material *can* be captured in a giant dub, and then restored to a new tape with different timecode as needed. Alternatively, when source material is digital and no online (or EDL) is expected, non-coded material can be captured and used; but the user must realize that media files from re-capturing the material will likely not correspond to the formerly-created sequences.

Digitizing is primarily in real time and often uses a video compression scheme to reduce the huge amount of data in the video frame into a usable and manageable size. Most compression rates range from 5:1 (for DV quality) to around 20:1 (common in Motion-JPEG compression for offline). With 3rd party hardware, uncompressed SDTV (and even HDTV) video can be captured and manipulated in many open architecture editing systems. Editors should determine the degree of compression required prior to starting a project, even if the system allows for multiple compression rates. Deciding this compression factor is based on estimates of the total required source material, and/or the amount of digital storage available at a given cost, and/or the desired resolution.

Audio is usually not compressed when digitized. CD-quality audio is 16-bit sound, sampled at 44.1KHz. Audio is frequently digitized at this quality, to allow for "online" type audio work along with the video; however, disk space can be saved by digitizing the audio at 22KHz or even 11KHz. For some specific applications, audio is digitized at 48KHz, which would of course utilize additional hard disk space—it is perceptually undifferentiable from 44.1KHz audio.

The idea of "appropriate" image resolution is a subtle one. Every type of digital compression has associated with it some kinds of inherent *artifacts*—visual anomalies—that are created in the images. If these artifacts are understood, certain problems can be avoided. For example, demonstrations of digital compression at trade shows tend to use demo materials with bold bright colors (azure blue sky at the beach, deep green water, bright yellow towels, *etc.*) and lots of close ups (eyes, faces, hands). While the average viewer might regard these as good measurements of digital compression quality, close shots and bold colors are actually masking many compression artifacts. A close shot may seem to be showing more detail, but only of the *subject* and not of the *video*. A far more difficult image would be one with lots of motion, lots of tiny details, and few patterns. A crowd in a football stadium. An autumn forest in New England. A bag of multicolored beads. A shadowy figure running through an alley. Particularly for MPEG, moving shots also demonstrate compression artifacts more than static shots. Examine dollies and pans when possible. Shots that show the familiar artifacts and are good litmus tests for digital systems. If you have scene-by-scene control over the digital compression during a project, and saving disk space is important, you might decrease resolution for medium and close shots, and increase it for wide or detailed shots. These are also key issues with the significantly higher compressions utilized for streaming web video.

Artifacts include such problems as *scintillation* (tiny sparkles in detailed regions of the frame), *quantization noise* (normally-stationary background elements may appear to be busy with small wandering colors), *color banding* (patterns of color that show up as bands in smooth areas), and the all-too-common *blocking* (anomalies created from the DCT mathematics and 8x8 interframe units that create hard edges in otherwise regular areas of the video).

Offline nonlinear systems usually digitize one field of video for every frame—creating images with half the effective resolution of the original videotape. In systems that utilize this method, most default to digitize field 1. While this creates good transfers for video-originated source, it causes some motion artifacts for material that has been 3:2 transferred from film. The only way these motion aberrations can be avoided is to digitize at 24fps, with an understanding of the type of 3:2 pulldown used to create the videotape. Dedicated nonlinear systems designed to work with film accommodate a real 24fps digitization—the preferred method for digital offline of film—avoiding the creation of motion artifacts and saving 20% of the disk space which is needed for the same material transferred at 30fps.

V I D E O T A P E   A S S E M B L I E S

The degree of nonlinearity is largely a function of how long you can edit without committing those edits to videotape. There comes a time, however, when it is either desired or perhaps essential that a linear assembly to videotape be performed. All nonlinear systems can record edits (output) to videotape. Sometimes a tape assembly is required when hard disk storage is insufficient to hold all the source material: in that case, an editor might choose to do a first pass, culling material down to smaller quantities before beginning a nonlinear cut. Other times, a system is performing offline work, and a high-quality assembly of the edits is required using an EDL and an online system. For whatever reason, an editor might need to interface with tape assemblies.

There are two primary ways in which a system can create a videotape of an edited sequence. The first and most common is to "print-to-tape." Less common on current nonlinear systems, but still prevalent in the online world, is the somewhat traditional "auto assembly." The basic concepts of the auto-assembly are applicable to any linear medium that must be managed or edited.

Printing-to-tape is a real-time assembly and wholly unlike standard auto assemblies. Since digital systems can preview long portions of edited material directly from the system (often an entire production), they have the ability to record it onto a "downstream" videotape. Most professional systems have machine control (to control a tape deck for digitizing), but some may not. With no machine control available, a print-to-tape option is the only way to get the video edits directly out of the system. The user would need to run a cable from the video/audio outputs of the system to the inputs of a video deck of some kind (FireWire makes this easier), and then manually start recording on the deck and playing of the edited sequence.

But even with machine control, there are issues of print-to-tape that involve the frame accuracy of the tape produced, especially when the project is to finish on film.

Unless the output is via a digital line and in a digital format (*e.g.*, DV over FireWire), the image quality of the print-to-tape output is going to be one generation worse than it was on the system in the first place (remember that it is a digital-to-analog conversion to get the output onto videotape). If image quality is an issue, there must be a digital output or a source tape-to-master tape auto assembly.

The traditional methods of videotape assembly—non-real-time tape-to-tape recording controlled by the editing system—can sometimes be employed to get an online master tape out of the offline editing system, but to do this the editing system would need to be able to control two or more video decks simultaneously. Regardless of the computer handing the assembly, whether for BetaSP or 24P, certain strategies have been developed that aim to make assemblies from linear source material more efficient. Some are standard (the major ones listed here); others can be used, usually facilitated by computers crunching on EDLs and strategizing the most efficient means to accomplish the particular goal.

A-mode assemblies record each shot in the order they appear in the edit decision list (EDL). These assemblies build the tape sequentially, as it will appear when complete, with no regard for the location of the source material. A-mode assemblies are often considered inefficient, though more flexible in online sessions should changes be required.

B-mode assemblies record shots in the order they appear in the edit decision list, just like in A-mode, except that if a source reel is missing, the shot will be skipped. Every edit that can be recorded from available source tape(s) will be. When this pass is completed, other source tapes will be "requested" by the computer; once loaded, the missing shots will be assembled. If more than one source is accessible at one time, systems will use all available sources. This kind of assembly is often called a "checkerboard" assembly since it leaves black "holes" in the recorded master to be filled in later by not-yet-loaded source tapes.

In a B-mode assembly, the first pass across the record tape assembles all the shots it can, based on available source tapes, leaving unrecorded "holes."

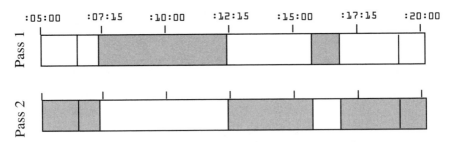

Then, when the pass is complete, the computer will ask for the missing tapes, and record those shots now available into the gaps. This assembly can only work if the computer doing the assembling (and the deck) are frame accurate and leave no errors. Even a 1-frame miss will show up and can ruin the entire assembly.

C-mode assemblies are very much like B-mode, except that the order of assembly is not determined by the EDL but rather by following the ascending location on the source tape (like a film pull list). This kind of assembly makes most sense when performing a tape-to-tape assembly with only a single source reel at a time.

As a general rule, you'll want to do a B-mode if the record tape is long and the sources are short (or fast); you'll do a C-mode if the record tape is short relative to the sources that are long (or slow).

D- and E-mode assemblies are slightly esoteric variations on B- and C-mode, respectively, and simply move any multiple-source transition events to the end of the EDL. This can be useful when you want to do the "easier" part of the assembly first, before dealing with lots of effects.

Linear auto assemblies take longer than real time. Depending on the length of the source tapes, the number of edits, and length of each edit, a videotape assembly can take anywhere from five to eight times the program's real duration. Because of this, the longer a project can refrain from assembling a videotape copy, the faster editing and re-editing can progress. In fact, this is one of the attractions of nonlinear editing—frequent tape assemblies considerably reduce the cost and time efficiency. Realize that linear systems, although they edit considerably slower and with greater difficulty, are building the videotape of the edited project as they go. Like film projects cutting film, projects cut via linear videotape systems are ready for delivery as soon as a editing is completed.

EDL Assembly Modes

Mode	Sorts by...	Typically used when...	Examples
A	Record TC	The record reel is much longer than the source reels; flexibility is needed in assembly, or the list isn't "clean"; any other uncertainty regarding the best choice of assembly mode.	Hour-long studio shows taped roughly in sequence, and with few sources
B	Source Reel, then Record TC	The source reels are shorter than the program.	Long-form: (documentaries, episodics, MOWs)
C	Source Reel, then Source TC	The source reels are longer than the program.	Short-form: (commercials, promos, music videos)
D	Source Reel, then Record TC; but puts transition effects at the end of the EDL	The same situation as B, but when there are lots of effects.	– – – –
E	Source Reel, then Source TC; but puts transition effects at the end of the EDL	The same situation as C, but when there are lots of effects.	– – – –

P R I N T O U T S

All professional systems have paper (or "hard copy") printouts as an output option from the computer.

The most common type is the **Edit Decision List** (EDL). You can designate a particular edited sequence and then the output format: Drop or non-drop; CMX, GVG or other type; a record start time, and EDL file name. This is an example of a CMX-type EDL:

```
TITLE: REEL 4          sc. 20 - 27        5,6,7

FCM: NON-DROP FRAME

001   BL    V    C          00:00:00:00  00:00:00:00  01:00:00:00  01:00:00:00
001   05    V    D     030  05:27:10:09  05:27:11:09  01:00:00:00  01:00:01:00
002   05    V    C          05:27:11:09  05:27:39:15  01:00:01:00  01:00:29:06
003   05    A    C          05:26:56:27  05:27:00:17  01:00:00:00  01:00:03:20
004   BL    A2   C          00:00:00:00  00:09:45:05  01:00:00:00  01:09:45:05
005   05    A    C          05:27:13:29  05:27:39:15  01:00:03:20  01:00:29:06
006   05    B    C          05:28:04:02  05:28:12:25  01:00:29:06  01:00:37:29
007   05    V    C          05:29:01:07  05:29:08:29  01:00:37:29  01:00:45:21
008   05    A    C          05:29:01:07  05:29:05:02  01:00:37:29  01:00:41:24
009   07    A    C          07:04:07:02  07:04:23:28  01:00:41:24  01:00:58:20
010   07    V    C          07:04:10:29  07:04:19:07  01:00:45:21  01:00:53:29
011   07    V    C          07:04:18:07  07:04:18:07  01:00:52:29  01:00:52:29
011   06    V    D     060  06:05:52:12  06:05:54:12  01:00:52:29  01:00:54:29
012   06    V    C          06:05:54:12  06:05:58:03  01:00:54:29  01:00:58:20
013   06    B    C          06:09:24:27  06:09:28:04  01:00:58:20  01:01:01:27
```

Next are the **film cut lists**. There are primarily four kinds: *assembly lists, pull lists, optical lists,* and *change lists.* For editing systems that work internally at 30fps (like video source material), they have within them algorithms for converting 30fps timecode into film numbers. For 24fps systems, the generation of film lists does not require such math. In both cases, a relational database will provide the link between the timecode numbers determined during the telecine session and the film numbers, allowing the system to "trace" (or match) back to the film. Usually this software is an integral part of the editing system, but more and more video editing systems require a separate computer program to complete these functions.

▶▶ An *assembly list* is made up of pairs of edge numbers, one for the in-point and one for the out-point of each shot in the cut sequence (like an EDL). These shots are listed in sequential order according to when they occur in the final edited sequence. On the following page are two different assembly lists:

Edit: R-1AB DZ-4TH SND   Project: **LADIES MAN-SCOTT**
Edit dated: 15:55, 6 Apr 2000   Format: 35mm
Report dated: 16:06, 6 Apr 2000   Cutting-copy version 5(CU13.2)
Assembly list

	Start	Length	Roll	Shot name	Start code	End
1	0+00	12+00	000-1	Academy Leader 000	1000+00	1011+15
2	12+00	27+07	850-1	Paramount Logo 850	1026+04	1053+10
3	39+07	26+06	059	FX1-1	800 1349+08	1375+13
4	65+13	1+15	252-1	4J-1	252 1253+04	1255+02
5	67+12	3+05	082-1	2-4	082 1154+15	1158+03
6	71+01	4+14	243-1	4A-2	243 1883+12	1888+09
7	75+15	2+13	082-1	2-4	082 1163+02	1165+14
8	78+12	2+03	252-1	4N-4	252 1771+00	1773+02
9	80+15	1+09	252-1	4P-2	252 1873+13	1875+05
10	82+08	1+14	252-1	4M-1	252 1508+06	1510+03
11	84+06	1+12	252-1	4M-3	252 1611+08	1613+03
12	86+02	2+14	082-1	2-4	082 1173+05	1176+02
13	89+00	3+09	252-1	4M-3	252 1618+08	1622+00

If a film-cutting list type is not specified by name, it is generally an assembly list.

➤ A *pull list* has the identical information as an assembly list, however it has been sorted not by sequence number (how they edit together), but rather by ascending film numbers. By organizing the shots in this way, an assistant or negative cutter can *pull* shots in the order they appear on the unrolling film. This facilitates the breakdown of film for future assembly:

It is important to note that for pull lists and assembly lists, there will usually be two options: *negative* or *workprint*. The lists are identical except that the negative list is built from key numbers and a workprint list is built from code numbers. Consequently, workprint lists have a further option of printing picture numbers or sound numbers.

```
CMX6000 WORK PRINT PULL LIST:        CUT MARGIN: 0 FRAMES      MEDIA: 35mm
PRODUCTION: THE SHELTERING SKY
VERSION #                          SPLICE LIST VERSION
   0004        REEL 4            sc. 20 - 27              5,6,7
--------------------------------------------------------------------------
             CODE NUMBER DUPLICATIONS IN THIS REEL
             SEQUENCE #   <   Overlaps with   >   SEQUENCE #
                  000                                  000
--------------------------------------------------------------------------
     LAB     SEQ   SCENE   TAKE        FIRST           LAST
     ROLL #   #                        CODE #          CODE #
--------------------------------------------------------------------------
     R-15    001    21     17    026       90.12   026      134.08
     R-15    002    21     26    026      247.11   026      260.12
     R-15    003    21     33    026      388.12   026      400.05
     R-15    008    22     15    028      218.06   028      243.12
     R-15    005    22B    11    028      370.15   028      377.15
     R-15    009    22B    11    028      420.08   028      434.02
     R-15    007    22B    15    028      651.04   028      665.07
     R-15    013    22B    15    028      730.05   028      740.14
     R-16    006    22     22    029      231.03   029      236.00
     R-16    010    22     22    029      288.11   029      292.04
```

Another factor in film list printing is the *cut margin*. Although film editors splice workprint with 2 to 4 sprockets of clear tape, negative cutters actually melt adjacent frames together with a "hot splicer." For 35mm film, editors leave a single frame clear at all splices to allow the negative cutter room to make the splice. Even though this hot splice only destroys a sprocket or so of film, allowing for a full frame is considered safer. This boundary frame, called the cut margin, must be accounted for at all splices. Prior to list printing, editing computers require a value for the designated cut margin. When a computer checks an edited sequence for use of duplicate or adjacent frames, the cut margin must be factored in to ensure that the negative cutting will go smoothly. Duplicate frame use will be part of a pull list or be in its own separate duplicate printing list.

▶▌An *optical list*, or optical count sheet, is a way to recreate on film those effects created and previewed in video. Many systems will locate effects in an edited sequence and generate a count sheet for an optical house to build the identical effect using film elements. These lists define the type of effect, indicate start and stop key numbers, and durations of the given elements.

```
Edit:          SC 50 VER 2              Project: MRS. DOUBTFIRE
Edit dated:    21:08, 15 Sep 1993       Format:  35mm
Report dated:  21:31, 16 Sep 1993       Cutting-copy version 4 (6.1)
Opticals list: printing handles 32 frames


     Start printing      Start out    Length   Full in             Stop

  4  50A-2 (roll A19)                  Dissolve 50B-1 (roll A19)
     KJ458822 5248+06    5258+08       3.00     KJ458822 5304+00    5307+10

  8  50-4 (roll A19)                   Dissolve BLK-         Cam (roll A19)
     KJ458822 5171+10    5179+06       3.00     KJ458822 5414+01    5420+09


End of opticals list
```

list courtesy of Lightworks

▶▌ A *change list* is the least commonly found list in general purpose nonlinear systems, but it is essential for electronic editing systems cutting film projects finishing on workprint. In the normal course of events, scenes are edited offline, electronically. Once scenes are cut, they will often need to be screened in a theater. This will involve the use of pull and/or assembly lists. However, once the cut film exists, changes made to the sequence on the electronic system will be somewhat difficult to re-create in the now-edited workprint. Because of this, a separate kind of list needs to be generated: one that will compare the original cut (which exists on workprint) to the current cut (created on the nonlinear system), and inform an assistant how to modify the workprint to bring it up to date. Change lists are also called "diff" lists because they highlight the *differences* between two edited versions.

```
Report: LADIES MAN 4TH CUT                    Original: R1AB-5AB DZ-2
Date:    6 Apr 100                            New:      R1AB-5AB DZ-4

Edit Reel No: 1
             One Pass List of cuts, inserts and dissolve changes
```

STEP	FOOTAGE START	FOOTAGE END	LOSE FTG	ADD FTG	SC.	CODE #'S DESCRIPTION	CUM CHG
1.1 *	408+07	413+02	4.12			205-1094+03-1098+14	
							-4+12
2.1	441+11	445+03	3.09			207-1218+02-1221+01	
2.2						207-1186+09-1187+01 Head Trim	
							-12+10
3.1	441+11	446+10			5.00	207-1253+08-1258+07	
							-7+10
4.1	449+08	452+10	3.03			207-1189+15-1189+15 Tail Trim	
4.2				.		207-1229+10-1232+11	
							-10+13
5.1	1006+08	1035+08	29.01			247-1217+06-1217+15 Tail Trim	
5.2						198-1135+14-1144+13	
5.3						065-1118+15-1124+00	
5.4						198-1355+09-1364+09	
5.5						065-1144+01-1149+04	
							-37+13

Many film projects only use electronic systems for *first* cuts; once they conform workprint, all additional changes are done on the celluloid workprint in traditional ways. This method is used because of the general difficulty in revising workprint from a change list, and because workprint editing is so widely known. On editing systems without the change lists feature, this method is the only option.

A change list will generally present the fewest modifications for converting one edited sequence into a new one. The user will have to select the two lists—presumably an old and a newer version of the same sequence, and then "request" that the system generate the change list. A change list is based on the fundamental property that only two actions ever happen when changing an original workprint into a new version: (1) a piece of film can be CUT out, or (2) a piece can be INSERTED in. Exactly how these operations are performed is critical. They can be done in two passes through the workprint (first to do all the cuts, then all the inserts), or in a single pass (cutting and inserting as you go). Neither method is better than the other, and *meticulous* attention to detail must be maintained throughout the entire process in either case.

Until digital projection of cinema from editing systems is mainstream, considerable labor and use of these printouts are required to maintain correlation between film and digital video.

▶▶ *Storyboards.* Some systems have the ability to laserprint cuts or dailies in storyboard formats. These printouts are particularly helpful when working with clients familiar with storyboards—as on feature films or commercial projects. The original storyboard print feature was offered in the Montage Picture Processor (below).

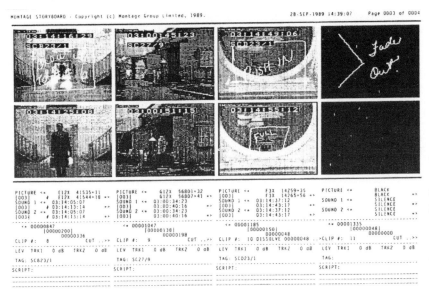

print courtesy of the Montage Group

▶▶ *System lists.* This is the catch-all category of hard copies of specific screens viewable on a particular editing system. These might include printouts of dailies listings, bins, logging screens, metadata reports and so on.

Chapter 5

Distribution

"Someday, some little girl in Ohio is going to make a beautiful film with her father's camcorder and become the next Mozart. Someday, the process will be demystified and available to everyone, and movies will become the artform they were meant to be."

—Francis Ford Coppola (1991)

Director: Apocalypse Now, The Godfather

A B R I E F R E V I E W O F B R O A D C A S T I N G

The average American household watches seven hours of television every day. Many people watch considerably more. Where do these images originate, and how do they arrive? It is difficult to have much perspective on the cable networks, and the Internet, and the distribution revolution, if we're unfamiliar with how broadcasting has already existed for decades and how it has evolved in recent years.

Programming

Someone, somewhere, creates a program. Sometimes it's original and sold to a network. Sometimes a network wants it created, and hires the people to make it. The prototype is called the *pilot*. Pilots are made, put on the air, and then either a number are ordered or the show is bagged. A typical order might be for 13 episodes, with an option to get nine more (the "back nine") if those work out. A full season is around 22 episodes. A one-hour episode of your run-of-the-mill TV show is shot on 35mm film, and costs about $2–3 million, on average, to produce and post. It is shot over the course of about a week and a half, and edited and posted over the next three. Sitcoms, shot on videotape (and only occasionally on film), get shot in about a day, and post-produced over the following week.

Once an episode of a show is cut by the editor, the director gets a few days to review it and revise it if needed; after that, the producer gets a few more days to review and revise (although they generally change pretty little). A show is cut precisely to "time," the contractually-agreed-to duration of programming. It would not be unusual for a one-hour show to contain precisely 44 minutes and 5 seconds of "show." When the producer signs off on it, the show is delivered to the network for approval. They bought it, after all, so they decide if it's done. If it was finished on film, it is telecined to a broadcast*able* master tape (it used to be 1", now it is D2 or Digital Betacam). In general, however, it is finished on videotape, and the offline creative work is regenerated in online, where the effects and color correction are done. This master tape has black slugs in the place of the opening credits, the commercial breaks, and the end credits. Once approved, a tape-to-tape color correction is done, and the online editor inserts titles, but leaves black slugs for ads. The show is microwaved to the network HQ, usually in New York.

The show is prepped for air by editors at the network, who insert the paid commercials (which arrive under separate cover, also usually on D2 or Digital Betacam), and then it is assembled onto carts, for broadcast.

Broadcasting

Broadcasting essentially consists of sticking a program tape into one of a series of cart machines—good VCRs with precise timing triggers—and hitting "go." This contiguous feed is beamed up at one of a handful of broadcasting satellites—basically radio relay stations in the sky.

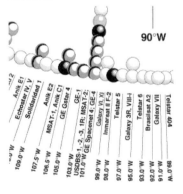

A map of broadcasting satellites in their geostationary locations around the earth. Below is detail showing names. Shading indicates operating frequency.

A satellite is covered with 16 (or 24) *transponders*—receivers/transmitters-that each can usually handle one channel of television (or 1,000 telephone conversations) at the same time. The company that owns the satellite rents out transponders to networks. CBS uses a bunch of them; HBO, a pioneer in using satellites, does too. There is an entire industry of people who build broadcasting satellites; they sell them to other companies who then pay NASA (or France or Japan...) to launch them into a geostationary orbit (a "GEO") around the earth, 22,300 miles above the equator—which means that from any vantage point on earth, they appear stationary. This makes it easy for a ground-based station to beam directly up at one, or receive signals down from one. A satellite circles for a dozen years before it dies. Today there are about 10,000 man-made objects of various kinds swinging around the earth, 500 of them working satellites.

A high-powered DBS satellite (Hughes HS-601). This big device (about 86 foot wingspan) is larger than the more typical types (like the HS-376 that looks like a 9 foot can) used for broadcasting.

Networks (ABC, CBS, HBO, *etc.*), like banks, have capital resources which they invest to make money. They use this capital to finance the enormous expense of producing television shows ("products") which they sell. They have arrangements with people to distribute these products, and because they have a substantial (and virtually guaranteed) audience, they can sell the rights to communicate with this audience, in mostly 30- and 60-second chunks, to advertisers who are thereby subsidizing some of the expenses of production. The networks make money both from advertising and from the asset value of their entertainment ("content"). A TV show can be sold on videotape or, if there are more than 100 episodes (about five television seasons), it can be syndicated.

Back to the Broadcast

Networks generate the programming and beam a microwave signal carrying the data up at their GEO satellite. Up there, it is received and retransmitted back down to receiving dishes on earth. Having a dish isn't enough. In many cases, you also need the codes to unlock and unscramble that signal to receive it. If you want, you can join the network's "club": buy the rights to pick up their programming on your dish and then redistribute the signals to your local audience—this would make you a network *affiliate*. Every network has affiliates. They have call letters (like radio broadcasters): KION, KCBS, WCJB. Affiliates pick up the microwave signals and convert them into two radio frequency (RF) signals; an AM signal to broadcast picture, and an FM signal to broadcast sound. Technically, television is just radio with pictures. The TV programming is encoded into AM/FM, boosted up in power with enormous quantities of electricity, and radiated from towers that send the signal out to anyone who can pick it up.

A typical broadcasting tower, with antennas strategically positioned to direct signals over the landscape.

There are many laws governing the rules of broadcasting, controlled by the Federal Communications

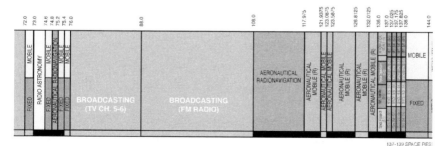

Detail of FCC spectrum allocations (from 72-114MHz), hinting at the full complexity.

◀◀ ER spectrum: p. 181

Commission (FCC). The FCC figures out how much of the electromagnetic spectrum can be used for television, and how much for other things, like radio broadcasting, cell phones, and navigation. They also make sure that stations using the same frequency for broadcasting are far enough apart not to bump into each other.

If I have a TV station in San Jose, I am granted use of a given channel in a territory around my station to broadcast. There is another station using the same channel in San Francisco, 50 miles north. The FCC regulates the overlap of our regions. If you live in Palo Alto, about half way between San Jose and San Francisco, you may be in the overlap area, and may get a jumbled signal. The FCC works to keep this to a minimum.

Let's say the San Jose and San Francisco channels are both affiliated with the same network, NBC. Since they are far enough apart they neither overlap signals nor compete with each other. But someone wanting to open a TV station in Palo Alto could not become affiliated with NBC since that area is already covered. But one *can* open a TV station without being affiliated with a network at all (as long as the FCC grants you a clear channel on which to broadcast). There are many *unaffiliated* TV stations. They are "independent." You can usually tell by their on-air graphics that they are independent. They tend to be proud of this. It means they don't automatically get to broadcast the programming made by (or brokered by) the big networks. They scrounge around and find programming on their own. They buy syndicated shows. They run lots of infomercials. They might go off the air at 2am. They have unique local programming.

At present, there are four *broadcasting* networks. Originally (in 1926), there was only RCA (and its TV division, NBC). NBC operated two networks, the Red and the Blue. In the 1940s they sold the Blue Network, which became ABC. CBS arrived later, first in radio, and by 1948 in TV. By the 1980s three networks, ABC, CBS, and NBC, carried most TV programming, as they had for more than three decades. In 1986 Fox Broadcasting Company arrived. Around the same time cable content providers started their own (non-broadcasting) networks. The original three networks began losing viewers as Fox and cable networks competed for their viewers. Hurt by the competition, eventually each was sold to a larger corporation. General Electric purchased RCA/NBC for $6.28 billion; Capital Cities bought ABC in 1985 (and then Disney purchased Capital Cities/ABC in 1995 for $18.5 billion); Westinghouse Electric purchased CBS for $5.4 billion, also in 1995. In 1998, after buying radio company Infinity Broadcasting, Westinghouse spun out CBS again. A year later, Viacom (once a division of CBS) purchased CBS for $34 billion. There is no end in sight to this mania.

There are 1,600 TV stations in the United States, spread out in 210 DMAs (designated market areas). Of these, 1,100 are affiliated with a major network (including PBS) and 450 are independent stations. Each station is a business. Some are owned in chains, by a bigger company. Some are independently owned. Some are owned and operated by a network itself (called "O&O"). The FCC has many rules about ownership. Newspapers, for instance, cannot own a TV station in an area where they sell papers. TV station owners can buy as many stations as they want, as long as they don't exceed 35% of the country. You cannot always readily tell who owns your local TV station.

Stations, like all businesses, try to make a profit. Broadcasters (whether of radio or television) generally belong to an enormous and powerful trade organization: the National Association of Broadcasters (NAB). Every year in April they have a trade show, which is one of the biggest conventions in the United States.

Networks like their affiliates to look good and professional. Like any other franchise, the parent provides the members with all kinds of paraphernalia to gussy up an otherwise-ordinary operation. You can usually tell the difference between a local mom-and-pop business and a national chain. Menus are slicker. Slogans are catchier. The Beatles are singing the theme music instead of someone's brother-in-law. Money has been spent. Networks produce copious graphics packages for their members, those fancy logos that fly by, the promotions for upcoming shows and "what's on this evening." These are produced by the networks and sold to the affiliates.

Channels

If you have an antenna on your house, it can pick up the broadcast signal from your local television stations. Signals beam through the air, the analog signals getting weaker the farther they radiate from the big transmitter near the station. Mountains screw up reception. Cities and buildings are worse. The signals for broadcasting television range from 54MHz to 890MHz. The FCC has divided this range of the electromagnetic spectrum into real estate that they sell. A "channel" of the spectrum that is wide enough to carry video is 6MHz; there are up to 83 channels in any given region of the country. Of the 83 television "channels," the first 12 channels (from 54MHz to 216MHz, with a gap in the middle for FM radio) are classified as very high frequency signals (VHF), whereas the remainder (from 470MHz to 890MHz), are ultra high frequency signals (UHF). There is no channel 1. In the grand scheme of things, these signals are not all that high in frequency, but when they were named the engineers seemed to think so.

If you're lucky, your television signal receiver (aka "TV set") can pick up local affiliates from all the major networks, public broadcasting, and a few independents; and if you're really lucky, you can get this whole array from two or three cities around you: the typical TV set receives all 12 of the VHF channels and a few UHF channels, all of varying quality, and representing a handful of programming sources.

Cable

Cable company promotion, circa 1969

In the late 1940s entrepreneurs developed "community antenna television" (or CATV), which wired local homes to a big shared antenna. Consumers who were connected to the network received clearer reception than they did on their own. Before long, the community antenna was receiving from a microwave relay station, and was able to pick up (and distribute) broadcast television signals from distant television stations. At first, cable filled in for areas with poor reception, but soon even areas with *good* broadcasting coverage enjoyed access to cable. By the 1960s the cables were able to deliver 20 channels, more than could easily be picked up on an antenna; by the 1970s, due to technological improvements in cable, it could transmit more than 100 channels. Cable still didn't catch on though, due to the high cost of laying coaxial cable: in rural areas around $10,000 per mile, and in cities where underground cable is required, between $100,000 and $300,000 per mile. As time went on, local businesses were started which ran coaxial cable from a central location to nearby houses. Over the decades, more and more houses were connected to this coaxial cable network. The central office is connected to a satellite dish that receives channels, just like a network affiliate. But the cable office is not an affiliate. It receives channels from (and pays for) a number of sources. Instead of rebroadcasting to its neighborhood (like a broadcast affiliate), it pipes its purchased signals over this coaxial cable to all who are connected. The central cable office then can buy channels (now a commodity) from anywhere and sell them to the connected network of homes. The advantage is that the signals are very sharp; even though they are usually analog, there is considerably less degradation of this signal than of signals transmitted through the air.

Broadcast television is free. It is paid for by the advertisers, who want you to watch the programming so you will see their ads. Cable is a subscription service. In theory, it doesn't need advertising because viewers pay to see it.

At first, all cable had to do to be better than broadcasting was deliver clearer pictures. No more fiddling with antennas. Broadcast stations that were possibly a little too far from your home could be piped in over the cable crystal clear. Originally, networks and network broadcasting affiliates didn't mind cable because their channel was still reaching the viewer, and was, in fact, better. But where a local broadcaster originally only competed with a couple other local stations for the attention of viewers, now distant stations were coming in clear, increasing competition for viewers.

But clarity of reception was just the beginning. When cable only provided local and regional TV channels, competition for the broadcaster was nominal. Before long, cable companies were looking for more and more channels to distribute in this pipe.

In the mid-70s HBO came along, selling movies via satellite—cable companies could sign up for commercial-free movies. Then other movie channels showed up (Cinemax, Showtime) that would license movies ("content") from movie studios and sell them to cable companies to distribute to households. Before long, this business model proved exceptional. Anyone could aggregate content, create a "channel," and sell it to a cable company to distribute. Cable companies grew. By the 1980s, cities were cable-wired or involved with battles over those rights. Networks that were designing product for a wide "American" homogeneous audience now had to contend with channels *targeted* to special interests.

Consolidation

On the other hand, cable companies were disparate and small. Even though there are almost 12 thousand cable companies, some have fewer than 100 customers. Once a local cable company hooks up everyone in a region, how does it grow? Answer: it buys the cable company in the next town over.

Over the 1980s, cable companies aggregated until they became fewer and fewer. A company that owns other local cable companies is called a "multiple system operator" (MSO). By 2000 almost a quarter of the nation was managed by Telecommunications, Inc. (TCI) the biggest MSO in the pond. At that time, almost 70% of American households had access to cable.

On the content side, cable "networks" proliferated. Sometimes a single parent would spawn multiple targeted channels, sometimes successful channels would get purchased by bigger companies to fill out some kind of viewer demographic portfolio. In 1999 there were a few hundred cable networks (from The History Channel, the Food Channel to ESPN2 and CNN). These cable *channels* compete to be selected by the cable *companies* as the included options in the 50 or so that will be piped through the coaxial cable into the homes. Content producers, like TV networks (ABC, CBS), and movie studios (Disney, Paramount), are motivated to buy cable channels (CNN, HBO), and distribution avenues (cable companies) as is any big company that is in the business of creating content (magazines, newspapers, *etc.*). Or the distribution channels might want to buy the producers. However it aggregates, the results are the same.

Today there are no pure "networks," just "media companies"—enormous conglomerates of content producers and distribution channels. One key to minimizing risk in producing content is in having a guaranteed outlet for distributing it. Not entirely different from the days when movie studios owned all the movie theaters in a region, they could make sure there was always an avenue to reach viewers. (And it is not entirely different from companies that build hotels in cheap, out of the way locations, and then build roads or railroads to move tourists to their hotels; this was how Miami and the Florida coast came to be what they are.) It's all about *content* (movies, products, hotels) and *access* (theaters, stores, Internet, trains).

DBS

There are a few places you can rest a satellite in a stationary orbit above the earth where it can "see" the entirety of the United States in its field of view. This is an extraordinarily good spot to be in. In 1982 the FCC auctioned off the spectrum associated with three of these great locations, and in 1993, the first direct broadcast satellite (DBS) was launched. The idea is that a satellite in one of these locations, with sufficient power, can beam down to anyone with a dish to receive it. Since DBS satellites are more powerful, their signals can be picked up with a relatively small parabolic dish, in most cases, less than 18" across. DBS is a broadcasting alternative to the networks and to cable. With about 10% of the market, two companies, DirecTV (mostly) and EchoStar (to a smaller degree), together own the three slots for national US direct broadcast.

Expansion

Researchers, at the urging of these companies, are trying to figure out how to get more channels into the face of consumers. They have limited options: (1) they can broadcast them to you, but there is still a limited spectrum and therefore limited numbers of channels; (2) they can add new cables between themselves and you, but this is phenomenally expensive and time consuming; (3) they can pipe them down the various cables already connecting everyone. When it comes to pipes into the home, there are really only two in place: telephone (*twisted pair*) lines and coaxial cable lines.

The phone companies put in all the phone lines. These lines, designed for teensy little voice signals, are not great for movies and videos, but they are virtually EVERYWHERE. The coaxial lines, connecting about 70% of all Americans, are owned by cable companies, and are great for movies, but are designed for analog one-way distribution of video/audio.

Of the three options for expansion, #1 and #2 simply didn't hold the promise of #3; but #3, pushing signals down existing pipes, poses challenges. The pipes are very limited, designed for other things. So scientists got to work to see if they could invent different and better ways to accomplish more with the same pipes.

Results

The phone company scientists invented DSL ("digital subscriber lines"). They want you to put a modem on your phone line, and connect it to something in your home for communication and entertainment. Phone lines are designed to move data in two directions (calls come in, calls go out). This is a key feature of phone lines. By compressing data, and by multiplexing signals (don't ask) in clever ways, the phone lines can move up to about 10Mbps—more than enough for audio and video. MPEG-2 now provides for broadcast television quality images at 4Mbps. One has to imagine this will even get better.

The cable companies have invented digital cable, which uses the same coaxial line, but pumps 100 channels into the pipeline. But cable television, like broadcasting, is designed to send information OUT and not get anything BACK. A coaxial cable can be set up to move 80Mbps of data through its pipe, but due to the way the networks were designed, everyone using the cable needs to "share"—meaning that if 10 people are using the same cable line, each only gets about 8Mbps; if 100 people are on the line . . . well, it gets pretty slow.

In designing a network, one important question is "how will the network be used?" If you don't know the answer to this with certainty, it is very difficult (and exceptionally risky) to build the specifications of the network.

Over-the-air broadcasting simply cannot compete with the cable networks in terms of quality and breadth. In an effort to revitalize the industry of broadcasting, new standards have been created for images that are kind of big for the telephonic pipelines—they pretty much need to be broadcast in the air or piped in through coaxial. This means that the broadcasting spectrum needs to be re-allocated. Channels are still 6MHz, but now they can be used for **one** compressed HDTV signal, or **four** compressed digital SDTV signals, or the like. People with TV receivers will once again get fewer channels, but the quality will be unparalleled. This is HDTV.

The debate about the rules and standards of this change raged for a decade. Clearly, broadcasters would want to maintain the high ground, but they couldn't deny the new power of computer users' and filmmakers' needs, and even the now-powerful cable companies wanted to have a say about it: the results are the DTV standards now in place. HD broadcasting has commenced, but will probably take the next decade to reach most homes.

The phone line network has been co-opted by the population to communicate with each other. Internet protocols invented 20 years ago have combined with personal computers to allow everyone with a computer to communicate with everyone else.

ISPs (Internet service providers) can vie for the role of the local cable companies of the new decade. By connecting homes to the Internet, they are the last gate to consumers, and therefore can position themselves in the content delivery networks. Streaming media can work its way over the Internet, but can also be broadcast by satellite directly to suitably wired ISPs (thus bypassing the IP router network), which pump the highband media over the formerly Internet-only pipes into the home.

While the pipes that connect people (and by definition, computers) will get bigger, the battle is more than just between the two concepts of broadcasting (the distribution of content from *few* to *many*) and narrowcasting (the distribution of content from *few* to *few*). More important, perhaps, is the role of one-to-one-casting, a broadening of the good ol' telephone. The pipes are in, and bandwidth is increasing. As the media conglomerates scramble to own the pipes and connect consumers with their controlled content, the individual is increasingly empowered to create and distribute content without their help.

THE ECONOMICS OF REVOLUTION

There is **production** and there is **distribution**. *Memorize this.*

Production is the inventing, creating, growing and manufacturing of a product. *Distribution* is how this product is moved from producer to consumer.

Before the Industrial Revolution, people made things in relatively small quantities; you'd bake some bread or hammer some horseshoes, and sell them to the people around you. Or to people who were a train/boatride away who wanted bread or horseshoes. This labor was reasonably skilled. You had to know a bunch about bread or metal to produce those products.

Then came the revolution.

In the Industrial Revolution, a hundred or so years ago, people realized that you could make things more cheaply if you made more of them at a time. And machines helped make more at a time. But the means of this mass production—big machines and factories—were prohibitively expensive to build. So some people who had money built factories, and hired other folks to work there.

The revolution in Russia in the early 1900s was partly due to the idea that a worker who didn't own the means of production was just a slave to the capitalist who did. And that capitalist would get rich from his investment; meanwhile the workers who actually made the products never got more than slave wages. The revolutionaries thought that the people doing the work should own the equipment so they could truly benefit from the fruits of their labors. It sort of makes sense.

Owning the *means* of production is a liberating, revolutionary, situation. One of the rules in the filmmaking and broadcast television industries is that the costs of production are high, and so it takes a big company to finance it. Similarly, when the specific tools of production are expensive, relatively few individuals are ever trained to work those tools, resulting in a sort of "priesthood" of the technically elite. How many people can you think of who have practical experience loading and running a 35mm motion picture camera? Fifteen years ago, you could count the number of people who had experience editing movies (it was a few thousand, only a

---

*This is not an economics handbook, so forgive the rather gross generalizations; the purpose will be clear later.

few hundred working consistently). There was a priesthood of video editors too. When the tools of production are expensive, a small number of people—the people who own those tools—get to decide what products are made.

A "Hollywood" movie costs tens of millions, sometimes a hundred million dollars, to produce. Even if you don't hire 20-million-dollar actors, or shoot for five months in Tierra del Fuego, it can be hard to make a full-length feature for less than a few million. That's more than most people have sitting around. But when the cost drops drastically, partially because the tools dramatically drop in price, access to filmmaking leaps. Think *sex, lies, and videotape*; think *Blair Witch Project*.

The widely discussed "desktop" revolution was indeed a *bona fide* economic revolution (and not simply a marketing declaration of everyone's excitement). The means of production for the publishing industry was now available to people with low-cost computers. Each of the desktop revolutions—in publishing, graphic design, video—began primarily as revolutions in the economics of *production*.

The desktop production revolution did not eviscerate the "high end" as some had predicted; rather, it opened up the "low end." The quantity of available goods increased, and not exclusively from the establishment. And because the costs to produce those goods was so low compared to the costs required to support the big industry, unit sales of products could be low and still reap profits for the producers. This happened with books and magazines and it is happening with video and film. The desktop video production revolution will open up the "low end."

But production is only half the battle.

The economics of *distribution* is the bottleneck for a complete revolution. As long as distribution—of pet food, cars, books, music, movies—relies on expensive or exclusive channels, Big Industry will maintain its hegemony. Distribution, and its expensive twin Marketing, are significant costs that need to be accounted for in the prices of goods.

For example, an independent CD publisher can produce a thousand copies of a disc for only a few dollars each; even selling 250 copies—an unmitigated failure for a big company—could result in acceptable profits for the independent. And the upside is high if all thousand copies should sell. Consequently the breadth of products increases for the public.

When pricing goods, a general rule of thumb says that each middleman who handles the product charges double what they paid. Eliminate one person or organization from the distribution chain, and product costs to consumers might be cut in half. When technology facilitates the marketing and distribution of goods, the effects on pricing is revolutionary.

Person-to-person selling online, whether it is MP3 music, books, or white elephants from your garage, has significant and positive economics (risk *vs* rewards). The upside is always the chance (though perhaps slim) that a product will have a wider appeal than the niche it was built for.

The Internet is the revolution in distribution that is the sibling to the desktop production technologies. By democratizing distribution in much the way that production is opening up, individuals are empowered to create products (books, designs, film, videos/TV, music) and sell directly to their audiences.

There is desktop publishing, desktop video, and now, ***desktop distribution***. It is the cornerstone for the revolution of a new century.

W H Y   T V   S H O W S   C U T   F I L M

There are a number of reasons why a production shot on film to be broadcast on television or distributed on videotape might want to cut the original negative.

In October, 1986, the New York Times published an article on the new revolution in editing. It referred specifically to the Montage, the EditDroid, and the Ediflex. The author described the evolving world of high definition TV and its incompatibility with current broadcast video. Film, it announced, was the only viable way to convert to high definition. The article acted as a catalyst that galvanized interest in nonlinear editing.

The second significant reason for having a film version of a television program is for foreign distribution. Europe and many other broadcast markets do not use the American NTSC video standard. They use PAL. Where NTSC broadcasts 525 lines of video at 30 frames per second, PAL broadcasts 625 lines of video at 25 frames per second. Although there are a number of very good "standards conversions" that will take an NTSC master tape and transfer it into a PAL tape, it is argued that only an original telecine from cut negative to PAL video is of sufficient picture quality to please European audiences.

Whether it is or isn't good enough is not the issue. The fact is that cut negative is as close to a super-high-quality universal standard media as the world knows. As distribution standards continue to change, film will always be an excellent universal source for transfer. Consequently, the need for projects to have a film cut is of ongoing concern to networks and studios.

HDTV and DIGITAL TELEVISION

What began as the development of a new broadcasting standard, called HDTV (or High Definition Television) evolved over the past 10 years to become a host of new digital video formats and standards. These will impact the production and distribution of moving pictures for decades to come.

The basic elements of HDTV broadcasting are as follows:

A system that provides wide-screen pictures with at least 2X the resolution (or sharpness) of ordinary TV, and CD-quality sound. Traditional (NTSC) broadcast video consists of 486 active lines of video, interlaced from two fields per frame, at a rate of 59.94 fields per second. Each line has 720 pixels; HDTV broadcasts video at significantly higher resolution—about 1080 lines on the screen, with almost 2000 pixels per line. An HDTV broadcast has about **six times** the visual information of NTSC. The aspect ratio for HDTV—the shape of the image on the screen—is unlike NTSC's 4:3 (kind of squarish); HDTV is 16:9 (wider, like a movie screen). The pixels are small and square (like a computer) and not big and rectangular (like a TV).

The term *high definition television* is actually considered limiting; the more general term for all these formats (including HDTV) is simply *digital television* (DTV) or sometimes *advanced television* (ATV).

Digital television impacts not only broadcasting, but has applications in the movie business, education, medicine, printing and publishing, photography, defense and manufacturing.

The issues of DTV are profoundly complex and impact many varied disciplines. For example, there are **production** standards which are different from **distribution** standards. Production standards have to do with the way HD is shot, recorded, and post-produced. Distribution standards have to do with the many, *many* ways this information can be delivered.

Delivering analog video is just what you think it is: you can deliver on a video cassette, you can receive it through the air on a TV. Delivery of digital information—including video—can be done with DVDs or telephone lines or coaxial cable, to name a few. So although broadcasting has been a significant part of traditional video distribution, it is only one (small) part of digital video distribution.

The broadcast standard (or transmission standard, which is the same thing) has to fit within constraints that have to do with the number of channels on the air, and the "size" of the channels (6MHz), among other things. Other distribution paths include coaxial cables, fiber optic cables, DVDs, and satellites. There is no way to adequately have a single distribution standard; all of these digital channels are somewhat unique. There could, however, be a single worldwide digital production standard; but determining what this would be would require an international effort of artistic, scientific and political cooperation.

Like the PAL, NTSC and SECAM transmission standards, developing a new US standard for high definition has been a high priority in the evolution of digital television. It was hoped that after 2006 the US will have phased out NTSC broadcasts, but this date proved to be too ambitious. Television sets, as we know them, will be virtually obsolete. The future will include boxes that blur the lines between televisions and computers,

---

◀◀ Pixels: p. 175

playing digital images and audio received as signals over the air, via wires and cables, or from other sources.

The FCC has insisted that their new standard must allow for a simulcast broadcast—at least during the transition years—where every TV station gets two channels (one for the new HDTV signal, and the second for a standard definition "low rez" version). This way, old sets can receive HD programs in NTSC, and new sets can watch NTSC *or* HD programming.

Why is this such a big deal? Because there are 93 million homes with one or more NTSC television sets in the US today. That translates into a lot of new television receivers that might be replaced—with new kinds of wide screen high-resolution displays that take signals from all kinds of digital sources. That's *billions* of dollars of revenue for new equipment.

There are a small handful of new videotape formats, and a number of them are what we might call "high definition." Some are *progressively* scanned (appended with a "P") and some are *interlaced* (appended with an "I"); their names usually reflect the horizontal line resolution (1080 or 720), and a frame rate (24 or 30). And while they all have CD-quality audio, 5.1 channels, and utilize MPEG-2 compression algorithms, they may be compressed differing degrees, with differing color sampling (the equivalent of 4:2:2, 4:1:1, *etc.*) and consequently have differing bandwidths. All of this dictates how they will be used.

For broadcasters and the public, there are many consequences to these changes. Using the proscribed MPEG-2 compression scheme, an HDTV signal, uncompressed at around 1Gbps, can be broadcast over the 6MHz channel inherited from the old analog TV days (about 19Mbps). But using similar compressions, those old standard TV signals (called standard definition, or SDTV) can be compressed into a *quarter* of their old bandwidth; thus, a broadcaster can play either 1 HDTV channel or their regular SDTV channel plus *3 additional standard-definition streams!* You can imagine the debates this might cause. The public often feels these valuable channels should be used for education or non-commercial broadcasts; the broadcasters want to use them for experimental work. PBS, for instance, utilizes the extra bandwidth for *Enhanced TV,* utilizing 19Mbps of channel alongside SDTV for various types of multimedia, educational materials, games, and other forms of data. The debate over the broadcast spectrum continues.

The highest qualities of DTV can produce surprisingly film-like results; shot properly and projected with a high-end digital projector, some form of HD video is expected to ultimately replace cinematic distribution (and eventually, production) of theatrical movies.

A   B R I E F   H I S T O R Y   O F   H D T V *

High Definition Television began as a new production standard that was first developed in Japan in 1970, and based on research done in the US in the '50s and '60s. By 1977 SMPTE had established a group to evaluate this new media, and by 1988, they sanctioned an analog standard called 240M, based on the Japanese's 1125/60 format. Although analog formats (like 240M, NTSC and PAL) require a broad base of acceptance and standardization, digital formats (for a variety of technical reasons) do not need the same degree of universality. The 240M analog standard was deemed to be too limiting for the future needs of entertainment and information technology.

In 1987, the Federal Communications Commission (FCC) established the Advisory Committee on Advanced Television Service (ACATS) to advise the FCC on technical and public policy issues regarding advanced television. The Advisory Committee consisted of 25 leaders of the television industry and hundreds of industry volunteers.

Initially, 23 different systems were proposed to the Advisory Committee. These systems ranged from "improved" systems that worked within the parameters of the NTSC system to improve the quality of the video, to "enhanced" systems that added additional information to the signal to provide an improved wide-screen picture, and finally to high-definition television ("HDTV") systems, which were completely new systems with substantially higher resolution, a wider picture aspect ratio and improved sound.

In the midst of this competitive process, a fundamental technological advance occurred when in May of 1990, General Instrument proposed the first all-digital high-definition television system. Within seven months, three additional all-digital HDTV systems had been proposed.

By 1991 the number of competing system proposals had been reduced to six, including the four all-digital HDTV systems. The Advisory Committee developed extensive test procedures to evaluate the performance of the proposed systems. From July 1991 to October 1992, the six systems were tested by three independent laboratories working together, following detailed test procedures.

The Advanced Television Test Center, funded by the broadcasting and consumer electronics industries, conducted transmission performance

---

*If most of this section sounds like a government document, that's because it is; this section is excerpted and edited from ATSC historical documents, submitted in December 1999.

testing and subjective tests using expert viewers; other national and international organizations tested aspects of the systems as well.

In February 1993, a Special Panel of the Advisory Committee convened to review the results of the testing process, and, if possible, choose a new transmission standard for terrestrial broadcast television to be recommended by the Advisory Committee to the FCC. After a week of deliberations, the Special Panel determined that there would be no further consideration of analog technology—an all-digital approach was both feasible and desirable.

Although all of the all-digital systems performed well, each of them had one or more deficiencies that required further improvement.

The Special Panel recommended that the proponents of the four all-digital systems be authorized to implement certain modifications they had proposed, and that supplemental tests of these improvements be conducted. The Advisory Committee adopted this recommendation, but also expressed its willingness to entertain a proposal by the remaining proponents for a single system that incorporated the best elements of the four all-digital systems.

In response to this invitation, in May 1993, as an alternative to a second round of intense competitive testing, the proponents of the four all-digital systems formed the Digital HDTV Grand Alliance. The members were:

AT&T (now managed by Lucent Technologies)
General Instrument
North American Philips
Massachusetts Institute of Technology
Thomson Consumer Electronics
The David Sarnoff Research Center (now Sarnoff Corporation)
Zenith Electronics Corporation

After a thorough review of the Grand Alliance's proposal, the Advisory Committee ordered a number of important changes, and the Grand Alliance companies proceeded to build a final prototype system based on specifications approved by the Advisory Committee. It was built in a modular fashion at various locations: the video encoder was built by AT&T and General Instrument; the video decoder by Philips, the multi-channel audio subsystem by Dolby Laboratories, the transport system by Thomson and Sarnoff, and the transmission subsystem by Zenith. The complete system was integrated at Sarnoff. Testing of the complete Grand Alliance system started in April 1995 and was completed in August of that year.

Another vitally important organization in this historic process was the Advanced Television Systems Committee (ATSC), a private sector organization founded in 1982. ATSC is composed of corporations, associations and educational institutions, developing voluntary standards for the entire spectrum of advanced television systems, including high-definition television. All segments of the television industry are represented within the ATSC, including broadcasters, cable companies, satellite service providers, consumer and professional equipment manufacturers, computer and telecommunications companies, and motion picture and other content providers. As well as documenting the proceedings, the ATSC developed the industry consensus around several standard-definition television (SDTV) formats that were added to the Grand Alliance HDTV system to form a complete digital television standard. Among other things, these SDTV video formats provide for interoperability with existing television standards and support the convergence of television and computers.

Following completion of its work to document the U.S. ATV standard, the ATSC membership approved it as the ATSC Digital Television Standard (A/53) on September 16, 1995. On November 28, 1995 the FCC Advisory Committee issued its Final Report. Accordingly, the FCC Advisory Committee recommended that the ATSC DTV Standard be adopted as the standard for digital terrestrial television broadcasts in the U.S.

After receiving the Advisory Committee's recommendation, the FCC released a Notice of Proposed Rule Making announcing its intention to adopt the ATSC standard. The notice stated:

"We believe that the ATSC DTV Standard embodies the world's best digital television technology and promises to permit striking improve-

vertical size	horizontal size	pixel shape	aspect ratio	frame rates supported includes both regular and 0.1% slower variations	scan type
1080	1920	■	16:9	24, 30	progressive
				30	interlaced
720	1280	■	16:9	24, 30, 60	progressive
480	704	■	16:9	24, 30, 60	progressive
			4:3	30	interlaced
	640	■	4:3	24, 30, 60	progressive
				30	interlaced

The infamous ATSC "**Table 3**"—its apparent simplicity belies its role as the epicenter of ongoing technical debates within the industry.

ments to today's television pictures and
sound; to permit the provision of additional
services and programs; to permit inte-
gration of future substantial improvements
while maintaining compatibility with initial
receivers; and to permit interoperability
with computers and other digital equip-
ment associated with the national information initiative. It was developed
and tested with the unparalleled cooperation of industry experts..."

On December 24, 1996, the U.S. FCC adopted the major elements of the ATSC Digital Television (DTV) Standard, mandating its use for digital terrestrial television broadcasts in the U.S. (The FCC did not mandate use of the specific HDTV and SDTV video formats contained in the ATSC Standard, but these have been uniformly adopted on a voluntary basis by broadcasters and receiver manufacturers.)

In 1997 the FCC adopted companion DTV rules assigning additional 6 MHz channels to approximately 1,600 full-power broadcasters in the U.S. to permit them to offer digital terrestrial broadcasts in parallel with their existing analog services during a transition period while consumers made the conversion to digital receivers or set-top boxes. The FCC also adopted a series of rules governing the transition to digital television, including a rather aggressive schedule for the transition. Under the FCC's timetable, stations in the largest U.S. cities would be required to go on the air first with digital services, while stations in smaller cities would make the transition later. Under the FCC's plan, more than half of the U.S. population would have access to terrestrial broadcast DTV signals within the first year, all commercial stations would have to be on the air within five years, and all public TV stations would have to be on the air within six years. Analog broadcasts would cease after nine years, assuming that the public had embraced digital TV in adequate numbers by that time. Part of the FCC's motivation in mandating a rapid deployment of digital TV was to hasten the day when it could recapture 108 MHz of invaluable (VHF) nationwide spectrum that would be freed up by the completion of the switch-over to more spectrum-efficient digital television technology.

In accordance with the FCC plan, digital television service was launched in the U.S. November 1, 1998, and more than 50% of the U.S. population had access to terrestrial DTV signals within one year.

VIDEO ON THE WEB

Evolving Business Models
"OPPORTUNITIES"

There is no more hotly debated topic than the implications of streaming web video. More and more individuals are acquiring "broadband" capability (the term is used for *anything* significantly faster than a 56K modem, usually >200Kbps cable modems or DSL). Along with this development has come the prospect of new business avenues for the establishment (big industry) to reach consumers—and the revolutionary notion that consumers themselves can produce and distribute video person-to-person without any need for an intermediary. Either way, the consumer's relationship to Media is about to change. Highlights include:

 New business model for the establishment—studios can release videos on VHS, DVD or stream it over the web. News organizations and television stations can have both broadcast and Internet (searchable and interactive) distribution.

Individual empowerment—anyone can stream video to anyone else, if they want.

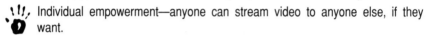 New advertising channel—studios (and individuals/small business) can display TV-like advertisements and trailers online. Where before the end of the century web ads were static or animated banners, in broadband they can be catchy video shorts, commercials, links to URLs, and more.

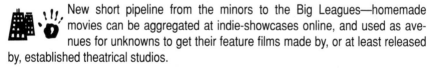 New short pipeline from the minors to the Big Leagues—homemade movies can be aggregated at indie-showcases online, and used as avenues for unknowns to get their feature films made by, or at least released by, established theatrical studios.

Internet Structure
"OBSTACLES"

Everyone knows that the Internet was developed by the Defense Department in the 1960s. The project from DARPA was to handle problems in communication that might occur if a nuclear blast took out a city. What they created was the "Internet Protocol," or IP. The key technology was the *IP router*. What they did was take information that previously was routed in a roughly continuous and direct way from place X to place Y, and break it into little packets of data that would get routed from IP router to IP router, working their way around the network until they eventually reached their target. Two packets going to the same place might not take the same route, and therefore might not even arrive in the order they were sent.

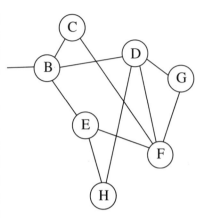

Data on the Internet doesn't necessarily take a direct route. It is broken into small packets, with some packets going B-D-F and others B-D-G-F. The system was designed so that if a bomb destroyed a node, like E, that data could still easily flow to its destination.

The Internet has worked this way for decades. Universities could communicate, and students and professors could use the network for both email and research. By the 1980s, businesses were springing up that would give consumers with computers access to this network—companies like AOL and Compuserve. In 1988, a reseacher invented HTML, a way to link articles to one another using keywords, thus providing layers and "depth" to otherwise flat journal papers. A few years later, Mosaic was invented, a graphical user interface that made HTML viewable and easy to interact with. Mosaic evolved into Netscape.

Every IP router has an IP address (like a phone number), a multi-digit (32-bit) number that looks like this: 12.0.44.254. Each group of digits (separated by a period) can be any value from 0 to 255 (each byte has 2^8, or 256, possibilities). When you have a mailbox at a service provider, it is *their* IP address that your mail first needs to find. "You" are just an account (like a folder) on their hard disk.

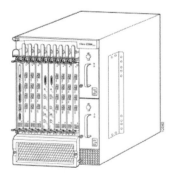

The plug side of a Cisco 12000 Series gigabit switch router

A word name can be assigned to an IP address. This is what a *domain name* is. A single computer on the Internet can have many domain names (and IP addresses) associated with it, but any IP address only points to one machine.

The Internet is an infrastructure. It is not owned by anyone, and its uses change as tools and technology change. Yet, the phone-line system connecting IP routers today is essentially the same as it was before the invention of the browser and the appellation "World Wide Web."

The more ubiquitous the Internet becomes, the less people who use it need to know in order to get by. As far as we are concerned, the most important thing that post-production people need know about the Web is its functionality with regard to audio and video. Getting a *downloadable*-style "movie" over the Internet is just like transferring any other large file. But because it was originally designed to chop up data into bite-sized packets, and ricochet them willy-nilly around the network (to be reassembled whenever it's convenient), the Internet provides some considerable challenges for data that really needs to be moved together, and quickly: like video and audio in a *streaming* format. Even if the bandwidth of the network were a non-issue (which it is *not*), the methodology of IP requires some fancy manipulations to maintain the appearance of continuous streams of data.

... and lest we forget, the other major problem with digitized video and audio data is that it's *huge*.

Coping With Limitations
"WORKAROUNDS"

As soon as post-production is completed, the question arises: *How do I get this onto the Web?* At this point, there are primarily two paths you can choose:

(1) Send a videotape (usually requested on BetaSP, DV or occasionally VHS) to a professional organization that manages web video. These websites (service bureau and ISP combined) will first determine whether they want your video (based on their own particular sets of parameters), and if they do, will manage the sophisticated compression, and ultimately the hosting of the video on their own site. They maintain the infrastructure for high-quality streaming to end-users. These include web businesses like *eveo*, *atomfilm*, or *ifilm*.

(2) The other path is really the promise of the Internet: that you can stream the video from your *own* website. Since "your own" website is likely hosted for you by a local ISP, this path would require that (a) you have the ability to compress your own video successfully, and (b) your ISP has both the technology and the desire to stream video onto the Web.

Streaming video is a particularly intensive process. The ISP needs to have sufficient throughput for the dozens (or hundreds, or thousands) of

◀◀ Digital video, size: pp. 154–155

people who might *simultaneously* request and stream dozens or hundreds of megabytes per user. They need special software and hardware products, and often need to manage licenses and royalties associated with using those products.

And the reality that the web really wasn't designed for moving continuous streams of video to anyone is yet another obstacle to smooth high-bandwidth streaming by a typical ISP. While most ISPs will struggle through this process, the aforementioned professional organizations (the *de facto* broadcasters of the Internet age) use what they call *edge networks*. These circumvent the body of the Internet, and only take advantage of the last mile of connection—from the local node directly to the modem of the consumer. To do this, the local node needs to receive the stream directly from a satellite managed by the streamer (and not from the IP router of another node). By bypassing the usual morass of IP routers, these *nouveaux* broadcasters can stream to the consumer much more directly, and only marginally using what might be called "the Internet."

If your ISP doesn't happen to have its own satellite, and must rely on *bona fide* Internet streaming to many simultaneous consumers, another workaround is through the use of "proxy servers." This is much like the use of mirror sites; the ISP duplicates source material at multiple strategic points within the Internet network and uses them to facilitate the caching of data more locally to the consumer. This is a method commonly utilized for the execution of live "webcasts."

But whatever method is ultimately used, the real key to streaming (at least for now) is highly-compressed video.

Compression: Your Main Squeeze

Creating video for the web means the conversion from some standard post-production format (DV, Digital Betacam, whatever) into a standard streaming format, a process shrouded in some mystery. If your video is analog, it must be digitized into a computer; though assuming you edited on a nonlinear system, your video is *already* digital. Ideally, the editing system itself will have an EXPORT function (or perhaps a third-party plug-in or stand-alone application that accomplishes the same thing) to do the conversion to a compressed format. Or perhaps it already uses an appropriate compressed format (M-JPEG, MPEG-2, DV, *etc.*), and so there's a file sitting on the hard disk waiting to go.

◄◄ Compression: pp. 156-164

Well, maybe not *a* file. Realize that the finished product in most editing systems is not a single digital ("media") file, but rather a *bunch* of digital files of the myriad original sources, connected together by another data file (the "edit decision list" or "sequence"). When this is the case, you can:

1. Use a software function to render the multitude of files that actually comprise your edited sequence into a single, all-inclusive ("flattened") digital file.

2. Output the finished sequence to videotape, and then digitize the result back into the computer, effectively "flattening" it in the process.

This second option is time consuming, but not detrimental to quality if the transfers are all-digital (as you would have with a FireWire connection). In systems with analog video connections, however, there would be two generations of quality loss: first, in the analog output/recording to tape, and then in the playback/re-digitizing into the computer. Assuming that quality is an issue, this is *a poor solution.*

Thus, the ideal is to have an editing system that generates the needed file directly as an Export option.

Depending on the edit system, though, these export capabilities vary. So it's possible that even though the resulting file might be (1) flattened, and (2) in a compressed digital video format, that format might not be nearly compressed *enough* for streaming on the World Wide Web. It's also quite likely that that file isn't one of the common highly-compressed file formats that web browsers and streaming media players expect to see.

So, even though it's possible to email or transfer high-quality digital files (in DV format, an AVI file, *etc.)* to the interested parties, consider that these are likely to be of enormous size: a 5 minute video in DV format takes up more than a gigabyte (see the Compression chart on page 174). And that's even assuming that the recipient's computer recognizes the file type, knows what to do with it, and then has the *chutzpah* to play it back smoothly, without stuttering.

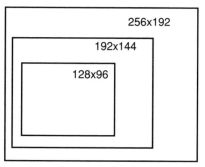

The same issues that apply to getting digital video to play on a computer also apply to getting it to stream on the web (see pages 154-155). Consequently, conversion utilities (and ideally, Export functions) generally

Three typical non-full-screen frame options for streaming web video. At a fixed bandwidth, larger frames equal worse looking video.

offer a range of compression and format selections. With an eye on your chosen distribution bandwidth (DSL? 56K modem?), you will need to select frame size, audio quality, player (QuickTime? RealMedia?), and ultimately codec (H.263? Sorenson? RealG2?).

How do you choose?

With respect to the latter: there are numerous codecs in the world, and more are invented all the time, each with its own specialties. H.261 was developed to compress videoconferences, so it handles video with talking heads, low motion, and generates correspondingly low data rates. The goal of the H.261 codec is to provide good quality with a small fixed window (176x144) at under 64Kbps. H.263 improved upon and largely superceded H.261, yet it's still optimized for video with little movement. Contrast these codecs with those like RealVideo G2, which are designed to adapt their data rate to the user's connection speed. Some codecs are better at *buffering* the data on the receiving end (a mechanism for coping with the haphazard "delivery" of IP packets), which on one hand, make video

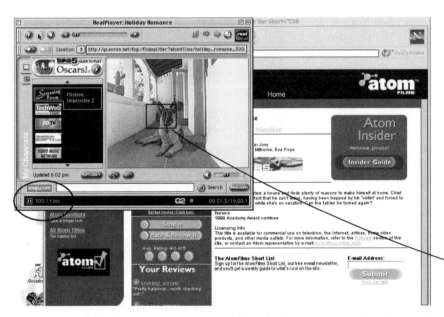

A streaming web "client" application (top left), playing a stream being fed to it by an independent film site's "server." In this example, the indicator shows a bandwidth of about 300Kbps; respectable, but nevertheless (in comparison to the original NTSC master videotape), a compression ratio of over 500:1. The picture thus suffers from compression artifacts, which are exaggerated even more when the window is enlarged to full-screen size (a piece of which is shown at right). A periodic "stuttering" of video frames is yet another manifestation of the bandwidth bottleneck.

take longer to start playing when selected, but on the other hand maintain smoother "high quality" playing streams. *Et cetera.*

As you can see, there are many considerations for determining an appropriate codec to use for your material. But what good is even the "perfect" codec, if none of your target audience is equipped to view it? Thus, selecting among these parameters is as much about brand recognition as it is technology. Corporate affiliations may impose "suggestions" that may or may not be ideal for your specific video. Thus, political and marketing factors, in addition to many subtle technological ones, are all to be considered in making such choices.

Once these decisions are made, however, the processing of the video file can begin. Currently these tend to be longer-than-real-time conversions, but faster processors (and/or hardware accelerator cards) will eventually make real-time, and even faster-than-real-time transfers more commonplace.

Better Compression

Regardless of codec, there are a handful of factors that affect video compression significantly:

- **Image "Noise."** The less noise the better. Better quality video has less noise. This is why you might shoot DV or BetaSP rather than VHS. Or use a 3CCD camera rather than a 1CCD camera. Even if the result will ultimately be teeny (240x136), high-end production tools will produce video with less noise.

- **Detail.** The less detail the better. Solid backgrounds. No wideshots of trees in autumn. No crowds in football stadiums.

- **Text.** Text is full of "detail," and compresses poorly. If you must use text, make it large, sans serif, good contrast with the background, and short.

- **Movement** (the on-purpose kind). The less movement the better. This means minimizing tilts, pans, dolly shots, and tracking shots.

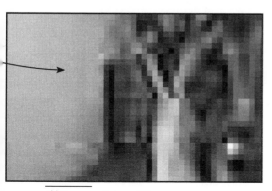

- **Movement** (the accidental kind). Camera shakes, from working without a tripod, are bad. Video from tripod-mounted cameras will not only compress better, but looks more professional as well.

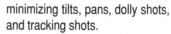

◀◀ Compression: p. 156

- **Edits**. Of course edits are necessary, but just know that more edits (fast cuts, MTV-style, etc.) compress worse than long scenes.
- **Effects**. The fewer effects the better. Both the transition kind (fades, dissolves) and the complex kind (keys) can compress poorly; in addition, they are hard to see in tiny windows.

These guidelines are not intended to suggest that your videos should be boring, just to help you understand the eventual implications of decisions that are routinely made in production and post-production.

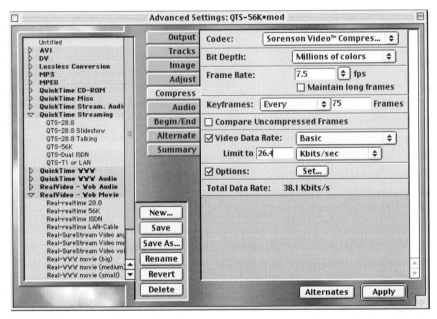

The advanced settings mode of a comprehensive encoding application (Media Cleaner Pro) hints at the potentially daunting task of choosing compression settings. Luckily, presets and "wizard" dialog boxes can simplify the decision-making process.

If you choose to bypass the professional Web streaming sites, and try to deliver video on your own (with your ISP's assistance), the main question is this: can your ISP stream? If they can, they may help you choose many of the parameters for compression. You will then only need to upload the properly-compressed file to the site. The end-user's browser, once seeing the format of the streaming video, should automatically load the appropriate plug-in to watch it live. If your ISP cannot stream, you can still deliver video, but it will be more like email or FTP. End-users will go to your webpage, select the file, and download it to their computer to watch later. The downside of this alternative is one of copyright and ownership. With streaming files, viewers never actually take possession of the file; with email or a download, they do.

While all the bandwidth limitations of the Internet are brought down full force on streaming video, the seminal days at the beginning of the 2000s give a hint as to the full potential of the medium.

S T R E A M I N G   M E D I A   F O R M A T S

Streaming video requires compression, which requires codecs. The codec software could be built into computers, it could be something the browser simply *does*, or it could be 3rd party downloadable software.

Historically, there are many codecs made by many organizations. There are codecs based on standard compression algorithms like MPEG-2, and there are unique codecs, like H.263, designed to be particularly good at compressing some particular kind of video (animation, head shots, long form, *etc.*). Whatever kind of codec compresses the video needs to be used to decompress it on the other side. Media players, like the ones outlined here, are decompressors that have some variety of codecs bundled into them. If you are compressing video, you need to think about what kind of player your end-viewer might be using. Microsoft makes a player that is particularly well-suited to Windows operating systems; QuickTime is comparably compatible with Apple products; and RealMedia stands independently, nicely cross-platform, with its own unique parameters that may or may not suit customer needs better than the alternatives.

QuickTime

In May 1991, Apple introduced a new format for playing moving digital video on their Macintosh computers. It is a file format (as are PICT, TIFF and EPS). It is also a method by which digital media can be moved, placed, and processed.

Following the announcement of QuickTime, many software developers began allowing for QuickTime "movies" to be placed in their applications—moving pictures could then be pasted into spreadsheets, scripts, flow charts, *etc.* By the late 1990s, QuickTime evolved into a streaming media

format. The QuickTime format is focused on maintaining the bit rate (and quality) of the streaming video. The player offers numerous codec options: H.263, Sorenson, MP3, and others. Unlike the alternatives, QuickTime streams by downloading some of a file to the user's computer, and begins playing once enough of the file is loaded. This is not as bad as downloading an *entire* file, but not quite as good as turning on a channel and watching what's on. The QuickTime format has become the basis for the MPEG-4 specification.

REAL Media

In 1995, Real Networks launched the REAL Audio player, the first company to provide this badly-needed application for the nascent World Wide Web. A few years later, when the web was "hot," and broadband made streaming video plausible, Real Networks' follow-up to the audio player was the video "media" player. Real utilizes its own proprietary codecs (*e.g.*, G2). These also encode the files with a series of internal settings which then tailor the bit rate to the connection of the viewer, thus making the format suitable for various (including low-baud rate) modem connections. In 2000, Real also added QuickTime compatibility to its server infrastructure. Real is presently the most popular player.

Windows Media

Not to be left out of any market battle, Microsoft launched its own player. Like Real, the Windows Media player stores multiple internal settings that make it work effectively over low-baud rate modems. When a file is to be played, the bit rate is optimized for the connection rate. The only downside of this method is that  the variable rate can degrade the video "playing" experience beyond the point that the makers of QuickTime happened to choose as acceptable for their codec. The Windows Media codecs include MPEG-1, MPEG-2 and MPEG-4.

Both Windows and Real have a fee structure that imposes charges on ISPs for streaming media (based on how much is streamed, and how often).

In some ways, the multiplicity of streaming formats is analogous to the videotape format wars of prior decades (VHS or Beta?), but because this battle is in software, the costs to distribute content in a handful of different formats is not extreme, and there doesn't need to be a "winner" in the consumer adoption of one over the others.

⊙ QuickTime

QuickTime is especially strong at broadband delivery. Movies may be progressively downloaded from a standard web server, or (with QuickTime 4) streamed from a QuickTime Streaming Server. The QuickTime Player and numerous other applications provide the strongest cross-platform playback. Alternate movies provide scalability.

⊙ RealVideo

Real files served from a RealServer are the most common true streaming media today. They may also be progressively downloaded from a standard web server., The RealPlayer provides playback on both Mac and Windows. Real's SureStream feature provides scalability.

⊙ Windows Media

Windows Media's MPEG-4 video codec offers a strong solution for modem delivery, especially for playback on Windows systems. Movies may be stored on a standard web server for 'HTTP Streaming', or truly streamed from a Windows NT/2000 Server. Intelligent Streaming provides scalability.

This "wizard" dialog (again, from Media Cleaner Pro) offers some useful insights into the subtleties of popular streaming formats.

Audio Streaming with MP3

"MP3" is the popular name for MPEG-1, audio layer 3 (there's no such thing as MPEG-3). The expression "layer 3" does *not* mean there are three tracks of audio, and that this is the third. There are three different audio coding schemes (*i.e.*, codecs) in the MPEG-1 format, called *layers*. Layer 3 is a particularly clever method, and uses a lot of scientific and psycho-acoustic strategies for throwing out tiny little subtleties that people may not really hear anyway. This sophisticated form of lossy compression means you can take digital audio data from an editing system, CD, *etc.* (at roughly 1.4Mbps uncompressed), scrunch it down by a factor of 10 or so, and still end up with a digital stereo signal that sounds awfully close to CD-quality.

The resulting bandwidth is small enough to flow nicely through some fairly small network data pipes. A DSL connection, for instance, will easily play MP3 audio with almost-CD quality. But like many compression and streaming schemes, MPEG (and hence, MP3) is "scalable." That means that if a connection isn't fast enough to stream smoothly at 128Kbps, or if that file is too big to download, there's always the option of encoding a lesser-quality version at a lower data rate—say, one that sounds more like FM-quality radio. In either case, with broadband modems a growing trend, the revolutionary implications for the music industry are a frequent (and heated) topic of discussion.

◀◀ MPEG compression: p. 160

ONGOING IMPLICATIONS
VIDEO ON WITH IN AND ? THE WEB

Let's take a trip.

Shoot some material: one hour (like a home video) or 100 hours (dailies for a big feature film).

It's in a computer. You have a database in the computer full of metadata about these hours of media that allows you to find any single frame, search for unique scenes or reorganize them into functional units. It searches and displays and gives you the ability to watch any video on demand, and modify any of the video in any way you want.

It's an editing system.

The video is on a terabyte server. It's connected to other computers in the building. It's a distributed architecture. A distributed workflow for the movie. You can edit the material you pull from the server while someone down the hall does sound effects and another person at another workstation creates visual effects. You preview her effects. You revise the cut. And so on.

You finish the movie.

You want people to see it.

You move the 100-minute compilation of the source material to another server, this one networked to everyone in the world. Maybe it's the same server you started with. When people want to see the movie, they just call it up and watch it.

But how do they find this movie? And when tens of thousands of other content creators are also placing their movies on the same network, how do people decide which they want to see?

They need workstations that allow them to search databases full of metadata about these hours of media. They want to be able to find any single movie in enormous volumes of material, search for scenes or organize them into streams of interest—channels—that can be watched end-on-end or popped through and sampled. They want to search and display and have the ability to watch any video (or "movie") on demand.

It's a television system.

What kind of television system can do this?

A television system that is like an editing system. This is precisely what editing systems are designed to do.

The network of the Internet, and the distributed architecture, creates a virtual petabyte server, or exabyte server (or maybe even zettabyte server) of media that is accessed by users everywhere. When the network is in a post-production facility (or in-house anywhere on a LAN), it might have expensive FibreChannel gigabit pipes for moving material. The Internet is cheap and huge and full of twisted-pair megabit pipes between workstations.

A screen from a leading videostream auto-logging software package, containing tools for scene detection, on-screen text recognition (OCR), voice recognition, and even face recognition.

News organizations already utilize mini-versions of the video-on-demand Internet, in-house, to build their news content for broadcasting. They have already developed workstations that are part network browser, part nonlinear editing system. Editing systems, once built with internal databases designed for a single user to access limited metadata about finite media, are now being built with enterprise-level databases, scalable to enormous magnitudes, able to be searched by many users simultaneously, on large networks, with ease. Logging source material into the metadata of these databases can be done with care, manually, by people who need to find every frame; or automatically, by sophisticated tools that parse endless streams of video and capture data based on the computer's "understanding" of the content in the stream. Less exacting, perhaps, but scaled to the task of logging vast archives of video. But all of this is about old media.

Let's meander a bit further …

Imagine that the material moving through the pipes of these networks is not only video, and not only webpages, but has begun to hybridize: traditional, dynamic moving images connecting with textural and other heretofore "static" content hiding in disparate parts of those networks.

They appear to interact in unusual ways, one with the other.

The nature of video "on" the Internet is only now forming. At first it will be an uneasy balance of static pages with "nested" video frames, or video streams with sidebar textural commentary; commercials with product ordering capabilities; databases with short video clips in fields; stories that flow passively or fork based on user input . . . but these aren't the *future* of video on the web. *That* hasn't been invented yet.

All we can say is that we are looking at the things that will evolve into what it will be; and it won't be like what have now. The media of the Internet has yet to be invented, but it will likely involve many of the things discussed here, and it will be created with many of the tools articulated in this book.

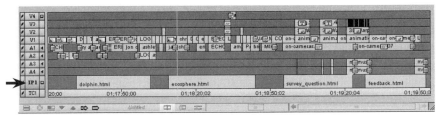

The convergence of television and the Internet creates the need for authoring capabilities for hybrid entertainment media. This Avid Symphony prototype (circa 1998) featured an "interactive programming" track, here labelled IP1. The functionality could be used to trigger web pages in sync with a television program (when viewed on WebTV-style set-top boxes), or with a video program (when seen in a frame on a web page).

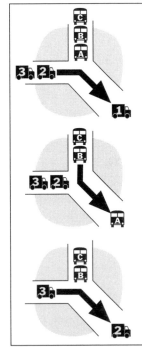

Multiplexing

If you need to move two or more streams of data and you only have one pipe, one common strategy is to multiplex the signals. Digital multiplexing ("MUX") is able to combine streams into a single wide stream, but return to the individual streams later on.

One analogy might be two lanes of cars that merge at a flashing light. One car from each lane is allowed to go when the light is green, and the two lanes alternate turns. This way, the two lanes can be merged in a controlled way.

In computers, the "flashing light" is driven by the "clock" of the CPU—the faster the clock speed (measured in megahertz), the quicker the multiplexer can alternate between the paths, and faster data can stream down the road.

Many types of data require multiplexing: DTV broadcast signals, DVDs playing, "packets" of data over the Internet, etc.

D V D

I t looks like a standard CD audio disc; it functions a great deal like a CD-ROM. The initials "DVD" originally stood for things: some people said *Digital Video Disc,* but also *Digital Versatile Disc*—now the initials officially don't stand for anything. DVD is the first distribution medium for digital, random-access and interactive video for the mass market.

The DVD format is backward compatible with the CD-ROM format, which means that a DVD player will play a CD-ROM (and CD-audio) disk, but not the other way around. Consequently, since CD-ROM players cost about as much as DVD players (which *do* a lot more), computers today are rarely delivered with CD-ROM drives; at the beginning of the new century, most PCs were built with DVD players in them. *Set-top* DVD players are for the most part a different industry, though, which unfortunately creates a dichotomy of DVD use and functionality in the medium.

DVDs in a *computer* are playing on high-resolution displays with progressive scanning, and are probably connected to the Internet (with a reasonably high-speed connection). The DVD can outperform web streaming video, if desired, and (in "eDVDs") can seamlessly augment this prerecorded disc with current metadata from the Internet. On the other hand, a DVD set-top box plays through a *television*, with lower resolution, limited screen area, and probably no Internet interaction.

Consequently, the design of material that goes on a DVD must either be targeted for one display method or the other, or downscaled to the lowest common denominator (the TV), as most are. This is one challenge of DVD creation and distribution.

Studios seem happy to release movies on DVD and treat it as comparable to making a dub for VHS video release. But in their focus on the disc as distribution medium, and closely watching expenditures beyond the bare minimum, they often miss the opportunity in the format. The disc is, after all, interactive. Should consumers so choose, and if given the chance, a great deal of interaction can be had.

Making DVDs

DVD *premastering* involves a number of steps, some of which are highly creative and others that are particularly technical. Simple interaction and menus are added quickly—often they're little more than a title card and some chapter stops. Like web or book publishing, DVD authoring is about preparing the materials for consumer interaction; designing a user's experience. DVD authoring tools are remarkably similar to Web authoring tools; an author of a complex site or DVD would begin by mapping out the tree and hierarchy of "pages" and links. Graphics, buttons, titles, *etc.* need to be created and integrated. Sophisticated DVDs have considerable content in addition to the baseline streaming of the film: interviews, script pages, commentary, and so on. DVDs are also useful media for non-entertainment content (*e.g.*, educational materials, corporate communications) which are all well-suited to the high-quality video and user-controlled interaction. In many ways, today's DVDs can model the Internet experience ideally possible with broadband (~4Mbps connections), and will serve to fill that niche for the many years before that broadband experience is widely available.

DVD Creation

Any video that is to play from a DVD needs to be encoded into a DVD-compatible format. Hardware or software is required to manage the video multiplexing and MPEG compression, and to generate the final "disc image" that ultimately finds its way onto a disc. This format conversion will increasingly be available as just one of a number of output options of any system managing digital video—in particular, from nonlinear editing systems. In much the way software like Photoshop works independently of file formats, but allows output in any number of options (*e.g.*, TIFF, JPEG, GIF), editing systems will output streams in DV, DVD, plain-ol' MPEG-2, QuickTime or RealMedia.

The most creative aspect of DVD creation is in the building of the user's experience. Elements for authoring must first be generated (edited sequences, title cards, any variety of graphics and content), and then a map needs to be conceived that moves users through the elements. Unlike most web links, DVD links have built-in intelligence that can make causal and temporal inferences from selections. For instance, a standard link might say "press this button and go here," but a link with intelligence would say "press this button and go here, unless you've already pressed it three times before, then go *here*; or unless you have already seen this other screen, whereby you should go *over there*." Increasing the sophistication of links means a more involved authoring process, but it creates a significantly more compelling medium.

As authoring systems continue to evolve, these tools become more streamlined, more graphical and intuitive, and more accessible to the non-technical creatives who will be developing innovative experiences.

DVD Anatomy

The various qualities that make DVDs unique fall into two distinct categories—**physical** and **logical**. Physically speaking, a DVD differs from a CD in two major respects:

- The bits are almost seven times as densely packed as on a CD.*
- Discs can be manufactured with *two* physical layers of information, and on both sides, yielding a total of up to 26 times the information on a disc that's the same "form factor" *(i.e.,* size and shape) as a CD.

Thus, DVDs can be manufactured in a variety of configurations; they are named for their approximate capacities:

Name	Description	Capacity**	
DVD-5	single-sided, single-layer	4.7GB	(37,600 Mb)
DVD-9	single-sided, dual-layer	8.5GB	(68,000 Mb)
DVD-10	double-sided, single-layer	9.4GB	(75,200 Mb)
DVD-18	double-sided, dual-layer	17.0GB	(136,000 Mb)

[Note: single-layer discs look silvery, like a CD; dual-layer discs have a golden appearance.]

In addition, the standard "single-speed" DVD drive is designed to spin the disc faster, and read about eight times more data off the disc per second, than a CD player or single-speed CD-ROM. Thus, it provides not just more capacity, but more "bandwidth" as well:

$$8x \text{ CD-ROM} = 1x \text{ DVD-ROM}$$
$$or \ 8 \times 1.3\text{Mbps} \approx 10\text{Mbps}$$

And that's it. The DVD disc itself is just a bigger, faster CD.

Of course, there are other features of the DVD format that make it unique compared to CDs or VHS tapes. The most significant is its ability to provide interactivity—it lets you make menu selections, and more. In addition, video can approach D1 quality. And there are multiple angles, audio soundtracks and subtitles available to select from. So what's involved in actually making a DVD of your finished, edited project?

On the logical side (that is, in terms of what's actually recorded on the disc), things get interesting. Load up a standard DVD-video disc on a computer's DVD-ROM drive, and you'll see some big, arcane-looking files with strange names. *It's this particular set of files, and the particular*

---

* You can see this for yourself just by looking: higher density means that a DVD reflects colors somewhat differently than a CD does when you hold it up to the light.

** In this unique case "1GB" refers to the metric (multiple of 1000), not the binary (multiple of 1024) number of bytes.

data that's contained within them, that gives the DVD its unique, interactive qualities.

This suggests an idea: couldn't you cram your own videos and menus into a set of files just like them, and then put those files on a hard drive, or burn them onto a CD-R, or maybe even email them? *You bet!* While this doesn't mean that a standard DVD player would read them (you'd need to create a *bona fide* DVD disc for that), it does open up a range of possibilities for anyone with access to those files and DVD player "emulator" software on their computer—which many computers now have. DVD is both an object (the disc) *and* a type of file format. This is an important distinction.

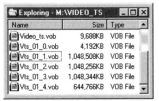

Part of a DVD file directory

The Nitty-Gritty

Creating a DVD means figuring out what you want the viewer to be able to see and do, and when, and then converting that into a form that the DVD player or software is able to recognize and play back. Here's an outline of the *premastering* process:

Asset management: Create and/or gather up all the different types of source files to be included on the DVD (video, audio, menu graphics, subtitles, stills, etc.).

Navigation: Figure out what menu screens you want to have, and what the viewer will have access to when clicking on each button. Much like designing the hierarchy of pages on a website.

Bit Budgeting: Tally up how much total "stuff" you've got, what capacity disc you're going to put it onto, and then figure out the highest quality video you can get, within those limitations. This is a little like playing "The Price Is Right"— you want to fill up the disc of your choice as close to capacity as possible, but without going over.

Encoding: Convert the video, audio, etc., into the types of "elementary streams" that DVD software and players expect to see. For video, this is some form of MPEG. For audio, there are several choices, including Dolby Digital, MPEG, and PCM.

Authoring: Import all the project's various assets into a specialized piece of software. (Menus and subtitles, generally created as TIFF or PICT files, are converted here into high-quality "MPEG stills.") Next, create links between the menu buttons and the individual videos. Add chapter points and other kinds of interactivity.

Proofing: "Beta" test the whole experience yourself and with the help of others. Debug stuff that doesn't quite work right.

Multiplexing: When everything's decided, export all of that digital data into the mysterious files that get written to the disc.

= a predominantly creative, aesthetic, "offline" activity   = a predominantly technical, "online" activity

"BIT BUDGETING" SIMPLIFIED

Bit budgeting can be a slightly tricky concept because there are not one, but *two* different limits to keep in mind: the total capacity of the disc you've chosen (*e.g.*, DVD-5, DVD-10), as well as the maximum speed the drive can "stream" that data off the disc.

Imagine a special water cooler that holds data instead of water. We can let the data flow out faster, which means a better quality picture, but there'll be less time before the supply runs out. Conversely, we can ration out the data slower and get a poorer quality picture, but it will flow for a longer period of time. (This is much like the usual quantity vs quality tradeoff on an NLE.)

But there's another catch: the bottleneck limits us to a maximum data flow of 10 million bits per second (10Mbps). That's the total "speed limit" at which a DVD player (or "1x" DVD-ROM) can read data from a disc. And this doesn't just include video—this one big *multiplexed* stream of data includes *all* the simultaneous video, audio and subtitle tracks that the viewer gets to choose from at any given moment.

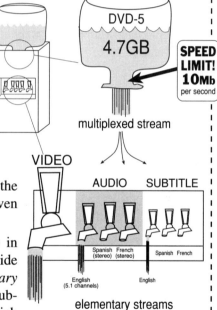

So, one of the first things to do in designing a particular DVD is decide how many of those individual *elementary* streams of data (picture, sound and sub-picture tracks) the viewer has from which to choose. She can switch between some of them or turn them off (kind of like TV channels). But the drive still has to read **all** of that data off the disc as it spins around in the drive (the one big "groove" on the DVD is that one all-encompassing multiplexed stream).

Thus, the challenge to bit budgeting is two-fold: (1) make sure all the simultaneous data streams don't exceed the DVD speed limit; and (2) make sure you don't run out of total disc capacity by the time the show is over.

Now, how much data do each of those elementary streams really use?

Video..up to **9.8* Mbps** (maximum)
         excellent quality................9 Mbps
         average quality.................4–5 Mbps
         lousy quality ....1.5 Mbps

Audio
PCM format ....0.768 Mbps (per channel)
Dolby Digital codec:
         surround (5.1 channels) ...0.384 Mbps
         stereo (2 channels) ....0.192 Mbps
         mono (1 channel) ....0.096 Mbps

Subpicture (i.e., subtitle) ....0.040 Mbps (per stream)

A Bit Budgeting Example

So how much video quality can we afford? We need to figure out the average "pace" we can read data off the disc (the total capacity divided by the total time needed), and then deduct the bandwidth needed for audio and subpicture streams. In this example, we'll allow 0.768Mbps for all the audio (English surround, plus two stereo foreign mixes), plus 0.120Mbps for three different language subtitle streams; that's 0.888Mbps reserved for stuff *other* than video.

Next, if we're planning on putting 60 minutes** of program material (3,600 seconds) onto a DVD-5 (*i.e.*, 37,600Mb capacity), then the average amount of data we can stream per second without running out of space is:

$$\frac{37{,}600 \text{ Megabits}}{3{,}600 \text{ seconds}} = 10.4 \text{ Megabits per second } (average)$$

That's even faster than we can physically read the data off the disc, so we'll figure on 10.0Mbps instead. But even after subtracting the audio/subpicture streams, we know we can use *very high quality* video (just over 9.0Mbps) and not have to worry about running out of data capacity for that entire hour. Now, instead, suppose we needed to encode 120 minutes of program material (7,200 seconds) onto the same DVD-5:

$$\frac{37{,}600 \text{ Megabits}}{7{,}200 \text{ seconds}} = 5.4 \text{ Megabits per second } (average)$$

Subtract the other streams, and that leaves us with only about 4.3Mbps for the video stream. *Hmmm*, does this mean we're stuck with "average" quality video for that whole two hours?

Maybe, but not if we get sneaky . . .

---

* This limit is somewhat less when using multiple picture angles.
** The running time also includes the video for any animated menus. Still frame menus take up negligible amounts of storage, though, and can essentially be ignored.

Encoding

One of the great things about MPEG is that many encoders can "throttle" the video data rate higher or lower as needed as they go along, depending on the complexity of the picture at a given moment. This is known as *variable bit rate* (VBR) encoding, as opposed to the real-time plain-vanilla *constant bit rate* (CBR) encoding:

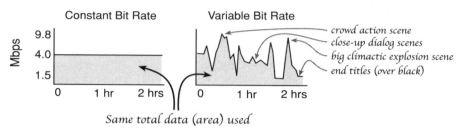

VBR encoding typically takes *two* passes to perform: once for the encoder to figure out a game plan, by tallying up how much quality each frame of video "deserves," followed by another pass to go through the actual encoding process. If there's too much program material to allow for a high CBR (like in the 60-minute example above), a VBR is your best bet for squeezing the best subjectively-looking picture into a finite amount of storage. It's like putting cash that you don't need right now into the bank, so that you can afford to splurge when vacation time comes around.

But what happens if you don't like the "decision" a VBR encoder made for the bit rate on a particular scene? Better MPEG encoders allow for a *segment re-encode*, which simply means that you can tweak a certain section by itself, without having to go back and encode the entire video over again from the beginning.

Authoring

Next, we bring all our various *media assets* (video, audio, graphics, *etc.*) into the DVD authoring software, and the fun begins. The essence of authoring is linking everything together in a way that hopefully makes sense to the viewer, and brings "life" and cohesiveness to a bunch of otherwise separate and static elements.

Like a web site, the menu screen that comes up when you initially start up a DVD is akin to a "home page," and each button operates like a "hyperlink" to another part of the disc. Authoring also entails deciding what the appropriate response should be at any given moment, when the user presses the Title, Menu, Return and other special-function buttons on the player's (physical or virtual) remote control.

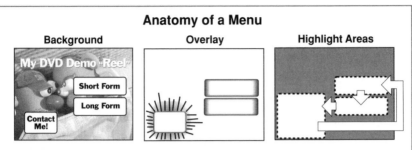

Anatomy of a Menu

Background	Overlay	Highlight Areas

Each DVD menu actually consists of a special set of three distinct layers:
1) **Background**: either a still frame or a moving video that includes a "picture" of buttons.
2) **Overlay**: a single graphic that contains something additional to display for each button, when selected (sort of like the JavaScript "rollover" graphics you'd see on a web page).
3) **Highlight Areas**: instructions that tell the player which portion of that all-in-one overlay graphic to display for any particular button, and how to create the illusion of navigation between them.

Multiplexing

Once you've finished authoring and proofing your DVD, it's time to *multiplex* (or "mux") all the concurrent video, audio and graphics files into an all-encompassing *video object* (VOB) file. This is the big "stream" coming out of the bottom of the water cooler bottle in our earlier analogy, which the DVD player or software eventually reads off the disc. By the way, that torrent of VOB data is actually split up into convenient 1GB chunks—and hence, is the answer to the mystery of what's in the files we saw in the DVD directory back on page 304.

Here are some other DVD authoring functions that the software may allow you to manipulate (either by name, or by using icons):

FP: First Play. What the player does automatically, if anything, when you first insert the disc.

GPRM: General Parameters. A set of registers that the player can use to "remember" things (temporarily, until the disc is ejected). Can be used to keep track of what's already been seen, "quiz" scores, etc., and thus provide a slightly customized experience each time the disc is played.

PGC: Program Chains. These are "EDL-like" pointers that tell the player what (portions of the) VOB "media files" to play back and when.

PTT: Part of Title. Technical name for a chapter point.

SPRM: System Parameters. Where the player keeps track of things like which angle, soundtrack, subpicture stream or menu button is the active one at the moment; region code, parental level, etc.

UOP: User Operations. Allows the author of the disc to disable certain functions (Previous, Next, Stop, etc.) at particular times. [This is where DVD "interactivity" can be actually worse for a viewer than plain old videotape, by preventing them from skipping over stuff they don't want to see.]

VM: Video Manager. The "root" menu which branches to any of the multiple Titles contained on the disc. Accessed via the DVD player's "Title" button.

VTS: Video Title Sets. Separate "sections" of the DVD, up to 99 per disc. Accessed via the DVD player's "Menu" button.

VIDEO-TO-FILM TRANSFERS

GETTING TO THE BIG SCREEN

There is perhaps no more satisfying moment in post-production than seeing your work projected onto a big screen. Film was meant to be projected, it captures the fine details of a scene—the subtle nuances and gradations in shadows, the glorious wideshots displayed on a panoramic vista of 16:9—and the impact of a close-up takes on a proportion unbelievable to TV producers.

There is a truism about having your face 30 feet big on a movie screen: cameras don't lie, and they capture every pore, every pimple, every blemish, and make them HUGE. This is true for human blemishes, and this is true for video artifacts. What is bad on the small screen, is unbearable when huge.

Let us start with the obvious: for the most part, video and film are incompatible: video is low resolution, low color fidelity, 30 frames per second, 4:3 aspect ratio; cinema is high resolution, high color fidelity, 24 frames per second and 16:9 aspect ratio. It is a professional challenge to convert film to video, but it is do-able. It tends to be easy moving from high-quality to low-quality. Going the other way isn't so easy.

But it is *NOT* impossible.

Over the years, methods of getting video onto film have been improved. Before there was videotape and video recording, archives of broadcasts were done by literally shooting a video monitor with a film camera. The device, called a *kinescope*, continues in use today, although its sophistication has improved. But the principle is the same, you need to shoot the video with a film camera . . . somehow . . .

The advent of HDTV improves the possibilities, and of course, the HDTV standard was developed with the conversion to and from film in mind. But this section is about standard definition video, and in particular, consumer and professional transfers. There is a pressing need to know about upconverting from the now-accessible DV formats to 16mm and 35mm film; it is the final breakthrough for the truly independent to hit the pipeline of the very exciting feature film world. Independents got exceptionally "hot" in the 1990s—they were nothing and in a few years they were everything. Winning Oscars like studios. Making stars from actors. DV is a crossover media, but video-to-film transfers are an important bridge. Every year since 1995 has seen an increase in the number of films shot on video (any assortment of formats), and transferred to film for the film festival circuit. A number of these films have gone on to win awards and launch careers of countless independent filmmakers. Quality issues aside, it is being done *all the time*.

The first thing to remember when performing this kind of transfer is simple: GIGO = *garbage in, garbage out*. Transferring to film will not make a lousy video look good. The ONLY way to have a good video-to-film product is to begin with good video. The better the video quality, and the more the video was created with this film transfer in mind, the better the end result will be.

The cost of video-to-final transfers varies widely, depending on a range of parameters (how much is being transferred, what format are you going to, how do you handle audio, *etc.*) but a rough idea for 35mm film is about $3/foot (or ~$425/minute for long-form) of finished material.

There are a couple of technologies that make video-to-film transfers possible; the predominant one utilizes an electron beam recorder (EBR). How these transfers are done is not as important, ultimately, as what you should know before going into one. Always check with the facility that will transfer your video for pricing and their specific instructions. Here are some basic guidelines to consider before shooting your video.

PICTURE

1. Aspect ratio. Film typically has an aspect ratio of either 1.85:1 (16:9) or sometimes 1.66:1 (a European standard). Shoot with safe-title and safe-action margins for the aspect ratio of your choice on your video-tape. Keep the action in the right area or you risk important scene material getting cut off in the transfer.

2. Include bars and tone for color and sound set-up at the head and tail of your video.

3. Shoot on as high quality tape as you can. The best transfers come from the best quality. Digital is better than analog; component better than composite; uncompressed better than compressed. Ideally, shoot digital component (D5, D1, DCT, Digital Betacam), although DV and MiniDV are reasonably acceptable. Also acceptable are digital composite formats like D2 and D3. Analog component formats (BetaSP and Betacam) and composite formats like (1" and 3/4") get more marginal, with consumer Y-C formats like S-VHS and Hi-8 acceptable, but not preferable. If someone offers you a HDTV 24P camera, take it. Remember: GIGO.

◀◀ Component/composite video: p. 120   ◀◀ Aspect ratio basics: p. 177

4. Focus. It seems obvious, but minute focus problems are particularly irritating on up-resolution. Consumer cameras often have lesser-quality lenses and optics and these minor video issues can be major film issues.

5. When shooting video, use the full dynamic range of the tape, avoid crushing the blacks or clipping the whites of a scene, and be as consistent as possible. If you use a scope, maintain chroma between 80–100 IRE units. Keep the light levels high enough to clearly separate the foreground subject and the background, rather than the traditional video look of full illumination. This is often referred to as shooting video *like film*.

6. Don't use the camera's electronic gain (low-light amplification); instead use cameras and lenses that can control shutter speed and f-stop. With higher light levels, stop down (*e.g.*, to f/32, f/64—the bigger the number the smaller the appeture), to increase depth of field. If all of this is new to you, take a class in photography.

7. If your camera has horizontal or vertical detail enhancement, turn it off.

8. If you camera has "soft" clipping circuits, turn them on. They will provide a better "film-look."

9. Avoid fast camera pans or shots with fast lateral (side-to-side) motion. One of the big differences between film and video involves frame rates and motion artifacts. Jerky motion or strobing will be exaggerated in such motion shots. Somewhat related to this, titles and credits should fade in and out (as opposed to rolling) so they don't strobe.

10. If your camera (or post-production system) can create *film-look*, **don't do it**. Don't add noise, dust, scratches, grain. Film-look software should only be used if you want a film look in your *finished videotape*. Film-look software is terrific for video releases (local TV spots, for instance) that you want to look as if they originated with a film shoot; not for projects that will actually finish on film.

AUDIO

Good audio can make mediocre picture work. Mediocre audio makes good pictures seem cheap. *Enough cannot be said about the importance of sound.*

While it might be preferable to record audio on a separate device (a Nagra, a DAT) and sync it up in post, it is sufficient to utilize the DV tape, as long as the audio is captured using better-than-built-in microphones, like lavaleers or shotgun mics. Have a separate person (utilizing headphones to monitor during recording) focusing on sound, levels, ambience and so on.

Also be aware if you are using any noise reduction circuitry—Dolby A or SR. The transfer facility will need to know all this kind of information.

ELECTRONIC CINEMA

Once a movie is completed, *release* prints of the film need to be duplicated and sent to all the theaters that will show it. Each individual set of prints is comprised of a dozen big reels, and can weigh hundreds of pounds—which means significant logistics and expense. True, all this effort is rather modest when compared to the dollars spent on the big production itself (famous actors, cars blowing up). But since it's not uncommon for a film to be released on 3,000 or more screens simultaneously, and for each set of prints to run $1,800 or so, that can still add up to an additional four or five **million** dollars *after* the film is done.

Clearly, this is a hurdle that has always prevented independent filmmakers from attaining wider distribution and exposure.

Enter electronic projection. While the projection of light through a positive celluloid film onto a silvery screen has long been considered an unparalleled experience, new technologies are now able to convincingly replicate (and even improve on) this. High-end video projectors—and high-definition 24P sources*—are immune to dirt, scratches, gate weave and other analog artifacts and deteriorations. Such a transformation in both acquisition and distribution will eventually have dramatic implications all throughout the "film" post-production process. Not only will celluloid release prints become a thing of the past, but all the conformation of workprint, pull lists and assembly lists will disappear, as will negative cutting. *Opticals* will no longer need be created optically, but generated electronically in post facility online suites or even desktop workstations, much as video is routinely postproduced today. Laboratories would not be needed to print circled takes (nor make release prints), but will be superceded by 24P telecine sessions (if the material was still shot on negative film), along with a down-conversion from 24P to offline qualities for editing.

Screening a rough cut on a large screen, once easy when everything was on sprocketed film, and difficult in the few-decade transition between film and video, will become as easy as patching the output of the computer into a video projector. It may even come to pass that editors can work in small rooms some days, or choose to sit on sofas in projection bays, editing and watching on a big screen *as they work*, seeing their project as the

---

* In an historic moment in 1999, George Lucas distributed a handful of high-def digital "prints" of his Star Wars: Episode I, and was able to project them digitally as well. Furthermore, a few select scenes in the film itself had also been shot using high-definition digital videotape. The experiment went well enough that he announced that Episode II would be shot almost entirely using 24P video, and receive even wider electronic distribution.

audience will ultimately see it. (Though they sometimes find it difficult to articulate, many editors attest to a difference in editorial pacing and rhythm that must be compensated for in order to make scenes "feel" right, when cut on a small screen but seen on a large screen.)

Also, it should be noted that so-called "E-Cinema" *doesn't* necessarily equate to *digital*, at least as far as the projector itself is concerned. High-quality projection can be accomplished by either digital or analog means, and experts are divided as to which method they prefer.

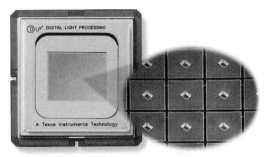

In any event, the new electronic cinema paradigm will mean a number of differences in how and what people see at "the movies," and raises a host of new issues and possibilities:

Technical: Projectors need to have sufficient brightness* and resolution for the screen size and aspect ratio used. Digital encoding must provide exceptional image quality, without

One of the competing technologies that makes electronic cinema possible is the DMD (Digital Micromirror Device™), utilized by Texas Instruments' DLP (Digital Light Processing™) projector system. A bona fide application of nanotechnology, each optical semiconductor chip has an array of over a million hinged, microscopic mirrors which operate as optical switches to project a high resolution, full-color image on-screen. A "refresh" rate of approximately 50,000 times per second essentially eliminates perceptible flicker.

noticeable artifacts, though at manageable bit rates. Server systems must be able to take punishing workloads with maximum reliability and fault tolerance.

Economics: Celluloid projection is old technology, and relatively cheap to install and maintain. E-Cinema, meanwhile, means formidable costs to be amortized over the lifespan of equipment that presumably will become obsolete more often. Business models are still evolving to cope with this eventuality.

Versatility (producers): Studios can wait longer before "locking" picture, and continue to make post-production changes up until—or even

---

* Brightness of image on a screen is measured in ANSI lumens. A typical movie or concert experience requires projector output of around 10,000–13,000 lumens to brighten those big screens. Midrange brightness, which trades projector size and power consumption for quality, runs in the 3,500–5,000 range; corporate presentations and small theaters might require 1,000–3,000 lumens.

after—opening weekend, with minimal added expense. That also means potentially more responsiveness (and/or indecision and pandering) in response to audience and critic reaction. Freed from the destructive nature of negative cutting, proliferation in numbers of different versions (*a la* software?) of a given movie is also conceivable.

Egalitarianism: While it has long been possible to transfer video-originated material to celluloid, the process is still expensive enough to keep even a limited cinematic release beyond the reach of many independents. Video projection dramatically lowers the barriers to entry in producing and distributing content for theatrical release, making the notion of "desktop cinema" increasingly feasible and even commonplace.

Distribution: "Films" can be shipped on physical media (such as multiple DVD-ROMs) rather than bulky spools of celluloid; but they are still subject to loss and theft, as well as the cost of duplication and logistics of delivery. Theatre chains may prefer digital transmission instead, via satellite, fiber optics, the Internet, *etc.* In either case, piracy and intellectual property rights remain an issue.

Versatility (theatres): Since the end "product" (*i.e.*, a digital bitstream) is no longer subject to the same physical limitations as film, theatre management will have unprecedented versatility to change programming at will, and at the last minute, according to demand. Independence from a cumbersome distribution medium could also mean more diverse and creative programming, perhaps from commercially-available HD DVDs. New business opportunities also include special non-cinematic events, such as live showings of the Superbowl, the Oscars®, *etc.*

Posterity: Celluloid (as well as sprocketed mag soundtrack) is still far more "archive-friendly" than video, with regards to both physical longevity and its immunity from technological obsolescence. Preservation of cinematic video content may thus prove to be an ongoing challenge.

Digital cinema will herald the end of celluloid in all but the art-house circuit. Ironically, as television becomes more cinematic in nature, cinema is itself evolving to become more like television. The competition between the two media for viewers will heat up as both HDTV broadcasting and digital cinema projection evolve. The end result—yet another chapter in the story of *Media Convergence*—presumably ushers in a new era in the democratization and aesthetics of the big screen.

◀◀ History and moving pictures: p. 39   ◀◀ 24P Note: p. 101

Chapter 6

Systems

"The overuse of *any* effect diminishes
its true worth."

—Edward Dmytryk (1984)

Director: The Caine Mutiny, The End of the Affair

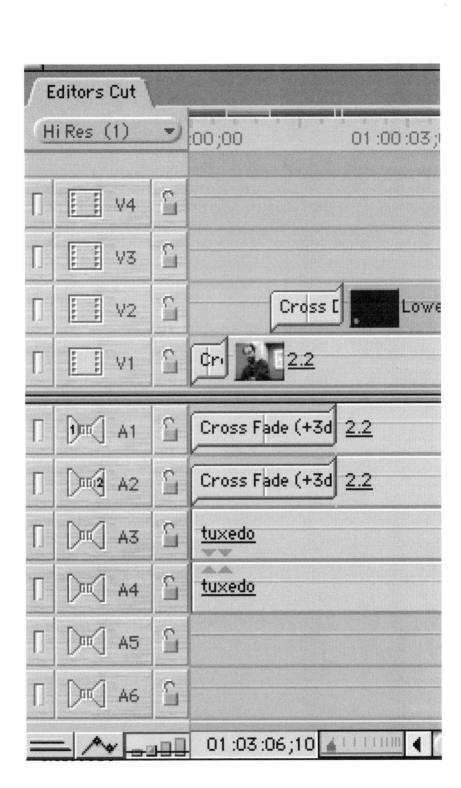

T Y P E S   O F   E D I T I N G   S Y S T E M S

IN THE BEGINNING, there was film. Film editing, which we will call TYPE I editing, reigned supreme for nearly a century, using uprights and flatbed editing tables.

At first, even the dawn of videotape did not stop traditional film editors from cutting tape as if it were film. For almost a decade, videotape was almost like another *kind* of film.

TYPE II editing is characterized by computer-controlled linear systems and the birth of timecode (1970); the media was usually videotape, but occasionally other things, like videodiscs. There was no database; there was one source reel for each set of dailies. There were the rudiments of nonlinearity in the form of a "preview" of the upcoming edit. These were accomplished by small recordings of bits of material to "B-rolls" that permitted short previews of certain edits and effects.

TYPE III editing systems began the Age of Nonlinear. Early generations offered multiple source tapes or discs to create random access and longer previews (extended nonlinearity). They had databases. Graphical displays. Late generation systems developed an important variation on the earlier systems: they offered the first *virtual masters*—extended previews, but with full user control. Because media was still analog, this functionality was only viable using laserdiscs. Computer-wise, it was not data intensive, but machine-control intensive: there was considerable scheduling of many source decks. To the editor, it was a dramatic departure from anything seen in the linear world.

The first generation of nonlinear systems debuted circa 1984–85: the Montage Picture Processor, the EditDroid, and the Ediflex. Each system

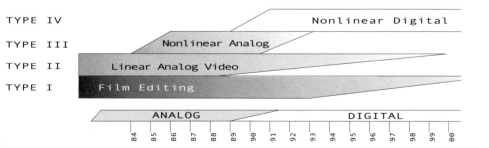

had to contend with a world not yet sure of electronic nonlinear. Film editors were being thrown onto these systems, and no one had a good idea of the best ways to use them, or for exactly what market the products were most suited. These things would come in time.

The second generation of nonlinear systems were a little smaller, a little cheaper, and sometimes designed with the perceived mistakes of the first generation in mind. The systems were the CMX 6000, the E-PIX, the TouchVision, the Link. The second generation was introduced to a community that was aware of the advances that had been made in editing. These systems were developed between 1986 and 1989.

TYPE IV systems were radically different from previous ones—source material was now digital. There was no scheduling of decks, no previews, no shuttling. Although perceptually it looked like some of the late-generation Type III systems, technologically it was a new entity. Like TYPE III, it was sometimes called *virtual editing*.

Type IV systems began with the EMC2 and the Avid. The Avid was introduced at roughly the same price as the Type III systems; the EMC, on the other hand, was the first system to offer *low-cost* nonlinear editing. Although these type IV systems debuted in 1989, for much of the professional world nonlinear editing began circa 1993–1994. The first years were solely nonlinear offline; by the beginning of the 21st century, nonlinear online had become commonplace. By the year 2000, nonlinear editing was an activity for consumers.

E V A L U A T I N G   S Y S T E M S

While digital editing systems are more alike than different, they *are* different; they look somewhat different, they have different computers using different interface styles, different functions and abilities, with strengths and weaknesses—usually related, at least in part, to *cost*. The following checklist is a partial summary of the questions that prospective users and buyers might want to have in mind when evaluating nonlinear systems:

INPUT/OUTPUT

All Systems

☐ What choices of image resolution are there? Can you mix levels of resolution within a project? Within a sequence?

☐ What types of digital and analog video/audio signals can be connected to the system? For video: Composite? Component? Y/C? SDI? FireWire? For audio: RCA jacks? XLR? AES/EBU? S/PDIF?

☐ What kind of VTR machine control is available? (Is extra hardware required?)

☐ Can you (optionally) "log" material before digitizing? Can that be done on a separate (non-editing) desktop computer?

☐ How well and *accurately* will the system handle timecode and/or control track breaks in the source material?

☐ Does the system place any limits on how much material can be used? How much material can be captured in one continuous "chunk"?

☐ Can the system import an EDL (Edit Decision List) from another editing system? In what formats? Can it then batch capture/digitize from that imported EDL?

☐ What formats of EDL will it output, if any (CMX, GVG, Sony, SMPTE, *etc.*)? In what "assembly modes" (A-mode, B-mode, C-mode, *etc.*)?

☐ What other file formats are supported for importing and exporting? Does the system support standardized interchange formats (such as AAF/OMF)?

☐ What types of features can be used to label and organize source material *after* digitizing has been done?

☐ How much work *might* be lost if the system crashes, or the power is suddenly shut off? What (and how time-consuming) is the recovery process? If there are RAID arrays, what kinds of fault tolerance are employed?

☐ Does the system have web-streaming output options? QuickTime? RealMedia? Windows Media? DVD? MPEG-2?

Film Cutting Systems

☐ What options are available for alternative film sizes and speeds (IMAX, 30fps film, 3-perf, 16mm, 8mm, 60fps, *etc.*)?

☐ If a mistake is made in the log-database (timecode-key number relationship), can it be corrected *after material has been edited*—without requiring re-editing?

☐ How "frame accurate" is the tape assembly?
☐ Can logs from telecine be input automatically?
☐ What format of telecine logs does it take?
☐ What kinds of film lists can be output? Pull lists? Change lists? Optical count sheets?
☐ Does the system account for re-using film material? Does it account for negative cut margins? How are these flagged and coordinated with potential requirements for reprints and opticals?
☐ If the system is editing via timecode with plans to convert to film edge numbers, what "rounding" corrections does the system make? Is it counting/editing at 30fps or 24fps? Are the implications of this acceptable?

EDITING FUNCTIONALITY

☐ How is most user input accomplished on the systems—keyboard? mouse? trackball? touchscreen? touchpad? *etc.*
Are the keyboard equivalents comprehensive and well-conceived?
Is the physical interface very complicated? comfortable? efficient?
Can third-party external controllers be used? mapped to functions?
☐ What methods are available for trimming? Numeric? Mouse? On-the-fly?
Can you examine the outgoing and incoming frames at an edit point at the same time?
Can they be adjusted together and if so, can they be interlocked and scrolled at play speeds? If so, what happens to audio? What control of audio scrub is there?
☐ How easily can the system perform a split-edit (also called an overlap or L-cut)? This is a very common function that should be effortless.
☐ What are the effects (wipes, fades, dissolves) capabilities?
How many simultaneous "streams" of video are there? One? Two? More?
What effects have to be rendered? How long will they take?
Can the direction of all the wipes be reversed (without having to rearrange all the shots onto different tracks)?
Does the system accept plug-in effects from other software products?
What about motion effects (like speed-up, slo-mo, fit-n-fill and freeze frames)? How are these represented in the EDL and film lists?
☐ Does the timeline force you to "checkerboard" material onto A/B tracks in order to create transition effects? How easily can you then make changes?
☐ Can you play the edited master in non-play speeds—variable fast, slow, forward, reverse, frame-by-frame, field-by-field? How does this affect audio pitch and can that be controlled? Are their options for audio scrubbing?
How smoothly can you vary the speed in the playing of material?
☐ How many audio channels can you edit with at one time?
How many can you input and output at once?
If fewer can be output than edited, can the multiple tracks be mixed down for output?
How many channels of audio can you monitor at once?
☐ Is the system capable of panning? EQ? Other audio effects?

☐ Can audio be easily moved from any track to any another track?

☐ Are audio levels remembered by the system? How are they adjusted? Physical console? On-screen? "Rubber bands"? Are rubber bands spline (Bézier) based?

☐ Can a shot or edited sequence from one project be easily moved into another project? What is the process?

☐ Can the digital source material be easily switched (*i.e.,* are the drives hot-swappable? Do you have to quit and shut down the system to do so)?

SYSTEM/VENDOR

☐ Is the system based on proprietary hardware, or open platform?
What happens when the hardware is updated?
Is there an upgrade path to other products, future updates, or future products?

☐ Is technical support available? What are the hours? Does it cost extra? What is its reputation among existing users? How long do they have to wait to talk to someone knowledgeable? How long does it take before problems are resolved?

☐ Are there diagnostic utilities for automatically isolating and troubleshooting problematic hardware/software/*data*?

☐ How easy is it to find a trained base of editors for this product? How many systems are currently installed?

☐ How much will it cost each year (per system) to keep the software and hardware up-to-date?

☐ Are there user groups and other third party resources active in your area? On the Internet?

☐ How easy or difficult is the system to set up and cable?
Is the physical configuration otherwise limited? Cable lengths, *etc.*?

☐ Is it available as a bundled hardware plus software product that can be installed on a customer-supplied platform (SGI, Mac, PC)?

Recognize that "more" isn't always better. A system that does MORE is usually more complicated, although not necessarily so. What you want is a system that is correct for the project(s) being edited.

From a facility standpoint, a more powerful system means more flexibility in the KINDS of projects that can use the equipment. From an editor's standpoint, there must be functional ease as well as compatibility between the system and the particular production.

◀◀ Computers: p. 202

TURNKEY SOLUTIONS

AND OPEN VERSUS CLOSED ARCHITECTURE

Occasionally, an editing system is just that: a "system"—a coordinated set of tools that produces a given outcome. A system made by a single company can be much easier to set up, use, and maintain. It would have precisely the right cabling and the specified monitors, the proper RAM configuration, and so on. The software and hardware would have been designed together in many cases. Such a company would be focused on the task of creating a well-integrated hardware/software combination that has **one** job to do, and is (presumably) designed to do it well. If that system breaks, you just call the company. In short, this system is a well-oiled machine that does nothing but edit, or render animation, or design 3D graphics. For the most part, this was the case until the mid-1980s.

But as computers dropped in price, and desktop computers gained extraordinary power, the idea of unbundling systems came into vogue. Closed architecture ("black boxes"—combinations of proprietary hardware and software) gave way to open architecture (so-called "white boxes"). In short, a single company only needed to develop software to do the editing or graphics, and their software would run on someone else's computer.

The power of open architecture allows hardware and software to develop separately. Since software companies can be small and often operate with low overhead (as compared to hardware companies), new small companies can proliferate and develop an exciting array of (often competing) products. This can be advantageous to consumers, who want product invention and market competition. Unbundling systems finally became feasible in the 1990s.

And unbundle they did. While companies now tend to specialize in either hardware or software, but rarely both, end-users are often left with the task of reassembling the systems from the parts: a computer from Compaq, a board from Pinnacle, RAM from a catalog, an accelerator from Matrox, an operating system from Microsoft, some application software from . . . whoever.

If the product doesn't work right, there are myriad places where the ball could have been dropped: incompatibility between the application and operating system, drivers, plug-ins, hardware configuration . . . a bug, well, *somewhere* . . . and just when support becomes the most critical issue, the proliferation of products, in particular inexpensive products, has made getting proper support challenging.

So with the flexibility of unbundled products comes a litany of potential problems. The alternative, then, is the *turnkey* solution.

"Turnkey" (which unfortunately looks like "turkey") is an old systems expression that means that you only have to *turn the key* and the thing goes on and works, as simple as starting a car. It's sort of like the expression "plug and play," although it refers to entire systems and not just individual components. A turnkey system is usually assembled by a third-party (a vendor, perhaps) specializing in the task. That third party will experiment with all the permutations of a product, determine what is the best set of compatible components at a given price, and then sell this complete (and hopefully reliable) solution to the anxious end-user. Bear in mind, though, that you might pay a premium for that service and convenience.

Lacking the services of such a *systems integrator* (or VAR, for "value-added reseller"), the best a consumer can do is check the trade press, read reviews, and talk to people who have gone before. Or just plunge in and take his chances.

But even shopping for turnkey systems can be confusing in its own way. While diligently doing your research at trade shows, it isn't always immediately apparent that seemingly-competing products are actually the *same* product in different guises, having been licensed, repackaged and rebundled by different companies, under different names. Untangling this web of products—some competing, some complementary, and some identical clones—can make comparison-shopping difficult at best.

If you are a sophisticated user, you may be comfortable with purchasing components (at least make sure they are recommended together), and assembling your own system. But if this gives you pause, look for integrated or turnkey solutions; these may be so completely configured that they include monitors, cables, drives and even pre-loaded software. You unpack, and plug it in.

System	Pricing	Reliability	Support
individual components	less?	least	iffy
turnkey: integrated (VAR)	more	more	more
turnkey: closed-architecture	most	most?	most?

◀◀ Nonlinear system reliability: p. 6

SYSTEM INTERFACE OVERVIEW

AN HISTORICAL PERSPECTIVE

This section will serve as a specific outline of some key products, placed in historical and technical perspective, in roughly chronological order. Unlike the more corporate discussions in Chapter 2 ("A Brief History of..."), this review will not be concerned primarily with economic success or failure, or realistic design strategy—only with fundamental form and interface.

Let us begin by looking at computers and a part of their evolution that involves us. In the 1970s and early 1980s, work was done at Xerox's computer research division in Palo Alto, called Xerox Parc. There, a number of innovative computer interface technologies and devices were pioneered—in particular, the bitmapped display, windows, the icon, and the mouse.

In 1981, when most computer companies or designers in the world were working from more standard PC-type monitors and technology, a few saw a different way for computers to interface. The first three nonlinear systems in our discussion are all extremely unique and innovative: (Lucasfilm's) EditDroid, (Cinedco's) Ediflex, (the Montage Group's) Montage Picture Processor.

The EditDroid (1984) shown with the "Death Star" console, as it was affectionately known. Note the central rear-projection widescreen display for the edited master, with the source monitor on the left and a SUN/1 workstation data monitor on the right.

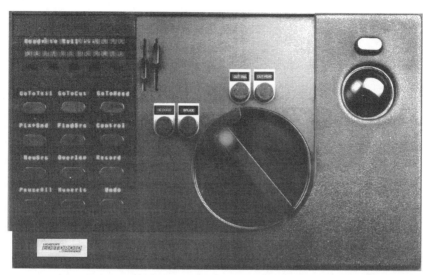

The EditDroid "TouchPad"—a stand-alone console for editing. On the right is a trackball and select button, on the left are "soft-function" keys that change as the editor performs different functions. In the middle is the metal "KEM-like" shuttle knob, and four buttons for editing. Editing took place largely on the data monitor via the trackball.

Although the early Ediflex prototype had been in some use prior to the release of the other two, all debuted as of NAB 1984, and tested on their first projects during the following year.

The first EditDroid worked primarily with six laserdisc players and two 3/4" decks. It got most of its random access from the discs, but could also make use of videotape for laying off parts of the cut, and for accessing source material not yet placed on laserdisc. The EditDroid ran on a SUN workstation, the highest powered platform of the available systems, and a bold step for 1984. (The SUN platform had actually been selected in 1981.) The system had a high-resolution bitmapped display, a graphical timeline that was used in editing (and not simply in representation), a highly designed editor's "TouchPad" that included a KEM-style knob for shuttle, a trackball, and some variable function buttons labeled with LEDs that changed as you worked. The cursor, as you moved across the screen, was a little book in the "Log Window," a scissors in the "Edit Window," and a tiny Darth Vader mask in the thin "Control Window" along the bottom of the screen.

Even with the laserdisc source, material needed to be scheduled and previewed—it was still not possible to roll material in a preview without recording. Effects were performed via a Grass Valley 100 switcher that was configured with the system.

The original EditDroid data display—the top portion of the screen is the LOG WINDOW, and below that is the EDIT WINDOW. The Log Window contains a listing of shots on any given laserdisc; the Edit Window contains the "cutting block" where picture and 2 channels of audio are cut, and a timeline (running right to left) where shots are placed. Below the timeline is a text display corresponding to the timeline, and along the bottom of each window are specific function buttons.

The Montage I controlled 17 Beta 1/2" hi-fi decks to achieve its random access. Material was broken into "loads" of up to four hours of source, and the metaphor was like hanging film strips in a film bin. In a complete departure from videotape editing, there was no "source" and no "master," *per se*; only big long strips that could be easily clipped apart into smaller strips, each hanging in the right order in a virtually infinite bin.

Thus the physical interface displayed seven of the shots in a bin at one time—the head of the strip on top and the tail of the strip on the bottom. This "head/tail" view proved exceptionally useful in identifying long pieces of film. The editor's console was designed like a stand-alone machine, echoing the vibe of the film editor's Moviola. It was tethered to the racks of beta machines, but it sat alone, upright, with a futuristic console face that included two large, eminently tactile and responsive black knobs, used for a number of functions, among them for rolling your view of the hanging shots. Probably no system since has created such a tactile physical space for editing.

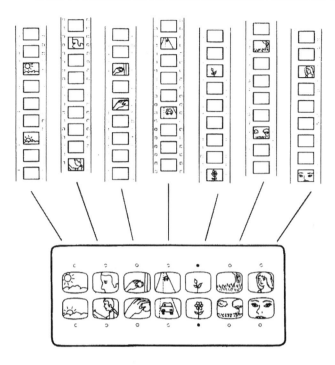

Above: an illustration from the Montage User's Manual showing the fundamental concept of Head-Tail views in a bin. The bottom part shows how those Head-Tail frames are displayed on the seven monitors from the editing console. Below: the original Montage console, with 3.7" b&w monitors for the bins (the middle section), the NTSC edited sequence (top right) and three 5" monitors (top left) for selected trims and splices, along with a status display. Highly tactile control/shuttle knobs are on the console, left and right, with a digitizing pad in the center for on-screen marks and notes.

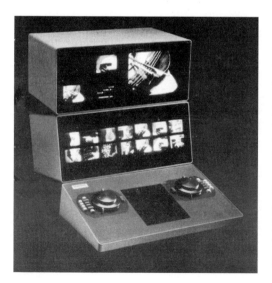

Also quite unique was the Montage's use of digital pictures; rather than wait for the Beta tapes to cue to any given head or tail, the Montage digitized the head and tail frames "labels" and stored them on a separate 10MB Bernoulli hard disk that went along with a project. When a shot was identified via the digital picture label and selected, the system then searched the tape for the identical frame and once cued, replaced the digital image with the videotape's color frame. This way, labels existed to represent all shots in all bins, even without relying on the videotape's ability to cue.

The cue time to play a preview ranged from a few seconds to a number of minutes in some cases.

Unlike other systems of the age, the Montage also allowed the editor to scratch notes on the digital picture labels with an "electronic grease pencil"—a pen and digitizing pad built into the console. All material input into the Montage was stored in the "source bin" where no editing could take place. From there, material was discarded from (to the "discard bin") or copied to. The editor built cuts into any of four separate "work bins" or in a "pull bin."

The Ediflex (below) was a third and still different kind of system. Tailored for episodic television editors, built by the only person with professional editing system design experience (Adrian Ettlinger), it was the only system from Hollywood. Where the EditDroid represented source material in a powerful configurable database (a *list*, in other words), and the Montage used its unique head/tail labels to represent everything, the Ediflex was based on the script. A lined script is the fundamental basis for source information—created mostly on the set during production, it is essential to most script-based editing—and a natural for the computer extension of source material. Although the process of getting the script into the system and correlating it to the source footage was a bit arduous for assistants, it

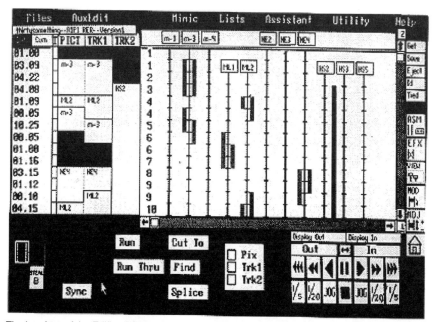

The interface of the Ediflex (1989): the timeline runs vertically at left, with the "script mimic" along the right. Takes run across the top, with numbers running vertically representing each line of dialog.

gave editors unusual functionalities. An editor could easily call up a given line of dialog from every take that covered the line. This was called the "Script Mimic."

All of these systems—so completely unique, as you might expect from three inventions made simultaneously with such different goals and hardware—were attempting to attract film editors to video, even though those editors had rejected the move during the previous decade. The systems were trying to re-create the film *style* with better computers and less expensive video technology. All were expensive for the time, with fully configured systems running from over $100,000 to $200,000. The source media for the Ediflex, like the Montage, was 1/2" videotape, and of inconsequential cost. The EditDroid required the use of laserdiscs, both expensive (compared to tape) and difficult to come by. The device used to record these write-once 30-minute laserdiscs was the Optical Disc Corporation (ODC) RLV recorder, itself costing more than $175,000.

The EditDroid, although arguably technologically superior to its competition, was labored with huge costs and barriers to viability (in the form of disc-dependence). The Montage began making inroads on long-form—television and movies—in various locations around the world (sales to big-name directors—Coppola and Kubrick—gave the Montage its biggest boost); the Ediflex settled into the Hollywood market as a rental-only system.

Although the interfaces were quite different, The EditDroid, the Ediflex and the Montage all scheduled previews, in much the way traditional videotape editors had previews: you chose a sequence or portion of a sequence, the machine would schedule the shots (either from the tape sources or the very fast laserdiscs) and then show you the preview. The editor had no control over the preview—it was shown in play-speed forward. The EditDroid enjoyed the luxury of quick laserdisc cueing, often a fraction of the time required by the tape-based systems.

In 1986-87, CMX released the CMX 6000, a laserdisc-based system that was considerably different from its predecessors. Although the simple

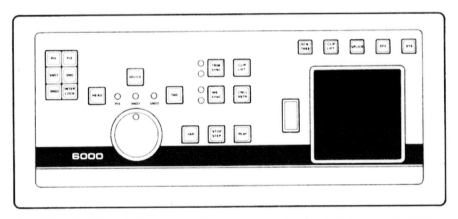

The CMX 6000 Edit Controller (above). Below, a system configuration (not shown are the banks of laserdisc players). Master monitor (left) and Source (right); In transition editing, the master thus displays the outgoing frame and the source shows the incoming frame. The source monitor also serves as the computer interface display when not running dailies, toggling between data and video.

style and dedicated console are often cited as the central features of the system, it was the **virtual master** that truly advanced the state of nonlinear editing. Laserdiscs cued forward or reverse as you shuttled the "master" around, which simulated an actual edited sequence.

The system had moved very far in removing the computer from inbetween the editor and the material; where the EditDroid was heavily computer dependent with its windows and icons, the Ediflex to a lesser degree, and the Montage the least, the CMX 6000 was very text list-oriented, but removed a great many computer functions (both to its credit and to a fault).

The greatest drawback of this system was ironically its defining characteristic—its solitary use of laserdiscs. The discs gave it the speed and interaction no prior system had, yet forced the user to be dependent on ODC disc premastering. This process was both time-consuming and expensive, and discs, once written, could not be erased or added to.

Still, the system was based on the very simple idea of locating source material in one monitor, and attaching it via a "splice" button to the master in the other monitor. Overlaps and inserts were particularly fast and easy.

Dependence on the computer interface is really not much of an issue until you have a bitmapped or icon-oriented display—the EditDroid was years ahead in this department, not seen again until the digital systems.

The Touchvision (1988), precusor to the D/Vision.

The BHP TouchVision had a technical architecture much like the tape-based Montage and Ediflex. But its interface again was unique. Instead of hiding the multiple sources that the computer relied on for random access, the TouchVision changed that perspective to a film-reel metaphor. Now, reels could be loaded onto individual decks, and controlled in much the same style as a multi-plate flatbed editing table. The system console was a dedicated tabletop with a KEM-style knob built in, further re-enforcing the filmic flavor. Although any of the tape-based systems could manage a multi-camera video or film editing style by dedicating a certain tape to a certain camera roll, multi-camera required a number of specific functions that would leave many systems un-utilized for these projects. Eventually, with later releases of software, both the EditDroid and the CMX 6000 could also manage multi-camera projects, but due to disc economics, they remained most viable for single camera (and ultimately) film rather than video source.

One system that was released with the original nonlinear products was an in-house system called the Spectra-Ace. Commonly known by the name of the facility, LaserEdit, the product was not nonlinear but only random access due to its use of laserdiscs for source instead of tape. Shots were accessed via timecode, manipulated much like videotape, and recorded to tape. The large, video-styled console generally did not facilitate the easy transition for film editors, so in time the system remained the domain of video-savvy editors, and became the predominant system in Hollywood for multi-camera television shows for many years.

Amtel's E-PIX sought to gain the advantage of the disc-based systems (fast access, high-quality images) while removing the dependency on the ODC laserdisc process. By creating a system with *recordable* videodiscs, this was solved, but only partially. The high cost of these WORM videodisc machines required that few be used on the system, which in turn required the evolution of clever new disc-use strategies. The basic premise was as follows: disc real estate is limited, expensive, and should not be squandered. Rather than transfer all source material to multiple copies of tape or disc, the E-PIX transferred selectively to the discs—in effect building "super select" discs of required shots. This concept was mirrored later, to a lesser degree, with early digital systems, as hard disk space was also somewhat limited and expensive. The E-PIX display had adapted a style not unlike the EditDroid—a cross between traditional videotape and a graphical interface. The timeline is a focus of the system, but so is timecode and timecode functionality. This makes the E-PIX video-like in

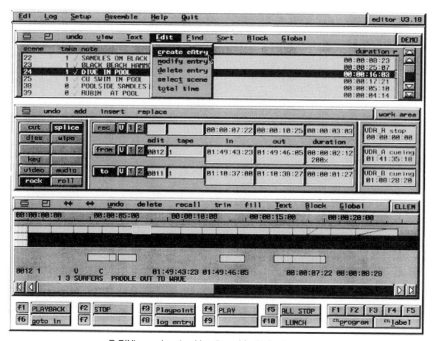

E-PIX's version 4, with a "graphical" display (1991)

its interface, like traditional linear systems or perhaps the Spectra System. Its console reflects this concept as it is neither particularly innovative nor simple. Its bitmapped monitor is busy with functions and timeline—but it was designed only to display information, and was not meant to be "active." The E-PIX was largely ignored in the small markets originally investigating nonlinear editing, but eventually found a serious niche. By the early '90s, professionals were not ready for the still-evolving functionality and poor image quality of the new digital system but were growing tired of the old nonlinear technology. The videodisc-based E-PIX took advantage of the gap between the analog and the digital worlds, and like the laserdisc-based CMX 6000, flourished to some degree during the transition years.

By the late '80s, a few companies began investigating the possibility of digital offline. Both the EMC2 and the AVID began as projects from frustrated computer graphics individuals dealing with tedious and slow linear tape editing. Evolved from the corporate market, rather than from (or for) Hollywood, these two systems looked to personal computer platforms and off-the-shelf technology for much of their design; their primary goal was to make the editing accessible, to liberate it from the "high

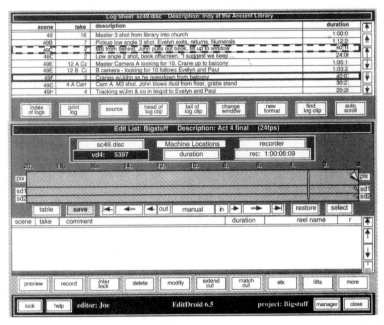

The EditDroid, re-released as the edtiDROID, version 6.5 (1990). Notice that the cutting block from the original system has been removed; and the System Window along the bottom has been expanded. The new editDROID lasted only briefly.

end." The EMC2 was positioned with a PC-based platform, common in American business; the Avid was initially developed on the Apollo graphics workstation but was ultimately introduced on the Apple Macintosh platform, less common for business, but growing in appeal with computer-friendly artisans, and leveraged with the massive religiosity of the Macintosh product followers. In a way, the chosen platforms set up the separation between the two products' thrusts: the PC for the low-end, and the Macintosh for the high-end—for the first many years of the two products, the EMC was consistently less than half the cost of the comparable Avid line.

Although the existing manufacturers had long expected digital video to make its way into offline, they were confident their markets would not accept the image quality possible at the economics required. Avid and EMC bypassed the resistance mostly by avoiding Hollywood, and the markets the other systems worked in. The response was unprecedented, at least since videotape and computer controllers were first introduced for offline in the '70s. Where the sum total of Montages, EditDroids, CMX 6000s, E-PIXs, TouchVisions manufactured and used from 1985-1990

Display from an early release of the EMC2. Note how event numbers are placed into the timeline display.

was perhaps under a few hundred, the numbers of EMCs and Avids sold in their first few years greatly surpassed that, and by 1993, their numbers were an order of magnitude beyond all other nonlinear offline products.

The EMC, the first digital nonlinear system, was careful to rely mostly on removable media, magneto-optical disks, for source. Although this limited the image quality in particular, it was the most realistic option available for a flexible system. The interface was primarily a timeline, and something like a slide viewer representation of source. Both EMC and Avid licensed the Montage head/tail image label patent for their systems. The EMC invented editing primitives that were perhaps more complex than required for the functionality, but the system was known as efficient and cost effective. The Avid moved editing the farthest into the computer style (the Macintosh style, in particular), by windowing not only all the data and graphics (like the EditDroid, and later the E-PIX), but also all source and master material. The Avid had the most highly-refined computer interface, well thought-out, and incredibly efficient for a software product. In spite of the company's efforts to offer various physical consoles for the Avid, it continued to appear that a mouse and keyboard were still the easiest and most streamlined ways to work with the system.

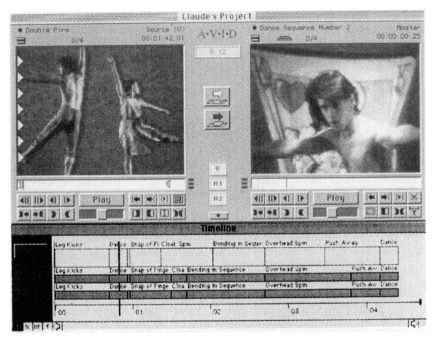

Avid/1 Media Composer, editing display, 1990.

By riding the multiple platform options of the Macintosh line, and by bundling their software in different ways, Avid was able to differentiate a product line far beyond the scale of any other nonlinear company. Avid's marketing expertise went far beyond any previous nonlinear product manufacturer; bold advertisement headlines like "the first digital nonlinear system to play at 24fps," relied on the public's still-naive views of nonlinear systems and positioned themselves as the originators of a type of product that was many years in development. Regardless, Avid succeeded where numerous others had failed—they proved that nonlinear video editing was both plausible and viable, and they introduced systems on a scale that was unseen since CMX.

In response to the immediate acceptance of the EMC2 and Avid, the earlier nonlinear system manufacturers began the quick transition to digital. Three companies in particular took their established styles, their markets and expertise and invented all-new digital systems—windowed, bit-

Avid's MUI console (1993); 12 user-configurable buttons and a shuttle "knob." The Manual User Interface was created in response to users' requests for a physical interface device.

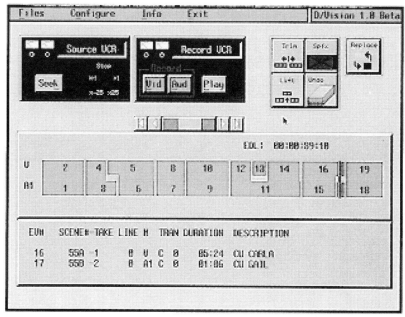

D/Vision editing monitor, system prototype version (1991). The D/Vision evolved into Discreet's edit* seven years later.

mapped, highly graphical. TouchVision produced the D/Vision, Montage produced the Montage III, and Ediflex produced the Ediflex Digital.

The D/Vision bore little resemblance to the TouchVision; it had a simple video metaphor that was easy to learn, and was reasonably functional. Competing with the lower cost systems like the EMC, the D/Vision was the first digital product truly replacing its analog product line. The

The Consequences of Open Architecture

As the 1990s progressed, and the digital nonlinear systems began to capture the editing market, the closed architecture systems began to wane, and with them, the work done in the interface integration of software and physical console hardware. New versions of even classic systems dropped their consoles for the economic advantages of software-only development, and therefore led to increased dependence on the ubiquitous keyboard and mouse. Ironically, one of the early attractions to nonlinear editing was the stylistic advantage of "film-style" editing over the keyboard-driven linear edit systems; by the time nonlinear editing became the norm, keyboards and computer monitors were as prevalent in edit bays as they had been a decade earlier.

◀◀ Open versus closed architecture: p. 202

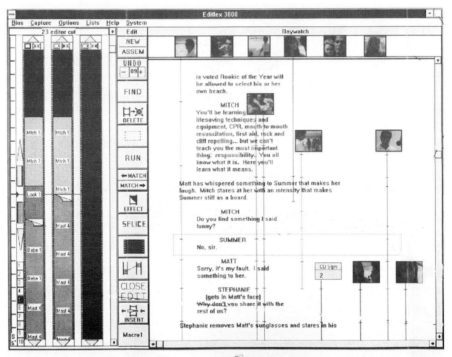

Ediflex Digital (prototype 1993). Areas of the display include the "Picture Gallery" (displaying the first frame of every take available at a selected point in the script), the vertical timeline along the left has scene names inside each shot, and up to 8 tracks of audio all in-sync to the script), and "picons" (or slate boxes of scene/take info) used to represent each scene in the script mimic.

Montage III removed the numerous video monitors used in showing the cut and the head/tail views, and windowed them into a display. The Montage Group pioneered a new kind of technology to break the logjam between fixed hard disk storage—with low cost per hour, good access and image quality—and magneto-optical storage—removable, archival. Montage developed a RAID magneto-optical striping technology that used four MO disks in parallel, increasing the effective quality possible with usual MO drives, but maintaining their removability. Numerous technical issues plagued the initial release of the system; it was shown at many trade shows for several years before its first use in 1993.

The Ediflex Digital was a re-invented system, with many of the flavors of the Ediflex "classic," but with the elegant addition of Link Editing System functionality (the project leader was one of the inventors of the Link system). The Ediflex Digital, like the Link, was based on the shooting script metaphor for source material. The "mimic" process had been re-

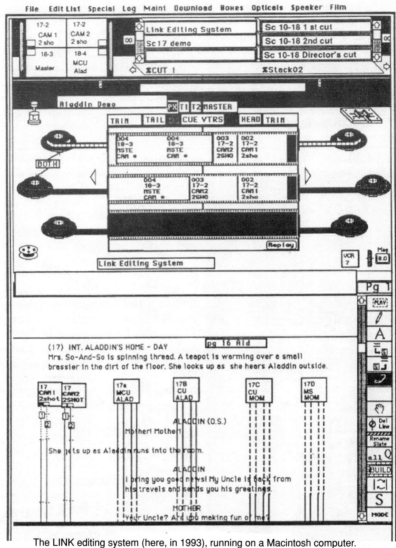

The LINK editing system (here, in 1993), running on a Macintosh computer.
Only a few were ever built. The system was only rentable in Hollywood.

worked such that assisting was considerably less arduous than it had been, and the graphical display harkened back to the Link as much as it did to the Ediflex. The vertical graphical timeline for the cut was unusual for picture editing systems (although more common in traditional audio workstations), but solved some of the "heads on left or heads on right" problems in timelined systems. While some of the graphical functionality associated with synchronization of tracks was quite unique, the system had adopted a screen "toolbox" that required mouse manipulation for much of the editing functionality. The Ediflex Digital was prototyped early in 1993, but the project was canceled and the company closed in the following year.

Unlike the other digital re-inventions of the analog systems, CMX began modifying its CMX 6000 system to capitalize on digital source rather than laserdiscs, but otherwise maintained much of the hardware and software for the release of their re-named CMX Cinema. The Cinema maintained the computer-low interface of the 6000 and the dedicated console while reducing hardware size, and adding, in effect, two giant digital "laserdiscs"—each capable of holding up to 26 hours with then-current hard disk technology. By adding specific tools for managing source material and a few much-needed functions to the architecture of the 6000, CMX augmented their analog system without actually inventing a new editing system. Of the re-designed analog systems, the Cinema was the only one to have evolved from its ancestor rather than to have been created new, in total. In spite of this, CMX terminated their digital nonlinear effort in 1994 and ceased development of the Cinema.

Once the window concept of editing was popularized by Avid, other software companies began creating their own nonlinear systems—manip-

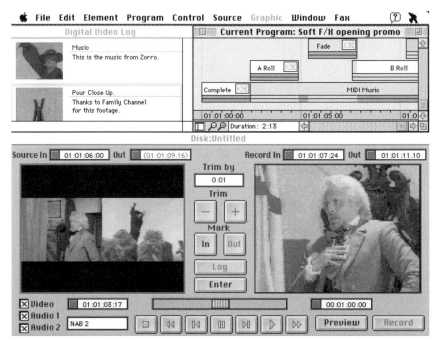

VideoF/X (circa 1992). Note the source log in the upper left, the timeline in the upper right, and the windows for source (left) and master (right).

ulating the various variables implicit in digital systems and applying cosmetic touches, both entirely driven by very specific markets.

D/FX was a company that had for years been producing high-end
video equipment—in particular, their award-winning graphics workstation, the Composium. In a move to enter the nonlinear area, D/FX introduced a low-cost editing package called Video F/X which in many ways
resembled an Avid interface. Their software-only option, Soft F/X was

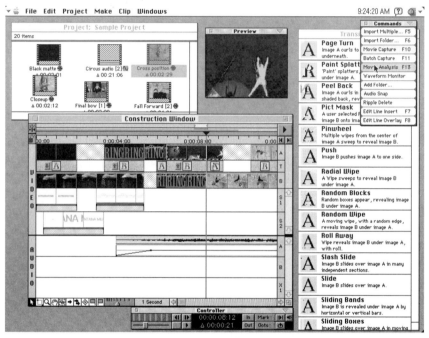

Adobe Premiere Version 4.0 (circa 1995) was the paragon of low-cost consumer video, but not yet
refined for professional users. Version 5.1 (circa 1998) was generally perceived to overcome that
barrier.

designed to open the way for post facilities to act more as "service
bureaus," where you would get source digitized for you and then take
it home to edit on your "satellite" system. The Video/FX was not fully
released and quickly evolved into the Hitchcock system in 1993. Soon
after, D/FX was closed and their Hitchcock system was sold to Aldus (pioneers of desktop publishing and inventors of such software as PageMaker),
where a small team continued its development as a high-end desktop
system. But after Aldus was purchased by Adobe (another desktop publishing pioneer, inventor of the PostScript printer language, and manufac-

turer of such popular graphics applications as Photoshop and Illustrator) in 1994, Adobe ultimately chose to stop development and disband the effort.

At the pure-software level, a number of Macintosh products have been introduced that can edit nonlinearly. The predominant entry here is from Adobe. Called Premiere, this software grew quickly and remarkably since it was introduced in 1992. Offered for under $1,000, Adobe's Premiere was based on a simple A-B video editor, but expanded on that idea to include multiple video and audio tracks. Although for many versions it was not a particularly efficient editor (displaying awkwardness at managing large volumes of source material), its strengths were its widespread usability for consumers and prosumers just learning about editing. Originally, the SuperMac Video Spigot made Premiere possible, but the release of the Digital Film card followed by Radius' VideoVision series of products made it viable. Premiere opened the door to "unbundled" editing software, allowing consumers to pick and choose appropriate CPUs, hard disks, digitizing boards, videotape decks, and so on. Prior to this point, nonlinear editing involved SYSTEMS, a combination of proprietary soft-

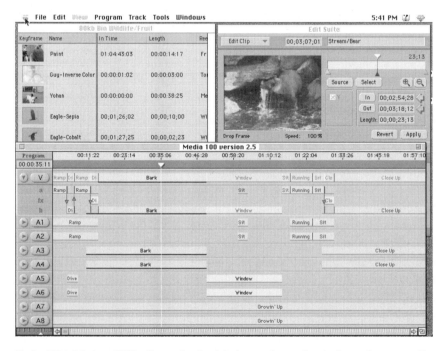

Media 100 (v2.5 circa 1995) with an attractive interface and exceptional price-performance ratio. Five years later, version 6.0 added significant functionality without the need for appreciable interface changes.

The classic Lightworks display (circa 1995) was largely maintained with the re-release of the system on the NT-platform, with highly augmented effects features (2000).

ware and hardware. Following Adobe's lead, a number of manufacturers began offering unbundled software for the lower-end users. By 1998 and Premiere version 5.1, the product made the transition from consumer video to professional feasibility. New tools for managing source material and timecode combined with the ever-increasing power of PCs raised Premiere to the level Adobe had enjoyed with Pagemaker for desktop publishing and Photoshop for image prepress. Resolution independence—the ability for low-cost software to *scale* in functionality based entirely on the power of the hardware on which it rested, and not on the software *per se*—became a cornerstone for new products. By 2000, Adobe Premiere 6 further targeted the software towards distribution independence—*edit once, repurpose many*—as the product fine tuned its market strategy towards the broadest base of users.

In 1991 an English group approached editing from a different point of view, and produced the highly stylized Lightworks. While it had many descendents as the company evolved on the corporate landscape, being used for multicamera work and being modified for news, its greatest success was in filmmaking. The Steenbeck-style shuttle knob provided remarkably sensitive control for moving-pictures, and sound scrubbing

Macromedia's original Final Cut, version 1 (circa 1998).

functionality appealed to the more traditional film editor. While many of the editing concepts were to be found in any number of digital systems, a few were unique and appealing. The architecture was designed to be unthreatening to technophobic editors, with databases organized into *rooms* with door-icons leading to them. The Lightworks was the first system to take advance of slip and slide clip editing features, where editors could open transitions in numerous ways, and push/pull material at will. It was years before competing systems offered the Lightwork's degree of editing sophistication.

Data Translation launched their Media 100 in 1993, and in a few years the product was a *bona fide* competitor to the Avid, offering often comparable performance at a fraction of the price. Eventually, the company renamed itself Media 100, and through a series of strategic alliances and mergers, positioned the product centrally as a host of media management tools.

Final Cut Pro was acquired by Apple Computer (from Macromedia) when the product was nascent, and it released in 1999. The system functions much in the style of the Avid and other digital systems, but is well-streamlined for DV and a seamless use of FireWire. These features immediately move

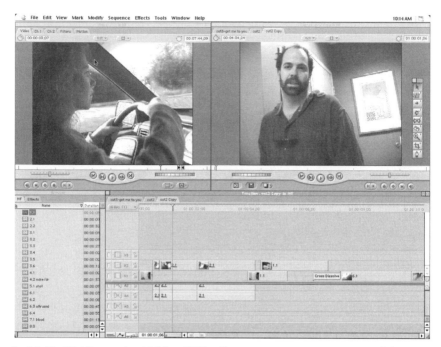

Apple's Final Cut Pro, version 1.2 (circa 2000), with a one-screen set-up: source upper left, master upper right, timeline lower right, and project bin lower left. Like many digital systems, virtually unlimited video and audio tracks can be managed in the timeline, although what can be played back together in real time is largely dependent on hardware.

the product into the role of pioneer in a new paradigm of editing system. Keeping with the extraordinarily smooth and modern designs of Apple Computer, the system was able to release with relatively high-end functionality in a low-cost package. In many ways, the product is the next generation of Adobe Premiere, particularly as it was developed by the same lead inventor as Premiere, Randy Ubillos. Where Premiere is truly an independent product, independent of platform, output formats, resolution, Final Cut is streamlined for the Apple, tailored for DV (although it works with High Definition Video as easily as MiniDV), and wed to Apple's vision for the future of high-end video production.

The first truly consumer-oriented nonlinear system, Apple's iMovie v.1 (circa 2000).

Also from Apple is perhaps the most revolutionary of their editing products, a truly consumer-oriented editing system. Called iMovie, and native on the consumer level-hardware from Apple, the system was developed by a completely separate team from Final Cut Pro. The system streamlines input from consumer/prosumer DV camcorders via FireWire, and allows users to trim in/out points of shots, manipulate them (in much the way the original digital systems like EMC2 and Avid/1 moved shots around as if they were slides for a presentation). While audio functionality is limited, titling is powerful with many built-in styles of titles made easy for consumers to add. Editing style is odd for a professional editor: originally only 12 shots could be kept in a source bin at a time (expanded in v2), with each removed as they are added to the simple timeline that runs along the bottom of the screen. While controls are minimal, it is quite capable enough for the videos of the average consumer.

The Eidos Judgement innovates in both architecture and interface. Vaguely reminiscent of the soft-edges and curves of Kai's Powertool products, the Eidos Judgement features an innovative timeline that represents the entirety of an edited sequence without needing to scroll off a page. The

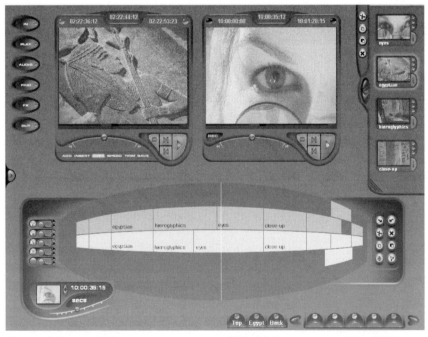

A new "twist" on an old idea: the system prototype of Eidos' Judgement system (circa 2000), with its novel timeline.

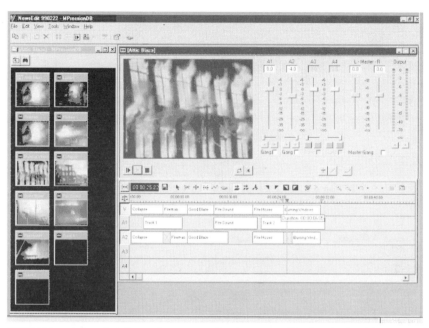

(2000) The evolution of media asset management system—a combination of network browser and nonlinear editor—tuned for news video production are exemplified by such products as the Vibrint (above) and Sony's ClipEdit (below). These kinds of tools in many ways descendents of the EMC2, which in 1995 became the text-based news editor, EditStar.

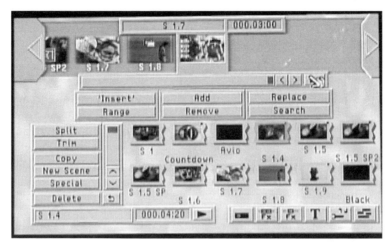

One of the newer generation of closed-architecture editing systems, the Casablanca/Avio from Draco Systems (2000).

timeline is viewed as if in a strong lens, enlarging the details at the now-line, and compressing superfluous information at the distant edges. From a structural standpoint, Judgement is built on an SQL server, implemented more like an enterprise level Internet database than a traditional editing system. This backbone has the potential to integrate well with network servers and users, and emerging metadata models. The video backbone is based on QuickTime 4; and although the initial release of the product is on a Windows PC, it is designed to be platform independent.

So Where Are We Now?

An historical perspective on editing systems of the past few decades suggests that the prevalence of open-architecture computers in the 1990s has yielded much similarity in user interface design. Even with their different icons and terminology, many systems are largely rearrangements of identical features. One advantage of closed-architecture systems (old and new) is that manufacturers have the freedom to design systems as they see fit, from the ground up, and therefore have the luxury of (potentially) greater leaps in design innovation.

It is possible that the next decade will bring us full circle, with the dynamic of ever-increasing computer power combined with a media-savvy computer-using public. Manufacturers may be able to overcome the conventional limitations of open architecture, and develop new media tools that capture the best of historical methodologies while innovating newer, more efficient, and more tactile interfaces.

T H E   B E S T   S Y S T E M

Bet you turned here first.

Okay. This is a trick, really. As you may have suspected, I'm not going to tell you what I think is the best system. Not because I think you might not agree, but because . . . *there is no "best" nonlinear editing system.* The question is actually kind of ridiculous. What do you mean by "the best"?

Least expensive? Best buy? Most versatile? Best for commercials? For drama? For theatrical films? Easiest to learn? Most popular? Most powerful? Most editors trained on it? Fastest cutter? Most reliable? Most likely to succeed? Best design? Most fun to use?

At least for these specific questions I have my own personal answers. But don't let anyone try to tell you which is best. Editors have many creative and technical preferences and vary based on their individual style of working and the types of projects they cut. Some editors swear by uprights; other editors prefer flatbeds. Every system—film, videotape, digital nonlinear—every editing system has fans and proponents, has owners making money, has clients totally satisfied, has success stories, has lists of credits. And of course, every system has detractors and drawbacks.

A few years following the introduction of a new system, the big features are pretty much working well, and the little problems are mostly hammered out. The systems by then are usually very stable, running well-tested software, and have a pretty large trained base of editors. The manuals are usually done by then, and customers are pretty happy.

If an editing system works—no matter how ridiculous its interface, cumbersome its editing style, or incredible its competition—if it works, someone will use it, and someone will love it.

If you have looked around, tried a number of systems, talked with working editors, and have located one system you like to use or are prepared to buy: you've done it. You've found the best system.

Congratulations.

Chapter 7

The Real World

"The editor is a storyteller. The editor restructures and rewrites and rebalances the story by juxtaposing images. What was in the script in the first place—and what the director's vision was at the beginning—is never quite the same reality of what comes out on the screen."

—Alan Heim

Editor: Network, All That Jazz

```
05:00    117-          KU 47                   455+0
08:00    117-5204+00   KU 47                 3455+0
02:20    117-3388+00   KU 46 0875-           0915+0
25:10    117-3606+00   KU 47 2534-0915+0
21:00    117-4000+00   KU 47 2534-0915+0
 0:00                  KU 47 2534-2211+0
04:20    117-4012+00   KU 47 2534-2028+0
58:20    117-4190+00   KU 46 0875-2028+0
58:00    102A-6000+00      10 2A -6000+0
 8:00    102A-6000+00      10 2A -6000+0
25:00    102A-6012+00   KU 27 5259-5268+0
20:00    102A-6054+00   KU 27 5259-5310+0
20:20    102A-6097+00   KU 27 5259-5353+0
38:20    102A-6155+00   KU 27 5259-5411+0
54:00    102A-6206+00   KU 27 5259-5462+0
 5:20    143-5000+00       14 3 -3000+0
 8:00                      14 3 -3000+0
52:10    143-3012+00       14 3 -5037+0
17:10    143-3038+00   KU 27 5259-5063+0
17:10    143-3064+00   KU 27 5259-5114+0
17:10    143-3090+00   KU 27 5259-5204+0
20:00    143-3120+00   KU 27 5259-5254+0
20:00    143-3150+00   KU 27 5259-5506+0
18:20    143-3178+00   KU 27 5259-5589+0
16:20    143-3203+00   KU 27 5259-5614+0
20:20    143-3234+00   KU 27 5259-5645+0
17:10    143-3260+00   KU 27 5259-5671+0
18:00    143-3287+00   KU 27 5259-5766+0
21:00    143-4000+00       14 3 -4000+0
 8:00                      14 3 -4000+0
21:25    143-4012+00   KU 27 5259-5800+0
         143-     +00   KU 43 3111-2929+0
                       KU 43 3111-3007+0
                       KU 43 3111-3082+0
```

P R A C T I C A L   E D I T I N G
M E T H O D S

This book promised to teach you *nothing* about the aesthetics of editing; there are far better guides to take you through those issues. Still, there are a few things about the practicalities of editing, in terms of general editing styles, that do seem appropriate for an introduction to editing and post-production. This section will provide a survey of those perspectives and, hopefully, make your editing sessions a little more productive in the process.

Depending on the type of project, you might be working most closely with the Director (in film), the Producer (in TV), or the Client (in various other situations). For the sake of simplicity, we'll use the terminology of film.

Principal Strategies

1. **The "Radio" Cut**—build structure, cut for sound, then fix.
2. **The "Art" Cut**—cut for picture, adjust sound as necessary.
3. **The "Linear" Offline**—build, build, build, until you're done.
4. **The Sculpted Sequence**—start big, and whittle down.

(Note: if some of these strategies sound mutually exclusive, it's because they are. The real trick here is in learning which approach is the best to apply in any given instance.)

1/ The Radio Cut

Basic philosophy: Cut for sound, don't worry about the images, certainly don't worry about the wallpaper, skip effects for now, get structure and/or timing right. Only then, once that's set, gussy up the picture tracks.

Edit in the most elemental way possible. Leave out L-cuts, cut-aways, effects, and filling in with room tone. Do as little fancy work as possible until you know you are on the right track. The more "vertical" editing you do (where picture has different edit points than sound), the more image tracks and/or audio tracks you need to adjust every time you modify the structure, and the slower and more painful any changes are going to be. This is particularly true when dealing with effects that require rendering.

Background: You impulsively buy a Christmas tree before your spouse gets home. On the one hand, you want to surprise and impress her with this

◀◀ Fundamental edits: p. 251

job well done. On the other hand, you have no idea whether she will want a tree at all, let alone THIS particular tree. But what do you do anyway? You pull out boxes of ornaments and trimmings, lavish them upon the little spruce, and after hours of hard labor, even manage to get the little crystal star on top. She walks in . . . She *hates* the tree. Common sense should have dictated that decorating the tree was dumb without first knowing that it was the right one.

Moral: Get the tree right before wasting time decorating it.

Mitigating factors: Editing is also about psychology. If you are the editor, part of your job is "selling" the cut to the director (*et al*). If you don't do some little things to make the cut look nice, the director might very well hate it, even when it's actually pretty good.

As an editor, you need to have a relationship of trust with your directors. They need to understand that you are sculpting something from their clay, and that it's cruder now than it will end up. But even a director who is your best friend will still like to see his material looking good, and if you don't even attempt to "sell the cut," there's always the chance that it simply will not fly.

The wife in this example may have hated the little tree when she walked in the door, but there was always the chance that the decorations would have made all the difference. This is a delicate trade-off, however, and a potential waste of time and energy if you guess wrong.

2/ The Art Cut

Basic philosophy: Cut for picture. *Images*, and timings of images, are the key element.

The more money spent on a production, on actors, on cinematography, the more appropriate this might be. It may mean stealing lines from other takes, it may mean looping or other post-sound work, but when the picture is the thing, cut it perfectly, and then figure out how to make the sound go with it.

Mitigating factors: Most people are forced to give up a good performance when the audio is ruined (*e.g.*, a poorly set up microphone, a jet flying overhead). Disregarding the sound and simply focusing on image is a luxury that few have. Still, some directors can be miffed if you disregard a stunning shot simply because the audio isn't quite right, but could have been saved with a simple little "cheat" from another take.

3/ The "Linear" Offline

Basic philosophy: Just because you're cutting on a nonlinear system doesn't mean that you have to work nonlinearly. You could still start at the beginning, build an area until it works (including the visual effects you like, music and sound effects, *etc.*), and once it's good, move on to the next part.

Younger editors, perhaps working on self-directed projects, may find this the most satisfying and "natural" way to work. In particular, when effects need to be rendered, an editor may want to see if they are working before determining what comes next. As great as nonlinear editing is, it's helpful to keep in mind that the end product is ultimately going to be viewed linearly. And editing in scene order, from beginning to end, is a luxury few directors have, or allow themselves.

Mitigating factors: The deeper you've dug yourself in, the more you tend to lose any remaining objectivity about what it is that you're doing, and why. And when you're through, if it doesn't work, it's also a serious challenge to unravel all those layers of work and then readjust them. Worse yet, if you're working FOR someone, they may be very unhappy that so much work was done prior to their "signing off" on the direction you opted to take. More than one editor has lost a job from pushing their own vision just a little too far, and not seeing eye-to-eye with the Auteur.

4/ The Sculpted Sequence

Basic philosophy: Editing is about culling large quantities of material into smaller quantities of material. For narrative projects, this requires a strict organizational structure (a script); but in many projects—whether documentaries, home videos, commercials or music videos, oftentimes a shot can go almost "anywhere" within reason. The process of selecting the good pieces, organizing them (like moving playing cards around in your hand), continually throwing shots out, making shots shorter, re-orga-nizing, and doing it again, until a structure emerges is a reasonable work-ing method.

There is a time-honored music video/commerical editorial concept called "beat the shot." It is where you use shot X until you find a shot more appropriate for that spot, then you swap in the new one. It is an appropriate tool in the "sculpting" of an editorial work.

If the project is video for a music track (either for professional or per-sonal uses), a rough cut without consideration of the music gets you closer to the shots you like, in a structure you like; adding the music track later to give you the rhythm and lyrics that will allow the refinement "to" the music.

Mitigating factors: This doesn't work for many scripted projects. You generally have an idea of how things go together based on the shoot.

Making Choices

Directors *want* to like the edit of their projects. They are predisposed to it. If you can't make their footage look good, they are more likely to believe it is your fault than theirs. At the same time, you need to build the structure of the project, set the pacing, choose good performances, and assemble a version of this story—in spite of the fact that it's only one of a vast number of plausible permutations of the given frames. Thus, there's a good chance that the one you choose isn't the one that the director had in mind.

But choose you do, every day; it's your job. And in a good creative working relationship, coming up with your own "take" on things can be one of the advantages (and even joys) of collaboration. So watch the material dispassionately, choose the good parts, and assemble a good product.

What happens if you need to show the director a very rough, early version of the cut? You'll need to *sell the cut.*

"Selling" the Cut

There are a few simple tricks that can help sell a first cut. Executing them takes extra time, and you will be making it a *little* harder to do subsequent changes, but sometimes that's the difference between pleasing your boss or not . . . between keeping a job and losing one. Remember, don't automatically do these things throughout the project, but only at those particularly rough moments where you sense the director might be irritated. Watch the edits through their eyes, and then . . .

1. Clean up bad sound *blips*. Lift out any particularly irritating moments when audio is clipped by the cut. If it works, adjust a frame out of picture AND sound to fix edits that miss the beat. Similarly . . .

2. Cross fade audio edits. Not always, and not when the sequence is dialog-driven; but in regions of high ambiance, or highly varying ambiance, a simple crossfade in audio can take the shock out of an edit. Most nonlinear systems can execute crossfades with very little fuss. *Alert:* since a cross-fade effectively lengthens the shots, you may *add* in a blip during the transition that was *cut out* when the cut was straight.

3. Drop a cut-away (of anything) over a jump cut. Jump cuts are remarkably jarring to viewers. Along with a bad sound cut, a jump cut can lead one to believe an edit isn't working, when you as an editor might know that it's actually easy to fix, as long as the timing of the dialog is right. (Another alternative, if you're still reaching a consensus on what basic direction to take, is to play just the soundtrack of a given section without the picture at all.)

4. Temp music. If a scene is going to have music, an opening sequence perhaps or a picture montage, get the director's favorite CD or (better) find a tune that is similar to the tone required for the scene, and simply drop it underneath. Temp music can be remarkably satisfying. And if you're not quite sure which cut to use, you can probably set up your audio mixing board to let you hear the editing system and the CD player at the same time, and try underscoring the picture "on-the-fly"—without lengthy digitizing—before deciding on a particular music cue.

5. 10 frame L-cuts. Otherwise abrupt edits can be "softened" simply by placing the audio and video transitions at different moments. These 10–15 frame overlaps can be a hassle to undo, though, especially when there are many of them, and particularly when you are only doing it to sell the cut. It can be a useful technique, however, to get you out of an occasional jam.

6. Leader at the head. It's nice to see your art placed in a nice frame. When running a sequence for the brass, give them a moment to settle in and prepare for the show. Don't start exactly at the head of the first shot you want them to look at. Back up. A slate or countdown is often appropriate. Get into the habit of slugging in "leader" or a little extra "pad" at the head of the sequence just as a matter of habit, if that's what it takes.

7. Temp (or real) sound effects. Nonlinear systems feature a multitude of spiffy DVE moves and other transitions, but visual transitions just by themselves can feel sterile. The right sound (a good "whoosh," perhaps) can be just the ticket to grabbing the viewer, by making that element of the production much more visceral.

8. Out-of-system acoustic effects. Don't be afraid to hum music, insert vocal "crashes," "pops," "pows," or "zowies" as the sequence requires, as long as the director and you are close enough to find this endearing. Doing so also communicates to the director how you see this coming together, and what your intentions are with regard to audio post.

Sometimes this is just the ticket to selling an otherwise bland first pass, and can be much more expedient in conveying your intent than actually hunting down and cutting in all those effects, which the director may or may not like anyway.

9. **Watch the director watching the sequence.** Sit back and see responses, what "hurts," what works, and what they are most proud of. Nodding isn't enough—make written notes on the director's feelings. It is important that they understand that you want to get it right, and that you want to process their input. (And immediately after you do, make sure to save a copy of the precise version of the cut that you both looked at, for future reference!)

PRODUCTIVITY TIP: The ability for a nonlinear system to use compiled "sequences" as source material for editing of a different sequence is a very powerful feature that some seasoned editors use to their advantage: Rather than delving repeatedly through long lists of dozens or hundreds of individual source clips to find their material, editors may opt to categorize footage into fewer and larger "virtual source rolls."

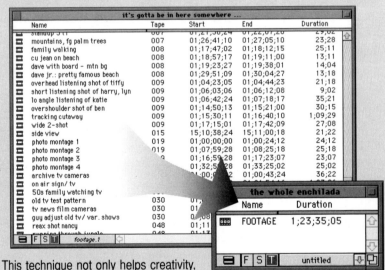

This technique not only helps creativity, but can generally be a more efficient way to work—it reduces the time spent searching for isolated snippets of material hidden away in bins, and also eliminates second-guessing as to which of those individual clips "deserve" to be reviewed.

D E C O N S T R U C T I N G
P R O D U C T I O N

The weaving together of picture and sound into a whole that's greater than the sum of its parts is integral to the "magic" of film. So, if you don't have an opportunity for on-site experience in production and post-production on professional projects, a fine alternative is to deconstruct a movie into its component parts: picture and sound.

Your homework: Take a film (one you've already seen before; you don't want to be distracted by wondering what's going on in the plot). Start with a videotape or DVD so you can control the play and freeze it if you want to. Select a single scene, watch it a few times. Once you are familiar with the action, plot and so on, (1) watch the picture *without the sound.* This makes it much easier to concentrate on the individual shots, which angles are used and in what order, the rhythm of the cutting, *etc.,* and develop an awareness of all of this independent of the soundtrack (which has its own competing rhythm).

(2) Now with *sound back on,* watch the picture again. Stop at edits; roll back and forth over them to examine the transitions. Notice how few special transition effects (like dissolves) most productions have. (3) Once you have a pretty good feeling about the scenes, put on headphones, turn your back to the monitor, and listen to the soundtrack *without watching the picture.* You can learn a lot and even begin to tell the difference between what was recorded on the set (*production sound*) and what was added after the fact (via *ADR* or *foley*); noticeable differences in sound quality, ambience or stereo panning are telltale signs to listen for when hunting for post-production footprints. And if you have access to a multi-channel surround system, you can even experiment with switching off or disconnecting individual speakers, to help isolate different parts of the soundtrack (dialog is usually confined to the center channel).

Important note: Don't restrict your education to A-run, Academy Award® nominees. In fact, never underestimate the educational power of watching truly bad, amateurish "B" movies. Most films we see are fairly well-executed, and tend to come together into a cohesive whole. Bad films are enlightening, by contrast. You can often get a peek "behind the curtain" as to the challenges involved in honing the various elements— picture editorial; cinematography and lighting; visual effects; dialog and sound effects recording, editing and mixing—into something that's ideally slick and unobtrusive.

◀◀ Introduction to production: p. 21

PRICING

1) How do you price equipment and rentals?
2) How much do you charge as an editor, post person?

Economics rule one: Things cost what the market will pay. *Period.*
If you're the only one in town who owns an editing system, you might make a lot more money than if you're in Hollywood, where everyone owns an editing system. Remember the key in negotiating rates is knowing the consumer's alternatives.

Aside from basics of *market-will-bear* economics, there are some guidelines for charging for things, regardless of what they are.

Equipment

In order to be worth purchasing in the first place, equipment needs to at least pay for itself over its expected "lifespan." *Broad rule of thumb*: computer-based equipment with a lifespan of 1 year will tend to rent weekly for 2% of the total purchase price; thus a $10,000 set-up including camera, system, monitors, and software might rent for $200/week. A higher-end professional system that costs $100,000 to set up is going to rent for somewhere around $2,000/week. This not only accounts for the depreciating value of the asset (your "stuff"), but also takes into account the "cost of money," also known as "interest." In high-interest times, when the cost of money is high, you get a premium for forking over cash to buy things that other people prefer to rent. Thus, when interest rates go up, charge more. (If equipment is expected to have a longer lifespan, like cars or furniture, a good approximation is to divide the above 1-year amount by the number of years you think you'll be renting the item.)

On the other hand, the best way to figure out pricing may be to look around. and see what other people are charging for the same thing.

Humans

Editing is a combination of technical know-how, aesthetic sensibilities and, hopefully, some practical experience. The more of any of these you have, in theory, the more you can charge. Experience tends to be a major factor.

There are unions that support and guide the editing profession. You don't have to be in a union to edit, but many professional projects (like Hollywood studio films and network television programs) require union affiliation. The Motion Picture Editors Guild in Hollywood and New York (both IATSE local 700), like all editors' locals, has negotiated minimum wages for people performing various post-production tasks. These wages change periodically. The rules for even getting IN the union are reasonably stringent, and the costs modestly pricey. Regardless, these wage listings should provide a starting place when thinking about editing income and relative value. Check with the union for the most up-to-date scale and rules of membership. The following amounts are rounded and are only guidelines:

JOB	min rate (per wk)
Editor ("picture")	$2,100
Assistant editor	$1,200
Apprentice editor	$900
Sound (or music) editor	$1,500
Videotape editor	$1,400
Video technician	$1,200

Remember, these figures are minimums that the union guarantees for their members (NB: some weeks are not 40-hour-weeks). Highly experienced film editors, including ACE members and Academy Award-winning editors, might make 2–3 times these amounts, or more. You, on the other hand, would probably be closer to "scale."

Freelance nonlinear editors, not associated with unions or union projects, make between $200 and $500 per day, depending on experience, the project (and its complexity), and expected time for the project (the more hours or days "guaranteed," the less you will make per hour or day—sort of a bulk discount). And, in some cases, the more graphical skills you can add to your expertise, the more this number may go up.

Professional post-production facilities, in the business of supplying top-quality, well-maintained equipment and having a bevy of superskilled artisans and technician on staff (or on-call), have their own loose guidelines for jobs and wages. These vary by city (New York and Los Angeles being, one would suspect, higher than Tulsa). The amounts in the table below illustrate the enormous ranges afforded by varying clients and budgets.

JOB	typical rates (per hr)
Offline editor	$20–85
Assistant editor	$15–25
Online editor	$25–75
Tape "op"	$15–25
Telecine operator/colorist (dailies/graveyard)	$20
Telecine operator/colorist (commercials)	$100+
Graphics compositor	$25–100+

T E L E C I N E   T R O U B L E

WHEN BAD NUMBERS HAPPEN TO GOOD FILM

While it's true that nonlinear editing systems themselves (especially those with a true 24fps editing mode) generally do a good job of keeping track of KeyKode numbers and other important film information—if that data wasn't generated accurately to begin with, it creates a classic situation of "garbage in, garbage out." Furthermore, the process of editing 24fps film via 29.97fps videotape involves a series of inherent workarounds, which means even more variables that can potentially mess up the "cut list" you'll eventually create on your editing system.

Film-to-tape transfers are an exacting art and science, and while expertise and attention to detail are critical, they can still vary widely. When booking a facility to do your work, remember that, as with so many other things, you often do get what you pay for.

Potential Sources of Error

Manual data entry. The telecine logging process requires manual entry of many detailed pieces of information, and hence is subject to human error. This includes camera roll and scene/take numbers, and audio timecodes read from often blurry-looking slates. Improper documentation of the film format used (*e.g.*, 3-perf instead of 4-perf 35mm) can also cause major headaches later on in the cutting room.

Equipment set-up. The KeyKode reader itself is located a short distance away from the "gate" where the film is actually scanned, which introduces an offset in the reading; in order to maintain accuracy, this delay must be precisely compensated for. The offset also varies depending on the particular path that the film is threaded through the telecine. Moreover, any electronic image processing in the chain (common for commercials and music videos, but generally avoided for "one light" dailies transfers) introduces an additional delay that would also need to be taken into consideration.

Equipment operation. There are several possible ways to record the very precise 3:2 pulldown pattern relative to the tape's timecode, and any deviations from the norm can conceivably create problems later on in the NLE.

Any needed shuttling back and forth of the film during the course of the transfer can conceivably throw off the counter sometimes used to gen-

---

◀◀ KeyKode: p. 98   ◀◀ Telecine/ 3:2 pulldown: p. 102   ◀◀ Film lists: p. 260-264

erate the videotape's "burned-in" window, and the information contained in the all-important *telecine log* file.

Any splices that might exist in the roll of film negative or workprint create discontinuities in the KeyKode numbering count, an obvious problem if such discrepancies go unnoticed.

Film production. When the camera assistant "checked the gate" after each shot was originally filmed, the camera could conceivably have been reloaded with a different perforation-to-KeyKode relationship than for the preceding shot. If so, this presents yet another possible complication in post.

What You Can Do

Make sure your nonlinear project was created for the proper film gauge/perf combination. Then, once you've imported the telecine log and digitized various shots, verify that the information in the system's database matches the KeyKode numbers on the "burn-in" window, and that [for 24fps systems] you see unique A/B/C/D frames, without stuttering or repeats. You'll want to do this near the beginning, middle and end of every shot.

Some projects take the process a step further, and transfer an already-synced workprint and mag soundtrack instead of separate film negative and Nagra sound rolls. Not only does this allow for more control and easier corroboration of results, it also eliminates time-consuming synchronization of sound (for every individual take) in an expensive telecine bay. Other cost-saving strategies include syncing up sound in the editing system; and avoiding workprint altogether. While it's possible to achieve perfectly accurate results using these latter methods, thorough familiarity with the particulars of each workflow is essential.

When all is said and done, however, there's no better way to assure accuracy than by actually comparing the telecine tapes against the film workprint/negative itself. This one-to-one verification (often using strategically placed "punch" marks at the head of each take or at the beginning and end of each roll) is an excellent way to be sure that the numbers in your nonlinear system really do agree with those on the film.

Lastly, if you're not already creating *Cut* and *Change Lists* as you go along, it's still prudent to test the procedure before your final deadline comes around. Any troubleshooting that's needed will benefit from the expertise of experienced and knowledgeable support personnel.

Overall, it's important to understand that every project is a unique and complex undertaking, and that your best first line of defense is to prepare for—and fully expect—a few "gotchas" along the way.

COLOR BARS

Better Living Through NTSC Monitor Adjustment

Everyone who works with video has had the experience of images looking one way on one monitor and another way on another monitor. Monitor calibration — making colors looks the same wherever you see them — is a challenge to consumers and professionals alike.*

To calibrate a monitor to its "correct" setting takes a few things. First, you need to have the tools (knobs) to adjust these settings; and second, you need a *standardized image* that gives you something consistent to set up.

Thus the invention of *color bars*. These are not just pretty colors, but a precisely engineered pattern of luminance and chrominance values. There are color bar *generators* (either in hardware or software) which output this pure color bar signal by which equipment can be adjusted. On computers, there are color bars image *files* (usually PICT or TIFF) that when viewed in applications also allow for connected monitor adjustment. There is an important distinction between these "pure" color bars and those that have been recorded onto videotape. "Recorded" bars may be distorted by numerous variables, and would only be used for *temporary* monitor adjustment, to view a specific video at its intended appearance.

Monitor Adjustment: Step-by-Step

Here are a few basics of adjusting your monitor, using the color bar signal, to achieve the desired professional monitor set-up required to make proper aesthetic adjustments to your program material. Remember, a video signal is made of luminance and chrominance. (If you don't remember, go back to page 120.) Though they may be labeled using slightly different terms, both professional monitors and consumer TVs have knobs that must be adjusted in a particular order:

1. Brightness (or "set-up" or "black level") controls the luminance part of the signal. To tweak luminance, set the monitor to monochromatic (black and white) if possible. While there are different ways to describe these values, the most common is an **IRE unit**:

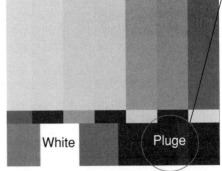

White   Pluge

---

* This section is about NTSC monitors. Computer monitors are an entirely different ball of wax.

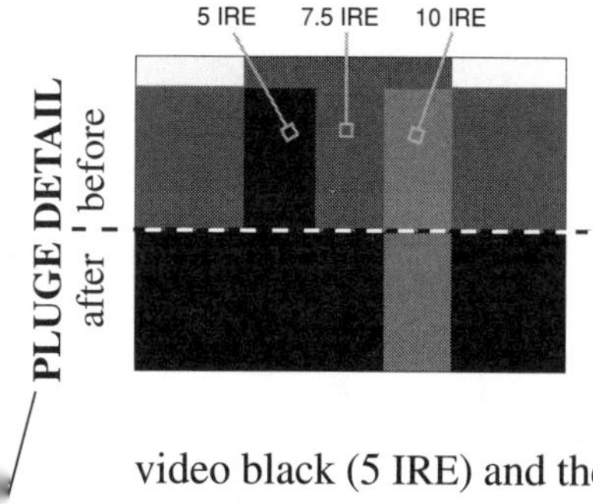

5 IRE   7.5 IRE   10 IRE

PLUGE DETAIL
after | before

a percent value from 0 ("black") to 100 ("white"). But 0 is too black for a video monitor (white is just right at 100); due to particulars of NTSC video signals, perfect *video black* is 7.5 IRE.

At the bottom right hand side of the color bars are three narrow bars, known as the *pluge.* * The bar on the left is blacker than video black (5 IRE) and the bar on the right is lighter than video black (10 IRE). The bar in the middle is perfect video black. Adjust the **brightness** knob until the left bar and the middle bar look the same, but you can still distinguish the bar on the right. *Simple.*

2. At the lower left side of the color bar pattern is a white bar. Turn the **contrast** (or "gain" or "white level") knob until this white bar overexposes. This will look like the white is spilling into the surrounding area ("blooming"). As soon as the white hits this point, back it off until it doesn't. You've now set the luminance properly.

3. **Saturation** (or "chroma saturation" or "chroma"): To get the colors set up correctly on the monitor, you don't actually want to see all the colors; turn on the *"blue gun" only.* The seven color bars represent various combinations of colors created by the electron guns in the monitor. Adjust the **chroma** knob until the bars on the *two ends* match the color and brightness of the little horizontal bars beneath them.

Use these for chroma adjustments...

...and these for hue adjustments.

Match these

4. **Hue**: Still with the blue gun only, turn the **hue** knob to adjust the *middle pair* of blue bars until they match the color and brightness of the bars beneath them. When properly set up, the bars should look like the illustration at left, with no break between the tall bars on top and the smaller horizontal bars beneath.

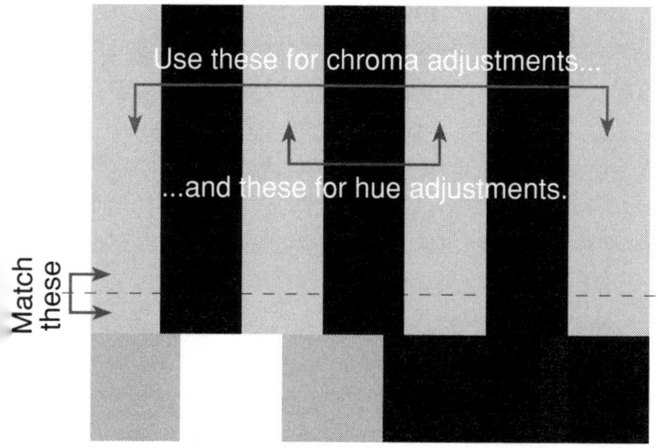

With only the blue-gun turned on (and thus, the red and green guns off), the monitor looks like this: bars of light blue and black bars. To adjust chroma and hue, you are concerned with the look of the tall blue bars, and the small horizontal bars beneath them.

---

* Geek trivia: PLUGE began as an acronym, standing for Picture Line-Up Generation Equipment.

D E M O   R E E L S
NOTES FROM THE FIELD

A good demo reel can make the difference between getting that great editing gig, or having to sit it out on the sidelines—so, understandably, a lot has been written on the subject. True enough, many rules are made to be broken ... but that still hasn't stopped our team of editing insiders from compiling this list, to help get you going in the right direction.

First Things First

Think about your viewers: *why are they even looking at this reel?* The truth is, they are not watching the reel to see your editing style. They are looking, mostly, at the **quality of the projects you have edited**. They want to know (1) the budgets of these projects that you were entrusted with, and (2) the style of the projects, as they relate to your capabilities and their needs. How will they judge this? They can surmise this info in a number of ways, and triangulate:

(a) *Film or video shoot.* Any professional can tell whether the material originated on film or video. Film costs more, and gets you more points.

(b) *Recognizable talent.* Are you editing public access or this year's Academy Award-winning feature? They can spot this in a second. Remember: they're probably not looking at the editing or the performances *per se*, but rather, the status of the project.

(c) *Lighting and Sound.* Again, pros can tell whether the project is of their caliber simply by being aware of the basic technical aspects of the production.

(d) *Breadth.* A reel of all-commercials will probably not get you a feature film. A reel of car chases may not land you a television sitcom. Think about making a reel that supports your claim that you can edit the things these people need edited. One of the great Catch-22s of the industry is that it is hard to get a job doing something until you've already done jobs just like that. (And once you have, there's an ever-present danger of being pigeonholed into that genre.) But good quality short-form success can lead to a long-form project faster than bad-quality long form.

Now, once you understand the foregoing caveat, you can see why the editing itself does have to be seamless — so it doesn't detract from the true subtext of the reel. So, with that out of the way, let's get down to the nitty-gritty.

Appearance

Much as we're not supposed to judge a book by its cover . . . you **will** be judged by your cover. It's only natural, though, given the brief window of opportunity available for conveying that crucial first impression. So make sure the physical reel itself doesn't detract from your chances even before your audience has a chance to put it in the machine.

Tape Stock—Splurge on fresh videotapes. There is little that turns off a recipient like getting a tape with the words "Seinfeld marathon" crossed out, and "Joe's Demo Reel" penciled in underneath. Also, avoid two-hour VHS tapes for a demo that lasts five minutes. Professional supply houses sell tapes in five or ten minute lengths that are ideal for this purpose (and are less expensive than the two-hour ones anyway). And don't forget professional-looking (and yet inexpensive) paper sleeves or plastic cases.

Labels—Make neat, legible labels for the cassette's face and spine, and if necessary, for the box itself. At a minimum, include your name, phone and pager numbers. It's also courteous to indicate the total running time (TRT), so your audience has an idea of what they're in for. Making those labels with a simple color inkjet printer can not only help demonstrate an aesthetic for graphics, but at the same time help your reel to stand out from the pile. Lastly, you might want to include your job title (presumably, some kind of "Editor")—while at first that may seem redundant, remember that a busy decision-maker may be reviewing the work of candidates for a number of different positions.

Format

The reasonable cost of VHS as a distribution medium does make it a common way of disseminating a reel; though for the same reason, be forewarned that it's also likely not be returned. (That's OK anyway, because it increases the odds that your tape gets passed along to yet someone else, later on.) For higher-end professional gigs, though (like commercials, episodics, features, *etc.*), a 3/4" videocassette is common (and more reasonable to expect to be given back to you afterwards). And if you're handing out your demo on DVD, just make sure that your potential client actually *has* a DVD player and *prefers* that format.

Duration: How Long to Make It

A ballpark answer is "as short as possible," which might translate to anywhere between, say, 2 and 8 minutes. But leaving someone with the impression that you just don't know when to pull the plug isn't a particularly good impression for an editor to give—imbuing video with suitable pacing is your *job*. Better to leave them wanting more rather than forcing them to shut it off before it reaches the end.

Creativity

A demo reel is a good overall opportunity to show off, creatively speaking. And that extends beyond just the clips you've chosen to show as the actual "content." An overall feeling of good production value helps sell yourself, and can even help make up for some elements (like graphics, animation, or sweetening) lacking in the body of the reel.

You might find a unique way to bring life to the "slate" (containing your name and other info) that begins and ends the reel. Or come up with interesting-yet-seamless ways to transition from one clip or section to the next. Above all, though, remember that flashy slates and clever transitions will not make up for content that is bad.

Content is King

Keep in mind a strange little twist of psychology: the decision for someone to watch your demo in the first place is an implicit acknowledgement that you *might* indeed be suited for the work. In that sense, the actual watching of the reel is partly an attempt to make sure there isn't any reason to **disqualify** you . . . so don't give them one! There's a tendency to judge you not by the best thing on the reel, but by the worst. Anything not up to the par of the rest of the material creates a disparity that might be nagging to the viewer ("Why that one lousy clip? Did she run out of decent stuff to show?"). While you may have included it for one reason (maybe it demonstrates that you know how to fly stuff around the screen with a DVE), the viewer may perceive it quite differently ("Boy, those effects sure look cheap!").

If you must include material that's a bit weaker, bury it in the middle. Put the biggest "splash" at the beginning of the reel, and save something good for the end—the first thing seen (the initial impression) and the last thing (the most recent impression) tend to have the longest-lasting effect.

Integrity

If you cut down a trailer for a film that was to air on your local TV affiliate (it started at 4 minutes, and you had to reduce it to 2 minutes), be careful that you are not implying that you (a) cut the original trailer, or (b) *gasp*, that you cut the original film. The odds are good that your target audience does know something about this original film, and you will not only *not* get the job, but possibly get blackballed (and thus *never* get a job) if word gets out that you are taking credit for material that you weren't really responsible for. Be EXCEPTIONALLY clear about what you are saying (and implying) that you are demonstrating.

Feedback

When you're putting together the reel and have questionable content or strategy, enlist the opinions of people you trust . . . and then take written notes. And even when a colleague flatly insists that "it's great!"—don't be afraid to probe a bit further: "What's the weakest part of the reel? What would you change, if you had to change something . . . ?"

Here's a good rule of thumb: if more than one person independently expresses the same reservation, give that feedback special attention. Collectively, they're trying to tell you something.

Specialist or Generalist

Since the goal of a great demo is to impress the viewer in the shortest amount of time, one approach could be to produce a single all-purpose demo that can be sent out unhesitatingly to anyone who asks the question: "Have you got a reel?" And even if that doesn't address precisely the style of editing that your prospect is looking for, that and a good resume (along with any verbal reputation that's preceded you, your overall list of credits, and a follow-up meeting in person) might be enough for you to land the job.

Or it may not. The more specialized the niche you're aiming for, the more likely it is that they'll want to see precisely the same type of work that their own project entails. To cover your bases, you might actually go to the trouble of creating two or three different styles of reel instead, each tailored for a particular type of editing. A newer option would be to use the multi-faceted nature of DVD to let the *viewer* choose what parts to see: the hot intro, or any of the short segments of more specific genre-work. Don't be afraid to take advantage of the interactive nature of this particular medium.

Good to Go

Offbeat advice: always keep a current reel (and resume) tucked away inside the trunk of your car. Strange as it sounds, many important hiring decisions are made at the last possible moment (either because of last-minute scheduling conflicts, budget meetings, or just plain procrastination). So if someone offers to send a courier over to where you're working today for another possible job for tomorrow, and the decision has to be made pronto . . . don't be surprised—be prepared instead.

Finally, in the end, remember that "stars don't audition." Meaning: you know you've made it when demo reels, like resumes, become unimportant. Once the world knows you are good, your name popping up at lunch is weightier than the fancy-shmanciest of reels. If you're reading this book, however, you probably will need a reel. Make it a good one.

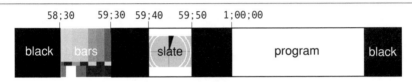

Tape Formatting

While a production may standardize any number of formats for delivery of finished elements or full productions, it is typical (and useful) to follow some basic guidelines for insuring smooth utilization of tapes. First of all, begin with black-and-coded stock, with a start time of 00;58;00;00 (2 minutes ahead of 1 hour) and use drop frame (DF) timecode unless told otherwise.

1. **Black**: 0–30 seconds at the head of the tape. Never start recording the actual program material from tape start (there is no way to pre-roll).

2. **Lay down color bars and tone**, to allow the person playing it back later to set up audio and video levels properly. Since these tweaks are done while the tape is playing, the rule of thumb is to provide 60 seconds. Start bars at 00;58;30;00.

3. **Black-Slate-Black**: after 10 seconds of black, record a 10 second "slate"— a title card of pertinent project information: title, date, run time, etc.—followed by another 10 seconds of black. If you need **Academy leader**, or other count-down, it's the last thing before program.

4. Leave sufficient **black at the tail** ("post-roll") after the end of the sequence, in case additional material needs to be added; you'll need a run of timecode for the machine to read during any subsequent pre-rolls. It looks better anyway.

AN EDITOR'S BAG OF TRICKS

If you earn your living as an editor, it's worth a modest investment in items that may help make your life easier, the quality of your work more professional, and even occasionally make you a hero with your Client. Admittedly, some of these items ideally "should" be supplied by your facility/rental company, but that's little consolation when you're under the gun rushing to meet a deadline, and a crucial item is just nowhere to be found.

Here, in no particular order, is a list of some odds and ends to consider having on hand:

• **Floppy disks** (Mac and DOS formatted), and at least a few of each applicable "density." (And don't forget single-sided double-density disks, if you need to prepare "RT-11" format EDLs for a linear online session. If you can pre-format a bunch of RT-11 disks ahead of time, so much the better.)

• **Zip disks** or the equivalent, for project backups; and also to "sneaker net" graphics files, *etc.*, from one workstation to another.

• **Plastic "2-pack" disk holders** or the like, for giving a copy of the final project files and EDLs to the client, when the project's done.

• **User "preference" or "settings" files** customize your editing system's keyboard, screen layout, *etc.*, to your individual taste. [Note: make sure you keep an archive of these files for each version of the editing software (and for each operating system platform, if applicable) since these files aren't always backward- or forward-compatible with previous/later versions of the software. Better still, also print out a "screen shot" that shows your personal on-screen layout, and have it to fall back on just in case.]

• **Headphones**: Invest in a high-quality pair (up to about $100), and don't be afraid to use them. Too many edit bays have a higher-than-desirable level of ambient noise (often from the cooling fans for the gear itself), or lousy monitoring or acoustics that prevent you from doing proper audio evaluation or troubleshooting. Because so much work traditionally done in online or sweetening bays is now routinely done at the "offline" stage, you'll want a way to cover yourself, especially before exclaiming to the next person in the chain: "But it was fine when it left here!"

• **System documentation** and other helpful reference materials. Even if you don't own the gear, it's still worthwhile to invest in your own copy of the manuals and other relevant documentation for the editing system(s) you use to earn your livelihood. Many of the little known "secrets" that help make an editor more productive or creative are actually there for the reading, right there in the book. Thankfully, one spin-off of the Internet era is that the documentation for many systems is now available on CD-ROM, or for download from the manufacturer's web site (and often for free).

In addition, if you have access to a CD/DVD-R "burner," you can assemble your own compilation of manuals, release notes, and a goldmine of other technical reference material from various web sites. While you're at it, you can also include your favorite useful software utilities (for data compression, font conversion, CD audio extraction, reading Acrobat files, *etc.*)

• **"Sharpie" markers**, in several colors; and a roll of white adhesive labelling tape: to marking and numbering videotape source reels and their boxes.

• **Dry erase markers** are occasionally useful for graphics and effects work. Sometimes, because of sluggish response times involved in creating multi-layered effects, it's useful to temporarily mark reference positions for effects on the monitor screen itself.

• **Basic office supplies** like pens and pencils (which clients love to "borrow"); sticky notes; a mini stapler; a notepad for taking notes during the session (gives the impression that you're actually paying attention); *etc.*

• **Small tools** such as a Swiss Army knife (regular or keychain size); a multi-function tool (such as a Leatherman or SeberTool); a small "greenie" screwdriver (for adjusting recessed calibration controls—but only if you know what you're doing!). And don't forget to pack a decent little flashlight (such as a mini-Maglite), helpful for sorting out cables in dimly lit equipment racks.

• **Small dictionary/thesaurus**: Avoid embarrassing and potentially costly spelling mistakes prior to delivering a finished program to a TV station or network (or worse yet, having it transferred to film stock). Since it's quite possible you'll be the one doing the typing in the first place, it's clearly better to get things done right the first time around. There's nothing more comical than an editor, a director, and a bunch of clients running around an empty facility on a Saturday night desperately trying to spell "ophthalmology."

• **Wrist rests**: cheap ergonomic insurance. One each for the keyboard hand (full-size) and mouse hand (mini-size).

• **Alcohol cleaning wipes**, for cleaning the mouse, monitors, *etc.* (A dirty mouse, in particular, can easily put a kink in your day.) A few Q-Tip swabs also help.

• **Small desk fan**: an occasional lifesaver, either for you or your gear. It's amazing how often an edit bay can be devoid of air conditioning, especially if you're working any nights or weekend shifts.

• **Calculator**: useful for tallying up timesheets, invoices, *etc*. And if it does time or timecode calculations, all the better.

• **Important phone numbers** for technical support (manufacturer, rental company, *etc*.); facility emergency numbers, other editors to refer clients to when you're busy, *etc*.

• **Invoices**, and other official paperwork. Create some professional-looking invoices. It's also helpful to carry copies of government forms (such as W-4, W-9, I-9, *etc*.)

• **Business cards.** If your client's happy with your work, they may ask to recommend you to others. Make sure it has your pager and cell phone numbers.

• **Pager and cell phone**: you'd be surprised at how "last-minute" many post-production hiring decisions really are.

Extra-Credit Materials

If you're still feeling wealthy after the little shopping spree above, here are some bigger-ticket items that can also be worthwhile investments:

• **A really good, ergonomic chair**: Post-production folks often spend long days sitting in lousy chairs. If you're thinking of spending years doing that, it's worth doing it with a minimum of discomfort. A decent chair (as well as lumbar support, foot rest, stretching exercises during your next "render," *etc*.) may help your creativity, your productivity—and not to mention, your you!

• **Portable Zip (or Jaz) drive, FireWire drive, CD/DVD-R** or the like, ideally usable with multiple operating systems. Larger formats can be the ultimate way to archive large project files (and even selected video/audio/graphics) for you and/or your client.

• **A modest collection of music and sound effects** is not a bad idea to have at your creative beck and call. To that end, there are many "buyout" production music CDs available from numerous sources at a fairly reasonable cost. It's also common to use existing movie soundtrack CDs for "temp" scores and screenings. Even a few good "sweetener" sound effects can be really useful in livening up otherwise sterile visual transitions. In any event, remember to investigate any licensing (*i.e.*, $) issues relevant to using the material in your finished production, particularly when it comes to music. At the least, be familiar with web-available libraries, like Beatnik (www.beatnik.com), which provide instant access to effects and music and will take care of appropriate licenses for you.

• **Portable TBC remote control**: One of the remnants of the old offline editing mentality was the luxury of ignoring video levels and leaving them at "preset" settings regardless, since all the video would be meticulously replaced in a subsequent online session anyway. But now that lots of video online work is actually being done in so-called "offline" bays, that luxury has been retired along with the old paradigm. Getting the best (or even acceptable) results means the ability to monitor the full video signal continuously in real time—and that usually means a physical waveform monitor and vectorscope that's separate from the editing system itself. But even so, you still need to be able to control the video levels that you're digitizing, which means having a set of timebase corrector controls available to you *at arm's reach*. Unfortunately, this is one of the most commonly overlooked accessories in many nonlinear edit bays.

• **Laptop computer**: useful for graphics manipulation, accessing email, researching on the Web, *etc.*, particularly if your system is busy rendering. Also sprinkle in appropriate software, clip art libraries, graphic textures, and so forth, as an additional value-added resource you can draw upon to enhance the finished project.

• **Camcorder**: to own, or even just to borrow, when necessary, for shooting your own swish pans and other "eye candy" elements. This doesn't have to be the latest DV model—a second-hand VHS might be fine, especially for abstract or "grunge" elements. Either way, it's ideal to "bump up" (*i.e.*, transfer) the footage to a professional, timecoded videotape format prior to digitizing. And when you're done, don't forget to create . . .

• **Your own personal compilation reel** (or CD/DVD) of really impressive material, including various eye candy elements that you've created, or encountered in your travels. Just the thing for novel transitions or quickie graphic treatments. (Remember that thousands of others have access to the same standard transitions that came with your editing system, so you may need to look elsewhere to come up with looks that are truly interesting and unique.)

Chapter 8

Theory

"[*Star Wars, Episode I*] is being put together more in the vein of traditional art, rather than the assembly line mode that Hollywood has adopted. I can write and shoot and edit simultaneously, and constantly upgrade what I'm doing. Like you would with a painting— you step back and look at it."

—George Lucas (1999)

Producer, Director, Nonlinear pioneer

INDIA superior

NO VVS

Terra florida

FR

Chamaho

Panuco Inf. Tornucarū

CVBA

Temistitan

Iamica

PARIA s abundat auro & margaritis

ORBIS

Insula Atlātica quam uocant Brasil & Americam

Catigara

Inf. infortu natæ

Regio Gigantū

Mare pacificum

Fretum Mag

H O R I Z O N T A L and V E R T I C A L N O N L I N E A R I T Y

This section was originally published in Nonlinear 3 (1995). It is reprinted here in its entirety, unedited, as the ideas covered continue to be useful in discussing nonlinear systems.

For this book, we define nonlinearity very specifically; in summary, it is a kind of perpetual preview. But let us delve further into the idea of nonlinearity, for it is central to understanding technological change.

An editing system is "nonlinear" as long as edits are not recorded to a master record videotape. Once shots are recorded from source to master, they lose their individuality, their distinctness. As long as shots remain distinct, it is quick and easy to manipulate them individually. Here are four shots, each from a different source reel:

1	2	3	4

As long as these shots remain separate, and can be previewed, I would say they are "nonlinear." I would go farther as to define this separation as HORIZONTAL separation and the preview of these discrete elements is HORIZONTAL nonlinearity. They are horizontal in this graphic; horizontal because they occur consecutively in time. Horizontal nonlinearity is concerned with temporal separation.

If I were to record this nonlinear sequence to a master tape, it would look the same, but I would have lost the horizontal separation: they would now exist as a single entity. Traditionally in the video domain this action is called an "assembly." Let us define the loss of any separation as an **assembly**.

Manipulations and modifications in the pure horizontal dimension are very facile and require little mental or computational work. Like shuffling cards, or trimming shots, or re-arranging slides, I can choose when each shot begins and ends, and the order in which they fall. I need to put shots in, take shots out, and ripple or not. I would go so far as to say that pure horizontal manipulation is mechanically simple. Unfortunately, there is little "real world" editing that looks like this.

Now, if time is frozen, and we are dealing with a single frame, there are still manipulations that go on.

Think of the final product: at any given instant, there is more than the single frame of picture—there is sound, and usually many tracks of sound. Often there are graphics: titles, effects, and so on. When these simultaneous events are depicted graphically, the most common method is by showing layers of tracks:

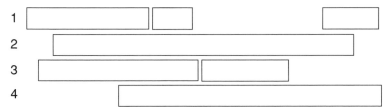

1
2
3
4

These layers show when events happen in relation to each other. These are vertical layers. As long as these vertical layers are maintained separately, let us define this as VERTICAL separation; if they can be previewed, let us define this as VERTICAL nonlinearity.

Vertical nonlinearity is not concerned with temporal issues, but with spacial issues. Like horizontal nonlinearity, vertical nonlinearity is not specifically for pictures, but for both visual and auditory information.

Vertical nonlinearity, like horizontal nonlinearity, exists for as long as the vertical elements are maintained as separate and discrete pieces. As long as vertical elements are maintained separately, they can be easily manipulated and modified. In the audio domain, a loss of vertical separation is referred to as a "mix-down"; in the video domain, the loss of vertical separation is sometimes called "flattening" or "rendering"; for our purposes here, let us continue to define a loss of any separation as an *assembly*.

It should be noted then that assemblies, mix-downs, and rendering are essentially the same kinds of processes—they combine separate elements into a less complex object—in doing so, separation is lost and nonlinearity is removed. And in all cases, this loss of separation is performed in order to view or preview a work.

It may be taken as fact that resources (energy, money, storage, *etc.*) must be utilized to maintain vertical or horizontal separation. Resources are saved as separation is lost. Ideally, elements would always exist with full separation, but reality dictates that at some point in the creative process, assembly is required. For the most part, even if separation can be economically maintained throughout the creative process, delivery of completed projects tends to follow a vertical and/or horizontal assembly.

The concept of synchronization, or "sync", is only fundamental to vertical processes. Clearly, editing as we know it today is a combination of vertical and horizontal manipulations. Viewing the interim stages or completed work is only possible when technology can simulate the assembly process. From a technology standpoint, it is easier to maintain separation of horizontal video elements and preview them (the common description of "nonlinear editing") than it is to maintain separation of vertical video elements and preview them (commonly called "real-time compositing"). But one can certainly assume that vertical nonlinearity will continue to evolve and proliferate as computational power increases.

For years, videotape editing was the editing of picture and two channels of sound, with three vertically separate tracks, but no horizontal separation except during the preview. Newer formats of tape made four channels of sound feasible. Additional or complex sound work is traditionally done on equipment designed to maintain vertical separation. A 24-track audio recorder does this. A digital audio workstation (DAW) uses digital storage to create a horizontal nonlinear system (remember that vertical separation existed prior to the "tapeless studio").

Effects and graphics have for years been synthesized and previewed in digital and analog domains, with some vertical and horizontal nonlinearity, but then mixed down and assembled to videotape.

As nonlinear editing systems evolved for picture, their evolution was primarily in the horizontal dimension. The unit of nonlinearity was the "load." A load consisted of a set of analog videotapes or videodiscs, and was limited. When the load became digital, on hard disks, resource allocation could radically change.

Digital storage brings up the issue of *reciprocity*: As some factors go up, others go down. Most commonly, as image quality (color, resolution, size) goes up, hours of storage goes down. If the goal is to store many hours of source material on-line simultaneously (*i.e.,* to build larger "loads") then clearly, you need to reduce image quality to meet this goal. This can be graphically shown; with image resolution (representing the quality of the image) on the vertical axis and quantity of source material on the

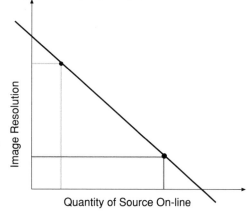

horizontal axis, a "utility" line crossing both axis sets up the inverse relationship.

At point 1 the image resolution is high and thus the quantity of source on-line is low; conversely at point 2 the on-line quantity is large and thus the image resolution is low.

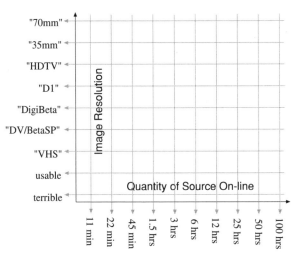

To make this graph more realistic, we can add some scales to the axis: for the image quality (y-axis) we can apply some generally understood degrees of quality; on the on-line source axis (x-axis) we can use a scale where each increment doubles the source of the prior value, beginning with 11 minutes of source and ending with 100 hours.

We will come back to this scaled graph in a moment.

It should be noted that as time and technology march forward, the graphical result of the increases in computer functionality will shift the utility line outward. This includes better and faster CPUs, larger and faster hard disks, faster bus and communications hardware, improvements in video and audio compression and so forth. There are many elements that are limiting factors in the evolution of computers in relation to these graphs. For example: larger hard disks, while they may allow for more source on-line, wouldn't alone allow for better image resolution; they are "bottle-necked" by limited data transfer rates. Similarly, once data transfer rates improve, and compression

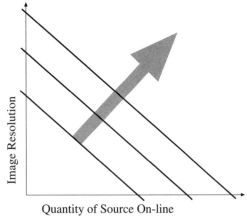

makes better images fit in smaller files, other factors (*e.g.*, the power of the CPU) may dictate the limit on how nonlinear (previewable) composited images may be.

This brings up two points. I would suggest that vertical and horizontal separation and nonlinearity are also inversely proportional. That is, with a fixed set of resources, to increase vertical separation reduces horizontal separation, and vice versa.

While from a technology standpoint this inverse relationship may not be appreciable due to the relative simplicity of horizontal separation when compared to vertical separation, I might go on to suggest that while the toolsets required for horizontal and vertical nonlinearity might be able to be codified and integrated due to many of their inherent commonalities, the tasks related to horizontal and vertical manipulations are mutually exclusive: the more vertical layers that are maintained separately, the more difficult any single horizontal manipulation will be. This is why traditional work in the vertical domain is usually restricted until after (or later) in the horizontal process. To put it simply, sound is done after picture for a reason. If you work a dozen tracks of sound and then recut the picture, it is more work than waiting to do the sound after picture is locked. The reality of film making and post-production tends to push this as far as it can go, but ideally, vertical work—both for visual effects and sound—is delayed until after picture is done.

Let us return to the evolution of computers, then, and the idea of computer complexity. In this case, complexity* is an abstract concept that ties together issues of computational complexity, required data bandwidth and transfer rate, and required storage. These features evolve at their own rate, but the net effect of this evolution is that different tasks become viable using computers of a given cost, with less complex ones first and more complex ones later. A simple diagram illustrates the point:

It is the complexity of computers that forms the bottleneck in the outward movement of the utility line on the earlier graphs.

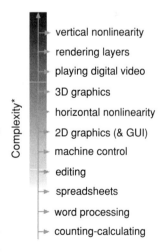

vertical nonlinearity
rendering layers
playing digital video
3D graphics
horizontal nonlinearity
2D graphics (& GUI)
machine control
editing
spreadsheets
word processing
counting-calculating

When we combine all of these observations, and look at the realities of post-production, the trends of the past decade are clear and a direction may become illuminated. First let's look at a set of typical post-production projects in terms of these values. Starting with quantity of source on-line, we can simply plot roughly how much source material is required for each type of production:

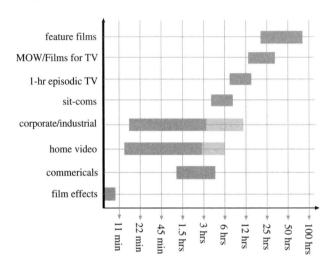

From the graph, we can see that to perform film effects we only need a few minutes of source material; to edit 4-camera TV sitcoms we need between 6 and 8 hours on-line; to cut a feature film we need between 24 and perhaps 100 hours on-line; and so on. So to utilize a nonlinear system for one of these types of productions, we now know the limiting value on the horizontal axis. For example, if the system can manage a maximum of 10 hours of source material, it will be difficult to edit a feature film, just about right (but on the low side) to cut an episodic TV show, and more than suf-

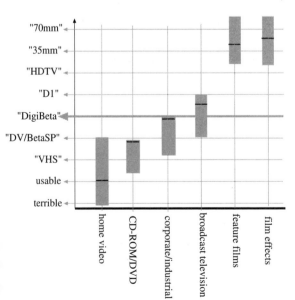

ficient for most commercials and home videos.

We can do the same thing for the vertical axis, image resolution. The image resolution of (standard definition) broadcast TV is less than that for feature films but much higher than that required for home videos or many corporate projects. Does this mean that you cannot edit a feature film on a system that only has the image quality sufficient for home videos? No, but you cannot

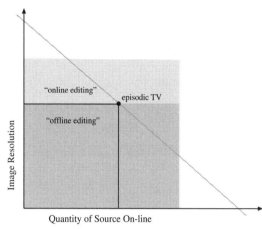

Quantity of Source On-line

deliver your product directly from the system. When a system's output image quality is less than that which is required for delivery, you have the natural division between online systems and offline systems.

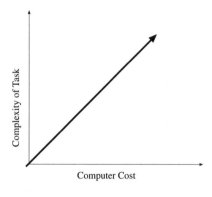

Computer Cost

Clearly, online systems are desirable in that they will require no other steps to deliver product; however, as long as economics prohibit the proliferation of online systems, offline may have to be sufficient. Affordability determines system complexity, and affordability is itself largely a function of prior expectations. TV post-production *expects* to pay thousands of dollars per week for equipment to edit; as time goes on, those productions have a choice—get more functionality for their money, or get the same functionality for less money.

What is happening in the post-production market is that as the line on the first graph shifts outward due to technological improvement, it will continue to move first in the domain that is most required until the maximum degree of value is reached, and then all future shifts in the line are immediately translated into increases in the less-required domain.

For example: feature film editing systems. They require a large quantity of source material (most-required). They will take whatever image resolution they can get. At first, image resolution is sacrificed to make sure all source can be loaded. But once that maximum threshold is reached (let's say 50 hours of source), additional improvements in compression and hard disk storage will be shifted towards improving the image quality. Since the *status quo* of film editing is offline, there is not a great deal of discussion about the reality of online film editing. However, you can be assured that at some point in the future, when film-quality digital video can be stored on a system costing a few thousand dollars per week, and the quantity of source hits the 18–25 hour range of our graph, people will want to print their digital video directly to film.

We can surmise that there will be a point in time prior to the above scenario where "broadcast-quality" video (lower quality than feature film) can be managed on a several thousand dollar a week system and the source on-line begins to rise. At first, short-form projects will utilize these systems for delivery: commercials, then longer projects like episodic TV. This point has already come, as would have been predicted, for news, home video and corporate video, and is one of the pressures on networks to implement a higher-resolution requirement for broadcasting: HDTV.

This discussion is oversimplified, but it can illustrate the kinds of issues in the evolution of nonlinear systems and nonlinear online.

FACETS OF POST-PRODUCTION TASKS

Horizontal tasks and vertical tasks are each unique, and ideally performed with specialized tools. It is certainly clear that issues of efficiency are often best handled by the separation of labors. To put this is today's terminology, I would say that traditional offline editing is mostly a horizontal task, best performed with horizontal tools—it is the building of the story, of the framework of a production, it is the process of culling down large volumes of source material into a small volume of relevant material; whereas traditional online editing is mostly a vertical task, performed best with vertical tools—it is the integration stage, it begins with a locked picture and then augments the look of the ele-

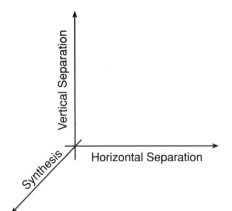

ments and integrates the cut with other elements, like titles and special effects.

Vertical systems, tailored for vertical separation and manipulation, are generally not well designed for "editing." Nor should they be. Horizontal systems, on the other hand, are only slowed down and complexified by the addition of many vertical elements (either for picture or sound); here it is perhaps most efficient when the verticality is kept to a minimum.

Equipment tends to have a strong bias towards either vertical non-linearity or horizontal nonlinearity. Note that horizontal and vertical distinctions have no direct relationship to online and offline systems. Recall that online and offline are relative terms related to output product: online quality can be defined as the quality required for actual product delivery; offline quality is any image or sound quality less than that. There is a tendency to relate vertically nonlinear systems with online quality and horizontal nonlinear systems with offline quality. Although this is a fair observation, it is important to understand that it is coincidental; because many complex effects are difficult to "describe" for re-creation, they tend to reside solely in the online domain, and thus verticality coincides with online quality.

Vertical and horizontal functionalities are not the only manipulations in today's systems. Another functionality is what I call SYNTHESIS. Graphically, it occupies no space, but produces the elements that are to be manipulated in the two dimensions. Shooting on location is a synthesis function (and is not readily managed on a computer). Creating 3D graphics is synthesis, as are 2D painting and text elements. Every system produced balances the functions of horizontal, vertical, and synthesis as required by the market for that product.

Not only can object-frames be created and maintained as separate frames vertically and/or horizontally (inter-frame), but synthesis functionality can also involve objects within a frame (intra-frame). The intra-frame objects—like object-oriented graphics (PostScript) and bitmapped objects—also can be manipulated with horizontal and vertical nonlinearity, *i.e.* they can be stacked and blended (vertical), and moved around the screen (horizontal). A film or videotape camera creates frame-objects of a single vertical layer, and at the intra-frame level are like bitmapped graphics.

The newest trend in nonlinear online systems is products that have vertical nonlinearity as well as horizontal nonlinearity, limited source capacity, and some synthesis functionality. Consumer-oriented products have a healthy mix of vertical separation (although complexity limits vertical nonlinearity), horizontal nonlinearity (the functionality is easy to deliver, especially at lesser image resolutions), and synthesis. More often than not, however, consumers and some professionals utilize the comprehensive synthesis functionality of products like Adobe's Photoshop and Illustrator, Macromedia Freehand, and Fractal Design's Painter—and simply import objects/frames into horizontal or vertical systems.

These theoretical constructs are not exclusive to video. The much-discussed desktop publishing realm is analogous as well. Writing and word processing are horizontal tasks, and we see proliferation of dedicated tools for the process. Publishing, on the other hand, is managed by specialized publishing tools. It is the integration and synthesis stage that begins with the locked manuscript, and then continues from there. The online assembly in the publishing industry is called *printing*. The real-time preview that characterizes nonlinear systems is generally referred to as WYSIWYG ("what you see is what you get").

In fact, with very little difficulty the tasks of publishing and video can be integrated where frames equal pages, TV safe lines equal margins, objects are imported, and output options are varied. While this integration may not yet have been perfected, this is the essence of multimedia authoring software.

There is no question that word processing software has many publishing functionalities; and publishing software has ample word processing functionalities; and yet the two tasks remain distinct, even by those who perform both. The phenomenal advent of desktop publishing, with its associated software products, has not eroded the word processing software market—in fact, reports seem to indicate that it has increased it. Professional writers and authors still tend to use the least complicated, most streamlined software for word processing.

This is not meant to indicate that traditional "horizontal" editors do not desire the additional power associated with verticality or with synthesis, only that they will desire it to the extent that it does not interfere with the efficient completion of their task.

I would suggest that the ideal human interface for horizontal manipulations is one that embodies rhythm and motion and is very physical, like a dedicated console. Vertical manipulations are perhaps best managed via graphical representations on computer displays and a corresponding mouse (or similar device). Can a graphical display manage horizontal elements? Absolutely. Can a physical console manage vertical elements? Somewhat, but perhaps to a lesser degree.

The bottom line is that even as traditional separations of labor—like online or offline—begin to erode, there may still be other rationale for a division of labor; the tasks associated with each arm (horizontal, vertical, synthesis) are specific and tools dedicated to those tasks will continue to thrive in the professional arena over the next many years. For consumers, generalization will be the trend, as systems go for high-profile functionality and diverse needs of a huge market, with less sensitivity to the subtle functions and styles of dedicated products. As will all products in all industries, there will be a continuum of product functionalities from the general (consumer) to the specific (professional).

THE FUTURE OF EDITING

AN INTERVIEW WITH THE AUTHOR

In the prior edition of NONLINEAR, this section discussed the future of offline; the discussion is still relevant (and can be found at the Nonlinear4.com website). But with the new century comes a new evaluation of editing technology in light of the much-discussed convergence of media. We caught up with the author at the recent NAB convention.

There are hundreds of nonlinear editing systems today; some are offline, some online, some involve DVD authoring, some output video for Internet streaming, some manage high definition compositing. What is happening?

Good question. More now than ever the "theory" I outlined about horizontal and vertical nonlinearity, and synthesis tools, is applicable. Regardless of your output options (either with regards to resolution or distribution venue), managing large quantities of source material, and cutting (and rearranging and pacing) that material into a stream—requires horizontal editing tools. Some software will do this easily. This is different from software that specializes in the generation of effects and the real-time compositing of multiple layers. And both of these are different from a "publishing" kind of application, which takes completed material and integrates it with interactive (*hyper-*) links, text and metadata, and fine tunes readability, color, layout, and so on.

To design those tools that make the media, it might help to think about how that media will be experienced. The following scenario is a vision of at least one way all this convergence could look over the next ten years. As you read this, think about the *tools* required to create the media . . .

It's 2010.

There are two general categories of home media: HDTV (the heir of broadcasting, now in 60% of homes), and iTV (the descendant of SDTV and the Internet). HDTV is high-resolution streaming media. It is the primary domain of the "broadcasters" and studios. High-end production and post-production is still relatively expensive and access to these channels is somewhat limited. HDTV is how we watch high-end "television" episodics, sporting events and movies. Feature films initially release on HD 24P, for theatrical (digital video) projection; but after a month of cinematic showings, they're available on premium HD cable networks and pay-per-view, though you can still see it in SD for less. To utilize this distribution channel, the first half of the decade was spent in a massive down-conversion of the film libraries of the world onto HD. By 2007 the bulk of filmed entertainment was converted, and the industries that burgeoned to mine those archives began to subside. [This is not unlike the years after the introduction of CD-Audio, where every recording in humanity was dusted off and re-mastered for disc. That "project" took from 1985-1995.]

itV is a combination of what we call "streaming media on the Internet" and late-20th century television. Like the massive down-conversion to HD, there was a comparable down-conversion and re-purposing of 601 video to fit within the formats of the old Internet. Material was logged (tagged with a bunch of metadata), and made marginally interactive and searchable. This also took the first two-thirds of the decade, then declined as less old programming remained to be converted, and as new material was designed for the medium itself. There are now thousands of "channels" (actually customizable playlists), many run by the old broadcasters; but most are high-niche providers offering material on-demand, 24/7, in SD. At any given moment, you can watch this iTV or pause it, freeze it, back up or scroll forward, search a different program, or change channels. Any particular stream may have embedded in it textural or meta screen information—from the script, to ancillary educational and production links, to e-commerce capabilities, as applicable. iTV is the *vox populi*, the person-to-person narrowcast that virtually anyone can offer. Some are supported by commercials and/or branded associations; some are micro-payment pay-per-views; some are independent, pushing forward any number of private agendas (*e.g.*, the Nature Conservancy Channel takes no ads, but continually reminds viewers about the importance of protecting the environment). Like old-fashioned "web rings," channels aggregate into self-selected bands, where accessing any one of the group facilitates finding the others. iTV can be searched by titles or content. New eight-character "content codes" are developed by the International Standards Organization, to facilitate searching over the massive Network.

In the home, there is one central computer, a server of sorts, which jacks into this cable/telecom network. This computer mediates the streams, records to its multi-terabyte storage array when desired, and distributes all the video/audio/data to the various monitors in the house. Individual appliances also have their own CPUs, and are connected (via UltraFireWire networks) to the central computer. Such is the home intranet. The monitors are flat-panel and 16:9. A super large display is in the family room, in front of the sofa. It is run with a handheld remote control, as TVs were a decade ago. It can show either HD channels or iTV channels—although the latter don't run full screen, but have a frame with banners and metatools on the sides, which are manipulated via the remote control. iTV shopping is executed in this way. Searching is done this way. [Petroglyph's patented nontextural search and display engine is now ubiquitous!]

There are 17" displays in the kitchen and in the den, both of which use more conventional wireless keyboards/mice, touchscreens and speech recognition. Most tasks which involve more-detailed text and numbers are done here, closer to the display, as is printing. But these terminals are networked to the same central system as are the entertainment displays. iTV and "found" video can accessed here as well, or saved for easy viewing later on the big screen.

PDAs (such as the Handspring Visor and the Palm XX) also act as highly-functional remote controls for the system, and can drive the channel-changing and iTV viewing/searching/recording. TiVo is a downloadable software plug-in, also driven from these remote controls, which manages all the storage, timeshifting, *etc.*, that video-on-demand entails. The handheld devices also contain portable versions of the home software (calendar, database, "check" book)—uploading, downloading and synchronizing to the central server when required.

Another common home device is a headset/earphone/microphone, which allows for dictation or voice recognition as needed, and also doubles as a telephone (with a small attachment, it can also go cellular). Any incoming call rings a bell and, on the iTV/HDTV displays, shows a pop-up dialog identifying the caller, and offering the choice of answering or routing to voicemail.

While all displays in the house can show data and video, some displays and rooms are better ergonomically-suited for passive entertainment vs active participation. This is largely a function of the body position and distance to the display/interface. So while a keyboard can easily be taken to the "TV" room, for typing and searching on the big screen, it may not be used this way often.

Overall, the home "computer" has evolved to not only compute but to act as a mediator of one- ("media") and two-way ("telephony") communications and entertainment.

WOW. What a tale. Do you really think this is the way it will play out?

It doesn't matter. What matters is that we need to recognize that the web experience today is a temporary state; that large economic and technological forces will create some important changes. There will be a merging of web development tools (like Macromedia Dreamweaver), DVD authoring tools (Sonic Solutions DVDit!), and other content publishing tools (Avid ePublisher, for instance). While these applications will also cross over into editing systems and graphical compositing systems, they will not all become one thing because the task of editing raw footage into bite-size chunks and full-length stories is NOT the same as the ability to take those finished chunks, fit them with metadata and hyperlinks, and lay them out onto 16:9 screens.

Can one person handle all that? Or will it always take a division of labor?

While I don't think all these tools will be found in one perfect product, many products will have functionality across a wide portion of the required gamut of needs, and *many people will be fully interested and capable of handling that wide functionality.* Still, there will be specialists, and teams working on projects will have certain advantages over those working independently. But it is important to note, as is true today in many fields, that a single person CAN manage the tasks and is therefore not *required* to have specialists and a multiplicity of products to manage everything.

Will there still be anything like "offline"?

Yes. If for any reason an individual cannot afford to edit and *finish* in the output quality he or she wants (film release, HD release, *etc.*), there will still be super-affordable tools that allow creativity on cheaper platforms (which basically means slower and/or lower-resolution), and people will use them. It may be what you do at home. It may be some people getting together with just enough capital to shoot in HD, but choose to post at standard definition, just to entice someone else into financing the more expensive HD finishing and distribution. That will be offline. Offline means *cheaper*; no matter what is invented, there will always be a cheaper path that some will want.

Why did you write this book?

I wrote it to answer my own questions about technology and these creative tools. I had an opportunity to witness the dawn of this age, and helped educate people who were forced to adopt technology when they didn't want to. They taught me a great deal about editing, and I helped communicate enough about technology to make the transitions easier. I don't teach engineers. This isn't a how-to book. This is the information I found useful or relevant, summed up in neat-enough packages to act as a launching pad and catalyst for those who are interested in learning more. And anyway, it was fun to do.

Appendix A

F I L M   F O O T A G E S

Just as metric numbers are "base 10" and binary numbers are "base 2," in a way, film numbers are "base 16." As you count, when you get to 15 you roll over another foot. Here is a small table that may help videotape editors and others unfamiliar with film footage counts:

For 35mm film, 4 perfs per frame...

1 foot	=	16 frames	=	2/3	second
1' + 8	=	24 fs	=	1	second
6'	=			4	seconds
15'	=	240 fs	=	10	seconds
30'	=			20	seconds
45'	=			30	seconds
90'	=	1440 fs	=	1	minute
180'	=			2	minutes
450'	=			5	minutes
900'	=			10	minutes

When counting film feet and frames, you must predetermine whether the first frame you start on is number "0" counting up to "15" or if the first frame is number "1" counting to "16." Zero counting (0–15) is the most common numbering scheme. You also must decide where you define your "0" frame in relation to each key/code number.

Kodak's introduction of KeyKode, their machine-readable edge numbers, solves these issues by standardizing and marking each *zero frame*. Adoption of this format will make the reading of edge numbers considerably more simple for all film handlers.

Appendix B

B I N A R Y   C O U N T I N G

Binary mathematics is a kind of counting in base 2 (it's no different from base 10, if you're missing eight fingers*):

Decimal		Binary			
Place 10	Place 1	Place 8	Place 4	Place 2	Place 1
	0				0
	1				1
	2			1	0
	3			1	1
	4		1	0	0
	5		1	0	1
	6		1	1	0
	7		1	1	1
	8	1	0	0	0
	9	1	0	0	1
1	0	1	0	1	0

Just as counting in base 10 is based on powers of ten, base 2 is based on powers of two. The powers of two, listed here, are familiar multiples to computers users:

$$2^0 \;=\; 1 \;=\; 0 \text{ bits}$$
$$2^1 \;=\; 2 \;=\; 1 \text{ bit}$$
$$2^2 \;=\; 4 \;=\; 2 \text{ bits}$$
$$2^3 \;=\; 8 \;=\; 3 \text{ bits}$$
$$2^4 \;=\; 16 \;=\; 4 \text{ bits}$$
$$2^5 \;=\; 32 \;=\; 5 \text{ bits}$$
$$2^6 \;=\; 64 \;=\; 6 \text{ bits}$$
$$2^7 \;=\; 128 \;=\; 7 \text{ bits}$$
$$2^8 \;=\; 256 \;=\; 8 \text{ bits}$$
$$2^9 \;=\; 512 \;=\; 9 \text{ bits}$$
$$2^{10} \;=\; 1{,}024 \;=\; 10 \text{ bits}$$

*with apologies to Tom Lehrer

2^{11}	=	2,048	=	11 bits
2^{12}	=	4,096	=	12 bits
2^{13}	=	8,192	=	13 bits
2^{14}	=	16,384	=	14 bits
2^{15}	=	32,768	=	15 bits
2^{16}	=	65,536	=	16 bits
2^{17}	=	131,072	=	17 bits
2^{18}	=	262,144	=	18 bits
2^{19}	=	524,288	=	19 bits
2^{20}	=	1,048,576	=	20 bits

The number of bits is the number of places a digit can have:

In 8-bits, the smallest number is $00000000 = 0$
and the largest number is $11111111 = 255$

The multiplication of two binary numbers always produces a result that has a total number of digits equal to the sum of the digits of the original numbers; for example:

$$10010 \times 011110 = 1001010100$$
(a 5-bit number times a 6-bit number equals an 11-bit number)

This might help explain why computers need many bits to do binary math—for digital audio and video manipulation (like 3-D graphics or music synthesis, *etc.*) Doing these computations in "real time" means a computer is doing millions of computations a second and moving large numbers (many bits) around the computer quickly. This takes more expensive computers with larger chips and faster processors.

For the digitizing of video, every pixel on the screen can have any one of a number of colors associated with it—the more choices of colors you have, the better the approximation of "reality." The number of choices for color are usually represented by the number of bits each pixel will be stored in. For example, a 16-bit system will allow each pixel to be one of 65,536 colors.

The greater the color choices, the better the images, but also the more memory needed to store an image. Many people consider "film-quality" color to be 8 bits for each of red, green and blue—or 24 bits total, or 16.78 *million* colors. The same image stored with 24 bits rather than 8 bits takes up more than 65,000 times as much memory.

Appendix C

T H E A T R I C A L   F E A T U R E S

AND THE REVOLUTIONARY DECADE OF NONLINEAR EDITING

For the historically minded, the ultimate adoption of electronic editing for film was determined by its use for theatrical feature films. Slowly over the course of the years from 1985 through 1995, more and more projects have been edited partially or entirely on this new equipment. Early adopters had to consistently face untested hardware and software in the harshest conditions and under extraordinary time and economic pressure. But it was from these directors' and editors' pioneering work that heralded in the acceptance of the nonlinear reality.

The following list follows the introduction of electronic systems for theatrical film work. This list is, of course, by no means the entirety of films edited electronically. Particularly in the years beginning with 1993, the number of films cut electronically has soared.

This listing will show only films with budgets over three million dollars. For some scale (and modest amusement), the *first film* any system cut is also included (in italics), regardless of the budget.

Year	Film Title	Director	Editor*	System
1985	POWER	Sidney Lumet	Andrew Mondshein	*Montage*
1986	SWEET LIBERTY	Alan Alda	Michael Economou	Montage
	MAKING MR. RIGHT	Susan Seidelman	Andrew Mondshein	Montage
	The Patriot	*Frank Harris*	*Richard Westover*	*EditDroid*
1987	FULL METAL JACKET	Stanley Kubrick	Martin Hunter	Montage
	Perfect Victims	*Shuke Levy*	*Jon Braun*	*CMX 6000*
	Garbage Pail Kids	*Rod Amateau*	*Leon Carrere*	*Ediflex*
	Mine Field	*Jean-Claude Lord*	*Yvles Lang Louis*	*TouchVision*

(1987 Academy Award, Scientific & Engineering achievement to MONTAGE)

Year	Film Title	Director	Editor*	System
1988	TORCH SONG TRILOGY	Harvey Fierstein	Nicholas Smith	Montage
	BIG TIME	Tom Waits	Glen Scantlebury	Montage
	COOKIE	Susan Seidelman	Andrew Mondshein	Montage
1989	FAMILY BUSINESS	Sidney Lumet	Andrew Mondshein	Montage
	EDDIE and the CRUISERS 2	Jean-Claude Lord	Jean-Guy Montpetit	TouchVision

1990	THE SHELTERING SKY	Bernardo Bertolucci	Gabriella Cristiani	CMX 6000
	GRAFFITI BRIDGE	Prince	Rebecca Ross	CMX 6000
	GET BACK	Aubrey Powell	Michael Rubin	CMX 6000
	ROCKY V	John Avildsen	Michael Knue	TouchVision
	THE OBJECT of BEAUTY	Michael Lindsay-Hogg	Ruth Foster	Ediflex
	THE DOORS	Oliver Stone	David Brenner	EditDroid
	ONCE AROUND	Lasse Hallstrom	Andrew Mondshein	Montage
	GODFATHER III	Francis Coppola	Barry Malkin	Montage
1991	TRUTH OR DARE	Alex Keshishian	Barry A. Brown	TouchVision
	HARLEY DAVIDSON and			
	the MARLBORO MAN	Simon Wincer	Corky Ehlers	CMX 6000
	DECEIVED	Damian Harris	Neil Travis	Montage
	LEAVING NORMAL	Ed Zwick	Victor duBois	Ediflex
	ALL I WANT for XMAS	Rob Lieberman	Dean Goodhill	E-PIX
	Critters 3	*Kristine Peterson*	*Terry Stokes*	*E-PIX*
	Let's Kill All the Lawyers	*Ron Senkowski*	*Christa Kindt*	*Avid*
	High Strung	*Roger Nygard*	*Tom Siiter*	*EMC2*
1992	KAFKA	Steven Soderbergh	Steven Soderbergh	EditDroid
	MEDICINE MAN	John McTiernan	Michael Miller	EditDroid
	WIND	Carol Ballard	Michael Chandler	EditDroid
	PATRIOT GAMES	Phillip Noyce	Neil Travis	Montage
	B. STOKER'S DRACULA	Francis Coppola	Nicholas Smith	Montage
	DISTINGUISHED GENTLEMAN	Jonathan Lynn	Tony Lombardo	CMX 6000
	STREET KNIGHT	Albert Magnoli	Wayne Wahrman	CMX 6000
	ARMY OF DARKNESS	Sam Raimi	Bob Murawski	EMC2
	KALIFORNIA	*Dominic Sena*	*Martin Hunter*	*Lightworks*
1993	LOST IN YONKERS	Martha Coolidge	Steve Cohen	Avid
	NEEDFUL THINGS	Fraser Heston	Rob Kobrin	Avid
	THE FUGITIVE	Andrew Davis	Dean Goodhill, *et al.*	Avid
	THE GETAWAY	Roger Donaldson	Conrad Buff	Avid
	ANGIE	Martha Coolidge	Steve Cohen	Avid
	LOVE AFFAIR	Glenn Gordon Carron	Bob Jones	Avid
	A BRONX TALE	Robert DeNiro	David Ray	Avid
	TRUE LIES	James Cameron	Conrad Buff, et al.	Avid
	TWO BITS	James Foley	Howard Smith	Avid
	NAKED GUN 33 1/3	Peter Segal	Jim Symons	Avid
	WOLF	Mike Nichols	Sam O'Steen	Avid
	YOUNGER and YOUNGER	Percy Adlon	Suzanne Fenn	Avid
	THE THING CALLED LOVE	Peter Bogdonovitch	Terry Stokes	E-PIX
	THE CONEHEADS	Steve Barron	Paul Trejo	E-PIX
	SLIVER	Phillip Noyce	Richard Francis Bruce	Montage II
	HEAVEN AND EARTH	Oliver Stone	David Brenner	Lightworks
	IRON WILL	Charles Haid	Andrew Doerfer	Lightworks
	BEVERLY HILLBILLIES	Penelope Speeris	Ross Albert	Lightworks
	MRS. DOUBTFIRE	Chris Columbus	Raja Gosnell	Lightworks

INTERSECTION	Mark Rydell	Mark Warner	Lightworks
WHITE FANG 2	Ken Olin	Elba Short	Lightworks
NATURAL BORN KILLERS	Oliver Stone	Hank Corwin	Lightworks
PELICAN BRIEF	Alan Pakula	Tom Rolfe	Lightworks
CLEAR/PRESENT DANGER	Phillip Noyce	Neil Travis	Lightworks
EIGHT SECONDS	John Avildsen	Doug Seelig	Lightworks
BLUE CHIPS	William Friedkin	Robert Lambert	Lightworks
DISCLOSURE	Barry Levinson	Jay Rabinowitz	Lightworks
BLOWN AWAY	Stephen Hopkins	Tim Wellburn	Lightworks
SPEED	Jan De Bont	John Wright	Lightworks

1994	ANGELS IN OUTFIELD	William Dear	Bruce Green	Avid
	FORGET PARIS	Billy Crystal	Kent Beyda	Avid
	DOLORES CLAIRBORNE	Taylor Hackford	Mark Warner	Avid
	RADIOLAND MURDERS	Mel Smith	Paul Trejo	Avid
	READY TO WEAR	Robert Altman	Geraldine Peroni	Avid
	TOMMY BOY	Peter Segal	William Kerr	Avid
	DROP ZONE	John Badham	Frank Morriss	Avid
	BRAVEHEART	Mel Gibson	Steve Rosenblum	Lightworks
	PULP FICTION	Quentin Tarantino	Sally Menke, et al.	Lightworks
	DIE HARD 3	John McTiernan	John Wright	Lightworks
	NELL	Michael Apted	Jim Clarke	Lightworks
	OUTBREAK	Wolfgang Peterson	Neil Travis, et al.	Lightworks
	CONGO	Frank Marshal	Anne V. Coates	Lightworks
	FRENCH KISS	Lawrence Kasdan	Joe Hutshing	Lightworks

(1994 Academy Award, Scientific & Engineering achievement to AVID)
(1994 Academy Award, Scientific & Engineering achievement to LIGHTWORKS)

1995	BRIDGES/MADISON CNTY	Clint Eastwood	Joel Cox	Avid
	WHILE YOU W/SLEEPING	John Turteltaub	Bruce Green	Avid
	SPECIES	Roger Donaldson	Conrad Buff	Avid
	VIRTUOSITY	Bret Leonard	Rob Kobrin, et al.	Avid
	THE NET	Irwin Winkler	Richard Halsey	Avid
	SHOWGIRLS	Paul Verhoeven	Mark Goldblatt	Avid
	NIXON	Oliver Stone	Brian Berdan, et al.	Avid
	JUMANJI	Joe Johnston	Robert Dalva	Avid
	ASSASSINS	Richard Donner	Richard Marks	Avid
	SABRINA	Sydney Pollack	Fredric Steinkamp	Avid
	BATMAN FOREVER	Joel Schumacher	Dennis Virkler	Lightworks
	NINE MONTHS	Chris Columbus	Raja Gosnell	Lightworks
	WATERWORLD	Kevin Reynolds	Peter Boyle	Lightworks
	MISSION: IMPOSSIBLE	Brian De Palma	Paul Hirsch, et al.	Lightworks
	MATILDA	Danny DeVito	Lynzee Klingman	Lightworks
	BOGUS	Norman Jewison	Stephen Rivkin	Lightworks
	VAMPIRE IN BROOKLYN	Wes Craven	Patrick Lussler	Lightworks

NO FEAR	James Foley	David Brenner	Lightworks
HOME FOR THE HOLIDAYS	Jodie Foster	Lynzee Klingman	Lightworks
LAST DANCE	Bruce Beresford	John Bloom	Lightworks
CASINO	Martin Scorsese	Thelma Schoonmaker	Lightworks

Although equipment manufacturers won't generally advertise it, many of these theatrical features were cut partially on electronic systems and partially on traditional workprint. The electronic cutting room, like the original film room, might have any combination of equipment: flatbeds, uprights, synchronizers, log books . . . similarly, there may be one or more electronic systems as well. There is no shame in only partially contributing to the process that post-produces a feature film—any way to efficiently, quickly, and creatively pore through 500,000 feet of film is an advancement to the film industry. A handful of directors even pioneered uses of videotape to assist in production and post-production prior to 1985. Inexpensive linear tape systems have been employed to rough-together scenes before cutting film. On rare occasions, entire movies have been edited linearly, on tape, and then conformed manually back to film.

Although all of the projects listed above are technically "theatrical" feature films, many of the lower budgeted films (like the countless ones not shown here) had only a contractual film release and then went directly to videotape distribution (called direct-to-video films). Regardless, all cut negative either directly from the systems or from a conformed workprint made (at least in part) from lists output by an electronic editing system.

This listing concludes with the films cut in 1995 as the principal equipment used in feature film work, the Avid Film Composer and the Lightworks, have both been distinguished with Academy Awards, approximately 10 years after the first films were cut electronically on the Montage. The Avid received the Academy's highest honor for technical advancement to the film industry, in 1998, signaling finally that digital nonlinear film editing was legitimized in Hollywood.

[For the most complete records of projects edited on any particular brand of nonlinear equipment, it is suggested that you contact the manufacturer directly.]

ACRONYMS and ABBREVIATIONS

24P: [video format] high definition 24fps, progressive scan

"601": [video format] formerly CCIR-601, now ITU-R601

AAF: [format] Advanced Authoring Format

ACE: [org] American Cinema Editors

A-D: [equip] analog-to-digital converter

ADB: [computer] Apple Desktop Bus

ADR: [post] automatic dialog replacement

ADSL: [network] Asymetrical Dedicated Subscriber Line

AES/EBU: [format, org] Audio Engineering Society / European Broadcast Union

ALE: [post] Avid Log Exchange

AM: [tech] amplitude modulation

ANSI: [org] American National Standards Institute

API: [computer] Application Program Interface

ARPA(Net): [org] Advanced Research Projects Agency (Network)

ASC: [org] American Society of Cinematographers

ASCII: [format] American Standard Code for Information Interchange

ATAPI: [computer] (IBM PC-)AT Attachment Packet Interface

ATM: [network] asynchronous transfer mode

ATR: [equip] audio tape recorder

ATSC: [video, org] Advanced Television Standards Committee

ATV: [bcast] Advanced Television

AVI: [format] Audio Video Interleaved

AVR: [post] Avid Video Resolution

BNC: [equip] bayonet Neill-Concelman (video connector)

CAS: [org] Cinema Audio Society

CAV: [storage] constant angular velocity

CBR: [video] constant bit rate (encoding)

CCD: [camera] charge-coupled device

CCIR: [org] Comité Consultatif International des Radiocommunications (now ITU-R)

CD-R: [storage] Compact Disc - Recordable

CD-ROM: [storage] Compact Disc - Read Only Memory

CDTV: [video] Conventional Definition Television (*i.e.*, NTSC, PAL)

CEMA: [org] Consumer Electronics Manufacturers Association

CES: [show] Consumer Electronics Show

CG: [video] Character Generator

CGI: [graphics] computer generated imagery

CGI (script): [web] common gateway interface

CK: [video] chroma key

CLI: [computer] command line interface (*e.g.*, DOS)

CLUT: [computer] color look-up table

CLV: [storage] constant linear velocity

CMOS: [computer] complementary metal-oxide semiconductor

CMX: [mfgr] "CBS-Memorex eXperiment(al)"

CMYK: [printing] Cyan, Magenta, Yellow and blacK

CPM: [business] cost per thousand

CPU: [computer] central processing unit

CSS: [web] cascading style sheets

CSS: [DVD] Content Scrambling System

CTDM: [video] compressed time-domain multiplex

CU: [prod] close-up shot

D-A [equip] digital-to-analog converter

DA: [equip] distribution amplifier

DAT: [equip] Digital Audio Tape

dB: [unit] decibel

DBS: [bcast] direct broadcast by satellite

DCT: [math] Discrete Cosine Transform

DDR: [equip] digital disk recorder

DF or DFTC: [video] drop-frame timecode

DGA: [org] Director's Guild of America

DHTML: [web] dynamic HTML

DIMM: [computer] dual in-line memory module
DLP: [distrib] Digital Light Processing™
DLT: [storage] digital linear tape
DME: [audio] dialog, music and effects (tracks)
DNA: [biology] deoxyribonucleic acid. (Just kidding. Seeing if you're paying attention!)
DNG: [bcast] digital news gathering
DNS: [web] domain name server
DOS: [computer] Disk Operating System
DP: [prod] Director of Photography
dpi: [printing] dots per inch
DRTV: [bcast] direct response TV (*i.e.*, infomercials)
DSK: [post] downstream key
DSP: [tech] digital signal processing
DT: [video] dynamic tracking
DTP: [computer] desktop publishing
DTS: [mfgr] Digital Theater Systems
DTV: [bcast] Digital Television
DV: [format] Digital Video
DVD: [storage] (officially stands for nothing)
DVE: [post] digital video effects
EBR: [equip] electron beam recorder
ECU: [prod] extreme close-up
EDL: [post] edit decision list
E-E: [video] Electronics to Electronics
EFP: [bcast] electronic field production
E-IDE: [computer] Enhanced Integrated Drive Electronics
EISA: [computer] Expanded Industry Standard Architecture
EM: [post] edited master
ENG: [bcast] electronic news gathering
EPS: [format] encapsulated postscript
ESS: [video] Electronic Still Store
FC-AL: [network] FibreChannel-Arbitrated Loop
FC-EL: [network] FibreChannel-Enhanced Loop
FD: [computer] floppy disk
FFT: [math] Fast Fourier Transform
FLOP: [computer] floating point operation
FM: [tech] frequency modulation

FTB: [video] fade to black
FTL: [post] film transfer log
ftp: [Internet] file transfer protocol
Gb: [unit] Gigabit
GB: [unit] Gigabyte
Gbps: [unit] gigabits per second
GHz: [unit] Gigahertz
GIGO: [computer slang] "garbage in, garbage out"
GOP: [MPEG compression] Group Of Pictures
GPI: [post] general purpose interface
GPRM: [DVD] general parameter
GUI: [computer] graphical user interface
GVG: [mfgr] Grass Valley Group
HDTV: [video] High Definition Television
HFS: (Macintosh) hierarchical filing system
HSB: [graphics] hue, saturation, brightness (color space)
HSL: [graphics] hue, saturation, luminance (color space)
HTML: [web] hypertext markup language
http: [web] hypertext transfer protocol
HUT: [bcast] (percentage of) households using television
Hz: [unit] hertz
IATSE: [org] International Alliance of Theatrical Stage Employees, Moving Picture Technicians, Artists and Allied Crafts of the United States and Canada
IBEW: [org] International Brotherhood of Electrical Workers
IDE: [computer] Integrated Drive Electronics
IEEE: [org] Institute of Electrical and Electronics Engineers
IFB: [bcast] Interruptible Foldback (or Feedback)
IN: [film] internegative
I/O: [computer] input/output
IP: [film] interpositive
IP: [web] Internet Protocol
ips: [video] inches per second
IRE: [unit] named after the Institute of Radio Engineers
ISA: [computer] Industry Standard Architecture
ISDN: [network] Integrated Services Digital Network

400

ISP: [web] Internet service provider

ITS: [org] Association of Imaging Technology and Sound

ITU-R: [org] International Telecommunication Union, Radiocommunication sector

JBOD: [storage] Just A Bunch of Disks (or Drives)

JFIF: [format] JPEG File Interchange Format

JPEG: [format, org] Joint Photographic Experts Group

JPG [format] JPEG format

kb: [unit] kilobit

kB: [unit] kilobyte

LAN: [computer] local area network

LFE: [sound] low-frequency enhancement channel

lpi: [printing] lines per inch

LTC: [video] longitudinal timecode

LZW: [format] Lemple-Zif-Welch compression

Mb: [unit] Megabit

MB: [unit] Megabyte

Mbps: [unit] megabits per second

MCU: [prod] medium close-up shot

MDM: [equip] modular digital multitrack

M&E: [post] music and (sound) effects track

M/E: [equip] mix/effects bank (on a video switcher)

MHz: [unit] Megahertz

MIDI: [format] Musical Instrument Digital Interface

M-JPEG: [format] Motion JPEG

MLS [prod]: medium long shot

MMC: [format] MIDI machine control

MO: [storage] magneto-optical

mob: media object

MOS: [film] supposedly "mit out sound" (*i.e.*, silent)

MOW: [bcast] "movie of the week"

MP3: [format] MPEG 1, layer 3

MPEG: [format, org] Moving Picture (Coding) Experts Group

MPSE: [org] Motion Picture Sound Editors

ms: [unit] millisecond

MS: [production] medium shot

MS-DOS: [computer] Microsoft Disk Operating System

MTBF: [computer] mean time between failures

MTC: [format] MIDI time code

MTS: [bcast] Multichannel Television Sound

mux: [signals] multiplex

NAB: [org] National Association of Broadcasters

NABET: [org] National Association of Broadcasting Engineers and Technicians

NATPE: [org] National Association of Television Program Executives

ND, NDF or NDTC: [video] non-drop frame timecode

NIC: [computer] network interface card

NL: [post] nonlinear

NLE: [post] nonlinear editing, or nonlinear editing system

NOC: [bcast] the Network Operations Center of a TV or cable network

ns: [unit] nanosecond

NTSC: [video, org] National Television Standards Committee

O&O: [bcast] (a TV station) owned & operated (by a network)

OC: [bcast] on camera (*i.e.*, a shot where the talent or actor speaking the dialog is seen on the screen)

OCN: [film] original camera negative

OEM: [business] original equipment manufacturer

OMF or OMFI: [format] Open Media Framework Interchange

OS: [computer] operating system

OS, O/S or OTS: [production] over-the-shoulder shot

P.A.: [prod] production assistant

PAL: [video] Phase Alternating Line

PB: [unit] Petabyte

PB: [video] playback

PCI: [computer] Peripheral Component Interconnect

PCM: [format] pulse code modulation

PD: [business] public domain

PDF: [format] Portable Document Format

PFL: [audio] pre-fader listen

P.M.: [equip] preventative maintenance

POTS: [network] "plain old telephone service"

P-P: [tech] peak-to-peak

PPP: [web] point-to-point protocol

PRAM: [computer] (Macintosh) Parameter RAM

PU: [production] pick-up shot

QA: [business] Quality Assurance

QC: [business, video] Quality Control

RAID: [computer] Redundant Array of Independent (or Inexpensive) Disks

RAM: [computer] Random Access Memory

RF: [tech] Radio Frequency

RFP: [business] request for proposals

RGB: [graphics] Red, Green and Blue

RGBA: [graphics] Red, Green, Blue plus Alpha (channel)

RLE: [tech] run-length encoding (compression)

R-MAG: [equip] removable-magnetic hard drive

RMS: [audio] root mean square

ROI: [business] return on investment

ROM: [computer] Read-Only Memory

RT: [equip] "real-time" effects

RTFM: [tech] "read the (freakin')
manual"

RU: [equip] rack unit

SAN: [computer] storage area network

SAP: [bcast] Secondary Audio Program

SC/H phase: [video tech] in NTSC, the relationship between subcarrier phase and horizontal sync

SCSI: [computer] Small Computer System Interface

SDDS: [film] Sony Dynamic Digital Sound

SDI: [video] serial digital interface

SDTV: [video] Standard Definition Television

SECAM: [video] Système Electronique Couleur Avec Mémoire

SGI: [manufacturer] Silicon Graphics, Inc.

SGML: [web] Standard Generalized Markup Language

SIMM: [computer] single in-line memory module

SMPTE: [org] Society of Motion Picture and Television Engineers

S/N: [tech] signal-to-noise ratio

SNG: [bcast] satellite news gathering

SOC: [org] Society of Operating Cameramen

SOT: [bcast] sound on tape (*i.e.*, sound bite)

S/PDIF: [audio] Sony/Phillips Digital Interface

SPL: [audio] sound pressure level

SPRM: [DVD] system parameter

SVGA: [computer] Super Video Graphics Array

TB: [unit] Terabyte

TBC: [equip] time-base corrector

TC: [video] timecode

T-CAL: [storage] thermal recalibration

TCG: [equip] timecode generator

TCP/IP: [web] Transfer Control Protocol / Internet Protocol

TCR: [equip] timecode reader

THX: [equip] Tom Holman's eXperiment (also, from the title of George Lucas' early film THX-1138EB)

TIFF: [format] Tagged-Image File Format

TLD: [web] top-level domain

TRT: [post, broadcast] total running time

UPS: [equip] uninterruptible power supply

URL: [web] uniform resource locator

USB: [computer] Universal Serial Bus

VAR: [business] value-added reseller

VBI: [video] vertical blanking interval

VBR: [video] variable bit rate (encoding)

VCA: [equip] voltage-controlled attenuator

VGA: [computer] Video Graphics Array

VHS: [video] Video Home System

VITC: [video] vertical interval timecode

VO: [prod] voiceover

VOB: [DVD] video object (file)

VTR: [equip] video tape recorder

VU: [audio] volume unit

WAN: [computer] wide area network

WS: [prod] wide shot

XML: [web] extensible markup language

Y: [video] luminance

I N D E X

1/2" tape   32
2nd unit photography   23
2:3 pulldown   *see* 3:2 pulldown
24P   37, 71, 101, 155, 172, 310-314, 388
3-perf film   26
3-point/4-point edits   231-232
3/4" tape, invention of   46
3:2 pulldown   105-107
   artifacts, in digitization   256
   naming conventions   xvi
   problems of   362-363
3D objects   236
3rd party hardware   203
4-perf film   26
4-point edits   224
4:2:2, 4:1:1, 4:2:0   165-166
4:3 and 16:9
   overview   178-179
   pixels   175-177
16:9   *see* 4:3
24fps and 30fps   102, 117
24P   37, 310, 312
25fps transfers   117
29.97   83, 113, 117, 362
30fps NTSC   117
409   36
60Hz power   117
1000, metric multiples   173

A

AAF   72, 99, 319
A-mode   258
Aaton, telecine logs   118-119
AC cable   137
Academy Awards®, Technical
      Achievement   60
academy leader   370
ACATS   283
access time, hard disks   170
accuracy   *see* frame accuracy
A-D conversion   149
additive property of color   181
address track timecode   86
Adobe
   After Effects   69, 175, 203, 234-236
   Premiere   41, 342-345

PostScript   189, 280
ADR   37, 359
affiliate, network   269
algorithm   210
alpha channel   234
AM/FM   269
Ampex   41, 43
   DVE   234
amplitude   147
amplitude modulation (AM)   186
analog systems   39
analog to digital converstion   149
anamorphic compression   178-179
animation   229-230
   software   230
ANSI lumens   313
anti-aliased fonts   188
Apple Computer   143, 295
array   *see* RAID
art cut, style   354
artifacts   255-256
ascender/descender   187
ascending timecode   90-91
aspect ratio
   basics   25, 178-179, 310
   of pixels   175-176
   in DTV "Table 3"   285
assemble editing   90
assemblies, videotape   257-259
assembly list   260-261
assembly, in theory   377
asset management, DVD   304
assistant editor, tasks   213
asymmetrical compression   160
ATSC   285-286
audio and video, storage   207
audio
   balanced/unbalanced   141-142
   hum   137
   resolve to mag   85
   rubber band   230
   sweetening   37
   tape   31-33, 85, 103
   timecode   86
   waveforms, in timelines   245
   waves   186
authoring, DVD   304, 307
auto-save   205-206

AUX, in EDL   93
Avid
    debut   58
    interface history   333-336
    OMF   *see* Open Media Framework
    Symphony   300
Avio *see* Casablanca

B

B-frames   161
B-mode   258
B-neg   27
B-roll
    effects   35
    offline   79
B-Y   165-166
backup   206
    RAID   167-168
bag of tricks, for editors   371-374
balanced audio   141-142
bandwidth   171-174
    RAID   168
bandwidth, table   174
Bargen, Dave   47
bars and tone   370
bars   *see* color bars
baseline   187
batch digitize   197, 253
baud rate   171
Beatnik   201
Bento   95-97
Bertolucci, Bernardo   1, 63
beta, invention   48
Betacam   32
Bézier curves   230
BHP   52
bins   215
binary vs metric, factoid   154
bit budgeting   304-306
bitmapped, fonts   188
bits per second   171
black   89-91
black and coded tape   90
black box   322
black level   364
Blair Witch Project   278
blank tape   89
blanking, vertical   87
blends   178-179
blocking, artifact   256
blue screen   234
BNC connectors   138

Brief History of Electronic Editing   41
brightness   132, 182
    monitor set-up   364
broadcast quality   14
broadcast spectrum   269, 274
broadcast, bandwidth   172
broadcasting, overview   267-276
browser/editors   299
buffering   292
burn in   *see* timecode
buzzwords   11

C

C   *see* chrominance
C-mode assemblies   259
cable, connectors   138-140
cable modems   173
cable TV   272-273
cables   136
Calaway, Jack   49
Caloway editing keyboard   33
camera rental   22
camera report   26-27
cancel/undo   226-227
capturing and digitizing   194, 253-256
Casablanca   348
Catmull, Ed   49
CBR   *see* constant bit rate
CBS Labs   41
CBS-Sony system   48
CCIR-601   170, 175, 204
CD-ROM   172, 301
celluloid, end of   312
center crop   178-179
CERN   62
CGI   236
change list   262, 363
channels, of TV   271
character generator   234
china marker   29
chroma key   234
chroma   *see* chrominance
chroma/luma effects   236
chrominance   120-123, 124-127, 132, 152
chrominance, sampling   165-166
cinema, electronic   312-314
circled takes   27, 213
Cisco   288
clip-based editing   225
ClipEdit   347
clock speed   202, 300

closed architecture systems   322-323
CMX   34, 45-66
  600   45
  6000   330-331
  EDLs   38
CMYK   132
coaxial cable   136-137, 272-273
code numbers   28, 98
codec   157, 292-293, 295-297
color   180-186 *see also* component
      video
  additive   181
  complementary   183
color and shading, in timelines   248
color banding, artifact   256
color bars   364-365
color correction   199
color correction, effects   236
color perception   185
color sampling   165-166
color space   130-135, 183
color television   40
color temperature   184-185
colors, illegal   134
complementary color   183
component/composite video   120-123
  in video-to-film transfer   310
  table of formats   128
composite video   *see* component video
compositing systems   229, 379
compressing images, concepts   154
compression   156-166
  anamorphic   178-179
  asymmetrical   160
  better   293
  for the web   290-295
  simple examples   163-164
  table of bandwidth   174
computer video, pixels   175-177
computer
  basics   202-206
  data sorting   214
  turnkey   322-323
cones and rods   181
constant bit rate   307
continuity notes   26
continuous and ascending
      timecode   90-91
contrast, monitor set-up   366
control track   87-90
Convergence   53
cooling, hardware   205

Coppola, Francis Ford   24, 49, 265
cost issues of nonlinear editing   6
count sheet   262
CPU   202, 300
C*r*, C*b   see* chrominance
creative indecision   8
cue points, in timelines   249
cut margin   261
cut away   251, 356
cutting block   29
cutting film using video   108-113, 363
cutting negative   30
cycle   146

D

D/Vision   66, 337
dailies   23
DARPA   287
DAT   32
Data Translation   67
data cables   137
data transfer rates   169
database   208-210
DAW   37
DBS   268, 274
DEC   38
deconstructing production   359
dedicated workstations   202
delete, by lift   225-226
deleting   220
demo reel   366-370, 374
Denecke   32
descriptions, of scenes   213
desktop   17
desktop distribution   279
diff list   262-264
digital audio
  capturing   255
  MP3   297
  signals   145-150
  storage issues   207
digital cinema   312-314
digital projection   312-314
digital subscriber line   275
digital systems, definition   17
digital video
  compression   156-166
  projection, flowchart   201
digitizing and capturing   148, 151, 194,
      253-256
disk rotation, bandwidth   171-172
disk, *see* hard disk

disks, RAID   167-168
dissolves   233
distribution   265-314
DLP technology   313
DMA   271
Dmytryk, Edward   315
Dolby Digital   306
dog legs, in splines   230
domain name   288
Doris, Robert   52
drop-frame timecode   83-85, 92
    why?   117
DSL   140, 275
DTV   280-286
dub method, assemblies   254
dupe negative   30
DV   201
DV timecode   88
DVD   301-308
    anatomy   303
    bandwidth   172
    color sampling   165-166
DVE   234, 240

E

E-cinema   312-314
E-Pix   332-333
E-PROM   205
edge networks   290
edge numbers   24, 98-99, 260-261
Ediflex   328-329,
    Ediflex Digital   338
    prototype   52
Edison, Thomas   186
edit mode, in EDL   93
edit sync guide   42
EditDroid   324-325
    interface   237-238
    timeline   244
Editec   43
editing
    assemble   90
    ergonomics   237-240
    flowcharts   196-201
    linear   3
    primitives   193, 216-230
    shot and transition based   220
editorial skills   9
editor's
    assistant   213
    bag of tricks   371-374
    wages   361

edits
    3- and 4-point   224
    fundamental examples   251-252
EDL   35-36
    3 and 4-point edits   231-232
    definition   92-94
    evolution   95-97
    print   260
    ripple   226
    telecine logs   118-119
EDLs, offline   79
eDVD   301
EECO timecode   44, 46
effects   233-236
    color correction   236
    in EDLs   93
    in web video   294
    motion   235
    motion paths   230
    optical lists   262
    paint   235
    switcher   35
    text   234
    timeline   229
Eidos Judgement   250, 346-347
electromagnetic radiation   180-186
electronic cinema   312-314
elementary streams, DVD   305
EMC2   58, 333-335
encoding, DVD   304, 307
enterprise level database   210
ER   see electromagnetic radiation
ergonomics   237-240
ESG   42
Ethernet   137, 172
Ettlinger, Adrian   44
evaluating systems   319-321
event numbers   93
Evertz, telecine logs   118-119
exabyte   173
export, web video   290

F

F/X   233-236
fades   233
fans, computers   205
fault tolerance, RAID   167-168
FCC   269-270, 280-286
FCM   92
feature film, editing schedule   195
fiber, optical   137
FibreChannel   95, 137, 173, 299

fields   208
  definition   87
  video   104
filament   186
file system organization   207-208
film cutting lists   260-263
film cutting systems   319-320
film editing, comparison   218
film finish, flow chart   198
film frames   25
film shoot, flow chart   197
film story   24
film style, insert   223-224
film, 3- and 4-perf   26
film, edge numbers   98-99
film-style trim   221-222
Final Cut   71, 344
Final Cut Pro   69, 72-75, 203 345
FireWire   137, 139, 143-144, 167, 204
first cut   222
first videotape broadcast   41
fit-to-fill   231-232
Flaherty, Joe   49
flatbed   29, 41
Flex files   118-119
floppy disk   36-37
flowcharts   196-201
  digital video projection   201
  HDTV finish   200
  MiniDV   201
  SDTV finish   199
  traditional film post   31
  web video   201
FM/AM   269
folded paper method, visualization   219
foley   359
fonts, bitmapped and vector   189
  antialiased   188
footage
  count   98
  numbers   24
format, tape   370
formatting, disks   169
frame accuracy   108-113
  definition   13
  film cutting with video   113-115
  limitations   116
frames, film   25
Framework, Open Media   see OMF
frequency   180
frequency modulation (FM)   186
frequency, audio   145

fundamental edits   251-252

G

gamut warning   133
gauges   136
generation loss   15
generating light   183
GEO locations   268
GIF   189
gigabyte   169, 173
Gone With The Wind   40
GOP   161
Grand Alliance   284-285
graphics applications, pixel
  issues   175-177
Grass Valley Group   34
green screen   234
Guggenheim, Ralph   50

H

H.261, H.263   292
halftone   134
handles   230
hard disk
  storage   204
  attributes   169-170
  file organization   207-208
  partition   207
  RAID   167-168
hardware   202
  third party   203
HDTV   30, 276, 280-286
  bandwidth and compression
    table   174
  flowchart   200
  future theory   388
  pixels   175-177
  quality   16
  video-to-film transfer   309
head and tail trims   29, 216-219
head/tail   216-217
header, EDL   92
heat sinks   205
Heim, Alan   351
hertz, definition   146
Hess Backup Methodology   206
high definition video   see HDTV
Hill, Dick   43
history and moving pictures   39-40
history, with undo   226-227
Hitchcock   341
horizontal editing, layers   228-229

horizontal nonlineariaty,
 theory 377-379
HTML 61, 288
hue 132, 182
 monitor set-up 366
hum, audio 137

I

I-frames 161
i.link 143-144
ICE, hardware 203
icons, picture 215
IEEE 1394 137, 143-144
illegal colors 134
image noise, web compression 293
image quality 14
ImMIX 63
iMovie 345
incandescence 183-184
index frames, database 210
insert editing 90, 220, 223-224
interframe/intraframe 161
interlaced video 87
Internet, historical 40 *see also* Web
interpositive print 30
intraframe compression, example 163
introduction to
 film post-production 24
 production 21
IP packets 292
IP router 287-288
IRE unit 364-365
ISDN, bandwidth 172
ISP 276, 289, 295
ITU-R601 xvi, 14-16, 174

J

jaggies 188
JBOD 167
jitter 36
JKL control 239
JPEG 159, 189 *see also* compression
Judgement, Eidos 250, 346-347
jump cut 356

K

Kelvin 184
KEM 29
 knob 238
key 234
key numbers 24

keyboard
 controller, cmx 50 47
 editing 33
keyframes 229-230
KeyKode 99, 362-363
Keylink files 118-119
KeyLog files 118-119
keys 35-36
kinescope 309

L

L-cut 222-223, 252, 357
LaserEdit 332
latent edge numbers 24
Laugh In 44
layers, of video 228-229
leader, slug 251-252
leading 187
letterbox 178-179
level, main 162
level, RAID 167-168
lift 220, 225-226, 252
light 180-186
light temperature 184
lightness 182
Lightworks 61, 237-238, 343
limitations of frame accuracy 116
linear editing 3
 problems 4
 offline, editing style 355
lined script 26-27
Link Editing Systems 339
lists, film cutting 260-263
"locking" the cut 8, 37, 198-201
log 208-209
log book 28
log/capture window 254
logging 194
logs, telecine 118
look-ahead preview, invention 44
lossless/lossy compression 157-159
LTC timecode 86-88
 accuracy issues 13
Lucas, George 49, 312, 375
Lucasfilm computer division 50
luma *see* luminance
lumen 313
luminance 120-123, 124-127, 132, 152
 key 234
 sampling 165-166
luminescence 183-184

M

machine-readable key numbers   99
magnetic film   85
mark-in/mark-out   231-232
match back (to film)   108-113
Matrox, hardware   203
matte   234
ME   *see* metal evaporate tape
Media 100   342, 344
Media Cleaner Pro   201, 295, 297
media assets, DVD   307
Media Composer, Avid   *see* Avid
media files   291
megabyte   169, 173
metal particulate tape   129
Memorex   44
memory   203
mentorship   7
menu, DVD   308
metadata   208-209
metal evaporate tape   129
metric vs binary, factoid   154
middle cut   178-179
MIDI cables   140
MiniDV, post-production
        flowchart   201
modem
    baud rate   171
    cable   173
modulation of amplitude,
        frequency   186
monitor
    adjustment   364-365
    pixels   175-177
    source/record   241-243
monitors and windows   241-243
Montage   51, 59, 326-327
Moorer, Andy   50
motion effects   235
motion path   230
Moviola   6, 29, 41
MOW   195
MP   *see* metal particulate tape
MP3   279, 297
MPEG   160-164, 275, 295-297
    compression table   174
    MPEG-2 family table   162
MSO   273
MUI   237, 336
multicamera mode   241
multiples of 1000   173
multiplexing   300

DVD   304, 305, 308
Murch, Walter   191
music video   195
MUX   300, 308

N

NAB   46
Nagra   32
naming scenes and takes   211-215
nanometer   181
narrowcasting   276
navigation, DVD   304
negative   261
negative cutter   30
negative, dupe   30
Netscape   66
network, edge   290
networks   267-276
NewTek, hardware   203
nomenclature   211-215
non-drop frame timecode
    definition   83-85
    in EDL   92
non-square pixels   175-177
nonlinear editing   4, 11
    comparison   218
    cost factors   6
    creative indecision   8
    historical perspective   39
    mentorship   7
    preparation   7
    reliablity   6
notes, in EDL   93
now line   249
NTSC
    30fps   117
    composite signal   122
    pixels   176
numbers, latent edge   24
numeric edits/trims   227-228
Nyquist, Joseph   149

O

object-oriented   189
OCR, in logging   299
ODC, founding   52
offline   79-82, 253
OMF   *see* Open Media Format
on the fly editing   33
online   79-82
Open Media Format   63, 66, 95-97
open architecture   202

consequences of   337
  systems   322-323
optical fiber   137
optical list   262
opticals   233, 312
organization, files   207-208
Oscar®   59-60
out-of-sync warning   232
outline, fonts   189
output, videotape   257-259
overlap edit   251-252

P

P-frames   161
Pagemaker   17
paint effects   235
Paintbox, Quantel   235
PAL   117
  composite signal   122
  pixels   175-177
pan-and-scan   178-179
Panavision   25
paper, folded method of
        visualization   219
parallel port   140
partition, hard disk   207
PCM format   306
PDP-11   38
perception of color   185
perceptually lossless   158
perforations   26
period   146
petabyte   173
phone connectors   139
Photoshop   189
  importing with layers   229
  plug-ins   236
PICT   159, 189, 304
picture icons   215
picture quality, nonlinear issues   9
pillarbox   179
pilot   267
Pinnacle, hardware   203
pitch, audio   145
Pixar   50
  Toy Story   69
pixel   151, 175-177
  color   165-166
plants, color   181
platform   202
play line   249
plug ins   236

pluge   366
point size   187
pointers, in editing software   217-218
post-production
  flow chart   196-201
  tasks, theory   384-387
PostScript   189, 230
prefix   24
premastering, DVD   301-308
Premiere   63, 341-343
preview   12
  historical   317
  look-ahead, invention   44
primary colors   120
primitives of editing   193, 216-230
print, interpositive   30
print, release   30
print-to-tape   257-259
printed takes   213
printouts   260-264
prism   181
production   21, 277
production, deconstructing   359
productivity tip   358
profile, main   162
programming   267
proxy server   290
pseudo-component   122-123
pull list   198, 200, 261
pulldown, 3:2   *see* 3:2 pulldown

Q

quality
  images   14
  videotape   129
Quantel, Paintbox   235
quantizing   134
quantum theory   180
QuickTime   63, 295-296

R

R-Y   165-166
radio cut   353
radio spectrum   180-186, 269
Rady, Bruce   55
RAID   144, 167-168, 207, 338
RAM/ROM   202-203, 205
random access   16
  nonlinear issues   8
RCA   141-142
RCA connectors   138-139
RC timecode   88

re-editing   220
Real Video   40 *see also* Real Media
real-time
    effects   233-236
    compositing   379
    preview   12
Real Media   40, 296
reciprocity theory   379
record in/out, in EDL   94
record monitor   241-243
reel names   211
reels, demo   366-370, 374
relational database   208-210
release print   30, 309, 312
reliability, nonlinear editing   6
rendering
    CPU   202
    theory   378
replacing   220
resolution   16, 80
resolution independent   189
resolve audio   198
resolve to mag, audio   85
retina   181
revolution   277-278
RGB   *see also* color space, component
        video
    cube   124-127
    discussion   182
    video signal   122
rippling   221-222, 226
RJ45   140
RLE   163
"rock n' roll"   237-240
rods and cones   181
Rolodex   208
rotational latency   170
rotoscope   235
RS-232/432   137, 140
RT-11   36, 38
rubber band, audio   230
rubber numbers   28, 98

S

S-video connectors   139
safe margin, TV   87
salary, editing   361
sampling rate   149
sampling, color   165-166
san serif   187
satellites   268
saturation   132

saturation   182
saturation, monitor set-up   366
save   205-206
scene descriptions   213
scene order,   23
scene/take labels, in timelines   248
scenes   22
scenes/takes, naming   211-215
scintillatoin   256
scope   127
screen, green or blue   234
Script Mimic   329
script supervisor   26
script, lined   27
SCSI   137,167-168, 170, 172-173, 204
sculpted sequence, editing style   355
SDTV
    flowchart   199
    pixels   175-177
SECAM, composite signal   122
segment re-encode   307
self-luminous colors   182
"selling the cut"   356
serif   187
set-up   22, 212
shielding, cable   136-137
shooting order,   23
shot identification, in timelines   248
shot-based editing   220
simo   197
Slate, Smart   32
slates, scene numbering   211-215
slide   225
slip   225
slug leader   251-252, 267
smart slate   32
Smith Splicer   42
Smith, Alvy Ray   50
snow   89
Sony   143
Sorenson   292
sorting scenes, computers   214
sound broadcast   186
source in/out, in EDL   94
source material, theory   379-380
source monitor   241-243
source reel ID, in EDL   93
space, letter and word   187
special effects   233-236
spectrum   16, 180-186
splines   189, 230
split edit   222-223

sprockets   26, 261
square pixels   175-177
SQL server   77, 348
Star Wars   312   *see also* Lucas, George
Steenbeck   29
storage   204
storage capacity, disks   169
storage, table of compression and
        bandwidth   174
storyboard printout   264
streaming
    media formats   295-297
    RAID   168
    video   289-290
striped timecode   89
striping data, RAID   167-168
SUN computer   50
superimposition   234
supply roll, on a flatbed   216
surface colors   182
sustained data transfer rate   169
sweetening   10, 37
    flowchart   198, 199, 200
Symphony, Avid prototype   300
sync point, database   210
synchronization   232
syndicaton   269
synthespian   236
systems
    historical overview   324-348
    how to evaluate   319-321
    open vs closed architecture   322-323
    printouts   264

T

tables
    assembly modes   259
    bandwidth of media formats   172
    composite/component video
        formats   128
    DVD types   303
    HDTV, Table 3   285
    MPEG-2 family   162
    RAID schemes   168
    video compression, bandwidth,
        storage   174
take-up reel, on a flatbed   216
takes   22
takes, circled   27
tape formatting   370
tape-start times   211
TBC   *see* timebase corrector

technical achievement, Academy
        Award®   60
Tektronix scope   127
telecine   32-33, 102
    log   118, 254, 363
    preparation   113
    troubles   362-363
temperature of light   184
temperature, computers   205
terabyte   169, 173
Texas Instruments   313
text, effects   234
texture mapping   236
third-party hardware   203
thumbnails, scene organization   215
TIFF   159, 189, 304
tilt-and-scan   178-179
time, spent editing   194
timebase corrector (TBC)   10, 35-36,
        374
timecode   83, 86-88
    address track   86
    audio   86
    continuous and ascending   90-91
    DV   88
    EECO   46
    invention   44
    reader   88
    window   32, 80
timeline
    animation   229
    cue points   249
    now line   249
    overview   244-250
    scale units   247
    shot identification   248
    zoom in/out   246
titles   188, 234
TLC, telecine logs   118-119
Touchvision   331-332
Toy Story   69
trace   36, 48
track enabling, timelines   246
traditional film post-production
        flowchart   31
transcoder, video   132
transfer, video to film   309-311
transition type, in EDL   93
transition years   39
transition-based editing   220
transponders   268
trim tabs   217

trim, numerically   227-228
trims, head and tail   29, 216-221
turn-around time, nonlinear editing   5
turnkey computers   322-323
TV channels   271
TV safe margin   87
type   187-190
types of editing systems,
        historical   317-318

U

U-matic   46
Ubillos, Randy   63
unbalanced audio   141-142
uncompressed video   174
undo/cancel   226-227
upright   28
USB   140

V

VAR   323
variable bit rate   307
vector, fonts   189
vertical blanking   87
vertical editing, layers   228-229
vertical nonlinearity, theory   233,
        377-379
VGA, pixels   176
VHS, invention   48
Vibrint   347
video
    black   365
    component-composite   120-123
    compression table   174
    compression, concepts   154
    desktop   17
    editing, comparison   218
    fields   104
    flowchart   199
    introduction to   32
    layers   228-229
    RAIDs   168
    storage   207
    streams   233
    titles   188
    transcoder   132
    uncompressed   174
    web   287-300
video-style
    insert   223-224
    trim   221-222
video-to-film

flowchart   201
transfer   309-311
VideoF/X   340
videotape
    assemblies   257-259
    broadcast, first   41
    quality   129
    reel names   211
Virage   299
virtual
    edits   218
    master   317, 331
    source roll, productivity tip   358
visible spectrum   180-186
Vistavision   25
VITC timecode   13, 86-88
VOB   308
volume   147
VTR invention   41

W

wages, editing   361
Warner, Bill   59
warning, out of sync   232
wave theory   180
waveform, digital audio   145-150
wavelength   146, 180
web video   201, 287-300
web, use of RAIDs   168
webcasting   290
WebTV   69
white balance   185
white box   322
Windows Media   296
windows and monitors   241-243
windows, pixels   175-177
wipes   233
wire frame   236
workprint   261
workstations, dedicated computers   202

X-Y-Z

x-height   187
Xerox PARC   324
XLR   139, 141-142
Y   *see* luminance
Y-C   122
YIQ, YUV   125-126
zero frame   98-100
zero-counting   99
zettabyte   173
zoom in, timeline   245-246

RON DIAMOND
www.nonlinear4.com/ron

Ron Diamond still fondly recalls his first "editing job." Sometime in his early teens, he recorded the off-air soundtrack of an entire week's worth of Johnny Carson monologues and skits, and on a whim, spliced them together into a satirical composite using some nearby scissors, adhesive tape, and a beat-up old reel-to-reel tape recorder his mother found

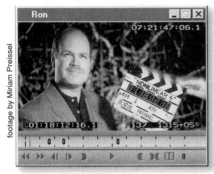

footage by Miriam Preissel

at a garage sale. His life was never quite the same again. He subsequently went on to become a radio engineer/ producer, audio engineer, technical director at Boston's CBS television affiliate, and video editor.

Ron and Michael first crossed paths during a demonstration of the EditDroid and SoundDroid proto-types at Lucasfilm in the mid '80s. At the time, *Nonlinear* was just the inkling of an idea for a series of handouts for Michael's prospective trainees. Fast forward seven years: *Nonlinear* is now in its second best-selling edition and has helped thousands (including Ron) to navigate the new editing paradigm. At a trade show, a chance encounter reintroduces the two, sparking a collaboration that continues to this day—the ultimate result of which is the book you're now holding.

Ron is currently a nonlinear offline/online editor and consultant residing in Los Angeles, specializing in the Avid Media Composer and Symphony. He has edited projects for most of the major broadcast and cable television networks, as well as for the occasional big screen.

NOTE: It only seems appropriate that the material new to this Fourth Edition of **Nonlinear**—the first major revision during the Internet age—evolved as a collaboration that was itself largely digital in nature. While **N3** made extensive use of penciled notes on the printed page, most of what's different about **N4** evolved via hundreds of emails, text documents, graphics and PDF files ricocheting their way back and forth through the packet-switching networks connecting Santa Cruz and Los Angeles.

Also, unlike prior editions of **Nonlinear**, this edition contains a number of original essays by Ron, including much of The Real World chapter, although his contribution is dispersed throughout these pages. Ron's efforts and insights continue to be influential to this work, as it exists both in print and on the Web.

—MHR

MICHAEL RUBIN

www.nonlinear4.com/mike

Michael Rubin is an editor, consultant, and entrepreneur, currently residing in Santa Cruz, California. Graduated from Brown University with an Sc.B. in Neural Sciences, he joined Lucasfilm's new division, The Droid Works, where he was instrumental in the development of training for the EditDroid, and in working with the Hollywood market in evaluating and adopting new electronic systems for film.

In 1987, Rubin joined the start-up Sonic Solutions (employee #3); and then moved on to CMX-Aurora as a staff editor (and later as Director of Philosophy) for the development and release of the CMX 6000, supporting its use in Los Angeles. He left CMX in 1989 to freelance edit.

Rubin has supported nonlinear editing systems on many projects: commercials and music videos; the first feature film edited on the 6000; considerable film-originated television programming and the mini-series *Lonesome Dove*. Rubin assisted Academy Award winning editor Gabriella Cristiani in her post-production of Bernardo Bertolucci's *The Sheltering Sky*, and was a principal editor on the Paul McCartney concert film *Get Back*. He was also an editor on one of the first television series to be cut on the Avid (1991). He is presently a fan of Apple's Final Cut Pro.

Rubin has lectured internationally on nonlinear editing: at the NAB conference, as a guest of the EFFECTS conference in Hamburg, Germany; at the International Television Symposium in Montreux, Switzerland, for the Directors Guild, for numerous university courses and conferences. He was also an advisor for the AFI-Apple Computer Center, and a teacher to hundreds of professional film editors. He is the author of a number of odd books and of a patent for new methods of Web searching and browsing.

Today, Rubin continues to shoot and edit on desktop video systems, and to write and consult. He is the co-founder and chairman of Petroglyph, Inc., an innovative private retail company in Northern California. Along with everything about his Macintosh, he enjoys writing, photography, films, travel, and reciprocating the love of his wife Jennifer and son, Jomo.

OLD-FASHIONED
ORDER FORM

Please send me _____ copies of **NONLINEAR/4** at $39.95 per copy, plus shipping and handling of $7 (first book); $1 (each additional); add 6% sales tax on orders shipped to a Florida address.

☐ Send information on the *forthcoming* **N4 Pocket Lexicon**.

SHIP TO:

Name: _____

Address: _____

City/State/Zip: _____

Phone/Fax: _____

Total amount included (check or money order): $ _____

Charge to credit card: ☐Visa ☐Mastercard ☐AmEx ☐Discover

Card number: _____

Expiration date: _____

Name on card:_____

Signature: _____

Foreign orders: Payable in U.S. funds drawn on a U.S. bank (write for shipping charges, and indicate surface or air) or by credit card.

MAIL: Triad Publishing Company
           PO Box 13355
           Gainesville, FL 32604

For credit card orders:

FAX: TOLL FREE to 1-800-854-4947
EMAIL: donna@triadpublishing.com

Contact Triad for bulk orders or special discounts.
Prices subject to change without notice.
www.nonlinear4.com

TRIAD

All orders shipped promptly.